AF389432

Catalysts for Syngas Production

Catalysts for Syngas Production

Special Issue Editor
Javier Ereña Loizaga

MDPI • Basel • Beijing • Wuhan • Barcelona • Belgrade

Special Issue Editor
Javier Ereña Loizaga
Department of Chemical
Engineering, University of the
Basque Country—UPV/EHU
Spain

Editorial Office
MDPI
St. Alban-Anlage 66
4052 Basel, Switzerland

This is a reprint of articles from the Special Issue published online in the open access journal *Catalysts* (ISSN 2073-4344) from 2018 to 2020 (available at: https://www.mdpi.com/journal/catalysts/special_issues/Syngas_Production).

For citation purposes, cite each article independently as indicated on the article page online and as indicated below:

LastName, A.A.; LastName, B.B.; LastName, C.C. Article Title. *Journal Name* **Year**, *Article Number*, Page Range.

ISBN 978-3-03936-595-1 (Hbk)
ISBN 978-3-03936-596-8 (PDF)

Contents

About the Special Issue Editor vii

Preface to "Catalysts for Syngas Production" ix

Javier Ereña
Catalysts for Syngas Production
Reprinted from: *Catalysts* **2020**, *10*, 657, doi:10.3390/catal10060657 . . . . **1**

Qinwei Yu, Yi Jiao, Weiqiang Wang, Yongmei Du, Chunying Li, Jianming Yang and Jian Lu
Catalytic Performance and Characterization of Ni-Co Bi-Metallic Catalysts in *n*-Decane Steam
Reforming: Effects of Co Addition
Reprinted from: *Catalysts* **2018**, *8*, 518, doi:10.3390/catal8110518 . . . . **4**

Anis H. Fakeeha, Siham Barama, Ahmed A. Ibrahim, Raja-Lafi Al-Otaibi, Akila Barama, Ahmed E. Abasaeed and Ahmed S. Al-Fatesh
In Situ Regeneration of Alumina-Supported Cobalt–Iron Catalysts for Hydrogen Production by
Catalytic Methane Decomposition
Reprinted from: *Catalysts* **2018**, *8*, 567, doi:10.3390/catal8110567 . . . . **17**

Yan Xu, Qiang Lin, Bing Liu, Feng Jiang, Yuebing Xu and Xiaohao Liu
A Facile Fabrication of Supported Ni/SiO_2 Catalysts for Dry Reforming of Methane with
Remarkably Enhanced Catalytic Performance
Reprinted from: *Catalysts* **2019**, *9*, 183, doi:10.3390/catal9020183 . . . . **33**

Ahmed Sadeq Al-Fatesh, Samsudeen Olajide Kasim, Ahmed Aidid Ibrahim, Anis Hamza Fakeeha, Ahmed Elhag Abasaeed, Rasheed Alrasheed, Rawan Ashamari and Abdulaziz Bagabas
Combined Magnesia, Ceria and Nickel catalyst supported over γ-Alumina Doped with Titania
for Dry Reforming of Methane
Reprinted from: *Catalysts* **2019**, *9*, 188, doi:10.3390/catal9020188 . . . . **42**

Xiaozhan Liu, Lu Zhao, Ying Li, Kegong Fang and Minghong Wu
Ni-Mo Sulfide Semiconductor Catalyst with High Catalytic Activity for One-Step Conversion
of CO_2 and H_2S to Syngas in Non-Thermal Plasma
Reprinted from: *Catalysts* **2019**, *9*, 525, doi:10.3390/catal9060525 . . . . **57**

Hae-Gu Park, Sang-Young Han, Ki-Won Jun, Yesol Woo, Myung-June Park and Seok Ki Kim
Bench-Scale Steam Reforming of Methane for Hydrogen Production
Reprinted from: *Catalysts* **2019**, *9*, 615, doi:10.3390/catal9070615 . . . . **70**

Bonan Liu, Liang Zhao, Zhijie Wu, Jin Zhang, Qiuyun Zong, Hamid Almegren, Feng Wei, Xiaohan Zhang, Zhen Zhao, Jinsen Gao and Tiancun Xiao
Recent Advances in Industrial Sulfur Tolerant Water Gas Shift Catalysts for Syngas Hydrogen
Enrichment: Application of Lean (Low) Steam/Gas Ratio
Reprinted from: *Catalysts* **2019**, *9*, 772, doi:10.3390/catal9090772 . . . . **84**

Andrea Fasolini, Silvia Ruggieri, Cristina Femoni and Francesco Basile
Highly Active Catalysts Based on the $Rh_4(CO)_{12}$ Cluster Supported on $Ce_{0.5}Zr_{0.5}$ and Zr Oxides
for Low-Temperature Methane Steam Reforming
Reprinted from: *Catalysts* **2019**, *9*, 800, doi:10.3390/catal9100800 . . . . **102**

Johnny Saavedra Lopez, Vanessa Lebarbier Dagle, Chinmay A. Deshmane, Libor Kovarik, Robert S. Wegeng and Robert A. Dagle
Methane and Ethane Steam Reforming over MgAl$_2$O$_4$-Supported Rh and Ir Catalysts: Catalytic Implications for Natural Gas Reforming Application
Reprinted from: *Catalysts* **2019**, *9*, 801, doi:10.3390/catal9100801 . . . . **121**

Abir Azara, El-Hadi Benyoussef, Faroudja Mohellebi, Mostafa Chamoumi, François Gitzhofer and Nicolas Abatzoglou
Catalytic Dry Reforming and Cracking of Ethylene for Carbon Nanofilaments and Hydrogen Production Using a Catalyst Derived from a Mining Residue
Reprinted from: *Catalysts* **2019**, *9*, 1069, doi:10.3390/catal9121069 . . . . **140**

About the Special Issue Editor

Javier Ereña Loizaga graduated in Chemistry (1991, Speciality: Industrial Chemistry) at the University of the Basque Country (UPV/EHU, Leioa, Spain). Since 1990, the year he joined the laboratories of the Chemical Engineering Department from the University of the Basque Country, his research has focused on the development of catalytic processes for obtaining fuels and raw materials from alternative sources to petroleum.

In 1996, he defended his Doctoral Thesis on a collaboration agreement between the "Catalytic Processes and Waste Valorization" research group (a well established/consolidated high-performance group in the Basque Country) and the Center for Chemical Reactors from the University of Western Ontario (Canada). The objective was to obtain gasoline from synthesis gas and CO_2.

Since 2000, his research work has mainly focused on the study of the following reseach lines (key to the industrial development of the concept of bio-refinery):

(1) H_2 synthesis by the steam reforming of dimethyl ether (DME) and ethanol.

(2)   Direct DME synthesis (STD process). The interest of this integrated process is based on both the product (clean fuel and H_2 source for fuel cells) and the raw materials (synthesis gas and CO_2).

(3)   The development of catalytic processes of interest from the perspective of sustainability.

Since 1996, he has been a Professor at the University of the Basque Country, where he integrates his research activities and teaching in subjects related to Chemical Engineering.

Preface to "Catalysts for Syngas Production"

Synthesis gas (or syngas) is a mixture of hydrogen and carbon monoxide, with different chemical composition and H_2/CO molar ratios, depending on the feedstock and production technology used. Syngas may be obtained from alternative sources to oil, such as natural gas, coal, biomass, organic wastes, etc. Syngas is a very good intermediate for the production of high value compounds at the industrial scale, such as hydrogen, methanol, liquid fuels, and a wide range of chemicals. Accordingly, efforts should be made to co-feed CO_2 with syngas, as an alternative for reducing greenhouse gas emissions. In addition, more syngas will be required in the near future, in order to satisfy the demand for synfuels and high value chemicals.

New research for syngas production is essential for reducing operating costs, improving the thermal efficiency of the process, and preserving the environment. Advances should be made in the following areas:

(1)   The development of new catalysts and catalytic routes for syngas production;

(2)   The optimization of the reaction conditions for the process;

(3)   The use of biomass, as a promising raw material for syngas production, due to its renewable character and potential for zero CO_2 emissions.

Further steps should be made to advance the catalytic processes for saving energy and capital costs, and for optimizing the quality and properties of syngas, such as H_2/CO molar ratio and absence of contaminants.

Javier Ereña Loizaga
Special Issue Editor

 *catalysts*

Editorial

Catalysts for Syngas Production

Javier Ereña

Department of Chemical Engineering, University of the Basque Country UPV/EHU, P.O. Box 644, 48080 Bilbao, Spain; javier.erena@ehu.eus; Tel.: +34-94-6015363

Received: 8 June 2020; Accepted: 10 June 2020; Published: 11 June 2020

Synthesis gas (or syngas) is a mixture of hydrogen and carbon monoxide, that may be obtained from alternative sources to oil, such as natural gas, coal, biomass, organic wastes, etc. [1–3] Biomass is a promising raw material for syngas production, due to its renewable character and potentially zero CO_2 emissions [4]. Syngas is an excellent intermediate for the production of high value compounds at the industrial scale, such as hydrogen, methanol, liquid fuels, and a wide range of chemicals.

This Special Issue on "Catalysts for Syngas Production" shows new research about the development of catalysts and catalytic routes for syngas production, and the optimization of the reaction conditions for the process.

This issue includes ten articles. Yu et al. analyze the performance of Ni-Co bi-metallic catalysts in n-decane steam reforming [5]. The addition of Co to the catalyst improves the hydrogen selectivity and anti-coking ability compared with the mono-Ni/Ce-Al_2O_3 catalyst. A synergistic effect between Ni and Co is observed, with 12% Co showing the best catalytic activity in the series Co-Ni/Ce-Al_2O_3 catalysts. In situ regeneration of a spent alumina-supported cobalt-iron catalyst for catalytic methane decomposition is reported by Fakeeha et al. [6] The main factors responsible for the catalyst deactivation are coke deposition and weak sintering of the metallic active phase (Co-Fe), which occur during the catalytic methane decomposition reaction and regeneration process. A facile fabrication of supported Ni/SiO_2 catalysts for dry reforming of methane is developed by Xu et al. [7] Due to the formation of much smaller Ni nanoparticles, this Ni/SiO_2 catalyst exhibits excellent coke-resistance performance and effectively suppresses the side reaction toward RWGS compared to that prepared with the conventional wetness impregnation method. The dry reforming of methane over combined magnesia, ceria and nickel catalysts, supported on γ-Al_2O_3 and doped with TiO_2, is investigated by Al-Fatesh et al. [8] The addition of CeO_2 and MgO to the catalyst enhances the interaction between the Ni and the support, and improves the activity of the solid. Liu et al. describe a novel one-step conversion of CO_2 and H_2S to syngas induced by non-thermal plasma, with the aid of Ni-Mo sulfide/Al_2O_3 catalyst under ambient conditions [9]. The optical and structural properties of the synthesized catalysts are significantly influenced by the Ni/Mo molar ratio. Moreover, the Ni-Mo sulfide/Al_2O_3 catalysts possess excellent catalytic activities for CO_2 and H_2S conversion, compared to the single-component NiS_2/Al_2O_3 and MoS_2/Al_2O_3 catalysts. The paper by Park et al. describes the effect that reaction parameters have on hydrogen production via steam reforming of methane, using lab- and bench-scale reactors to identify critical factors for the design of large-scale processes [10]. The temperature at the reactor bottom is crucial for determining the methane conversion and hydrogen production rates when a sufficiently high reaction temperature is maintained (above 800 °C). However, if the temperature of one or more of the furnaces decreases below 700 °C, the reaction is not equilibrated at the given space velocity. Liu et al. study a novel sulfur tolerant water gas shift catalyst (SWGS) developed for the applications under lean (low) steam/gas ratio conditions [11]. The adoption of the lean steam/gas SWGS catalyst significantly improves the plant efficiency and safety, and remarkably reduces the actual steam consumption for H_2 production, decreasing CO_2 emission. The paper by Fasolini et al. summarizes the synthesis, characterization and catalytic behavior of Rh-based catalysts, obtained by using the Rh_4 $(CO)_{12}$ neutral cluster as the active-phase precursor [12]. The preparation method allows the deposition of the cluster on the surface of $Ce_{0.5}Zr_{0.5}O_2$ and ZrO_2 supports, which are synthetized by the microemulsion technique, being the catalysts active in the low-temperature steam reforming process for syngas production. Methane and ethane steam reforming over $MgAl_2O_4$-supported Rh and Ir catalysts is analyzed in the paper by Lopez et al. [13] The Rh- and Ir-supported catalysts exhibit higher activity than Ni catalysts for steam methane reforming. Catalyst durability studies reveal the Rh catalyst to be stable under steam methane reforming conditions. The results of this study conclude that a Rh-supported catalyst enables very high activity and excellent stability, for both the steam reforming of methane and other higher hydrocarbons contained in natural gas, and under conditions of operation that are amendable to solar thermochemical operations. In the paper by Azara et al., iron-rich mining residue is used as a support to prepare a new Ni-based catalyst for C_2H_4 dry reforming and catalytic cracking [14]. The deposited carbon is found to be filamentous and of various sizes (i.e., diameters and lengths). The analyses of the results show that iron is responsible for the growth of carbon nanofilaments and nickel is responsible for the split of C-C bonds.

In summary, these ten papers clearly show the relevance of obtaining syngas for further applications, such as the production of hydrogen, methanol, liquid fuels, and a wide range of chemicals. Nowadays, efforts are being made on the co-feeding of CO_2 with syngas, as an alternative for reducing greenhouse gas emissions. I would like to thank all the authors of this Special Issue.

I am honored to be the Guest Editor of this Special Issue. I would like to thank the reviewers for improving the quality of the papers with their comments. I am also grateful to all the staff of the *Catalysts* Editorial Office.

Conflicts of Interest: The authors declare no conflict of interest.

References

1. Gao, J.; Guo, J.; Liang, D.; Hou, Z.; Fei, J.; Zheng, X. Production of Syngas via Autothermal Reforming of Methane in a Fluidized-bed Reactor over the Combined CeO_2-ZrO_2/SiO_2 Supported Ni Catalysts. *Int. J. Hydrog. Energy* **2008**, *33*, 5493–5500. [CrossRef]
2. Rezaei, M.; Alavi, S.M.; Sahebdelfar, S.; Yan, Z.F. Syngas Production by Methane Reforming with Carbon Dioxide on Noble Metal Catalysts. *J. Nat. Gas Chem.* **2006**, *15*, 327–334. [CrossRef]
3. He, M.; Xiao, B.; Liu, S.; Hu, Z.; Guo, X.; Luo, S.; Yang, F. Syngas Production from Pyrolysis of Municipal Solid Waste (MSW) with Dolomite as Downstream Catalysts. *J. Anal. Appl. Pyrolysis* **2010**, *87*, 181–187. [CrossRef]
4. Molino, A.; Chianese, A.; Musmarra, D. Biomass Gasification Technology: The State of the Art Overview. *J. Energy Chem.* **2016**, *25*, 10–25. [CrossRef]
5. Yu, Q.; Jiao, Y.; Wang, W.; Du, Y.; Li, C.; Yang, J.; Lu, J. Catalytic Performance and Characterization of Ni-Co Bi-Metallic Catalysts in n-Decane Steam Reforming: Effects of Co Addition. *Catalysts* **2018**, *8*, 518. [CrossRef]
6. Fakeeha, A.H.; Barama, S.; Ibrahim, A.A.; Al-Otaibi, R.L.; Barama, A.; Abasaeed, A.E.; Al-Fatesh, A.S. In Situ Regeneration of Alumina-Supported Cobalt-Iron Catalysts for Hydrogen Production by Catalytic Methane Decomposition. *Catalysts* **2018**, *8*, 567. [CrossRef]
7. Xu, Y.; Lin, Q.; Liu, B.; Jiang, F.; Xu, Y.; Liu, X. A Facile Fabrication of Supported Ni/SiO_2 Catalysts for Dry Reforming of Methane with Remarkably Enhanced Catalytic Performance. *Catalysts* **2019**, *9*, 183. [CrossRef]
8. Al-Fatesh, A.S.; Kasim, S.O.; Ibrahim, A.A.; Fakeeha, A.H.; Abasaeed, A.E.; Alrasheed, R.; Ashamari, R.; Bagabas, A. Combined Magnesia, Ceria and Nickel Catalyst Supported over γ-Alumina Doped with Titania for Dry Reforming of Methane. *Catalysts* **2019**, *9*, 188. [CrossRef]
9. Liu, X.; Zhao, L.; Li, Y.; Fang, K.; Wu, M. Ni-Mo Sulfide Semiconductor Catalyst with High Catalytic Activity for One-Step Conversion of CO_2 and H_2S to Syngas in Non-Thermal Plasma. *Catalysts* **2019**, *9*, 525. [CrossRef]
10. Park, H.G.; Han, S.Y.; Jun, K.W.; Woo, Y.; Park, M.J.; Kim, S.K. Bench-Scale Steam Reforming of Methane for Hydrogen Production. *Catalysts* **2019**, *9*, 615. [CrossRef]
11. Liu, B.; Zhao, L.; Wu, Z.; Zhang, J.; Zong, Q.; Almegren, H.; Wei, F.; Zhang, X.; Zhao, Z.; Gao, J.; et al. Recent Advances in Industrial Sulfur Tolerant Water Gas Shift Catalysts for Syngas Hydrogen Enrichment: Application of Lean (Low) Steam/Gas Ratio. *Catalysts* **2019**, *9*, 772. [CrossRef]

12. Fasolini, A.; Ruggieri, S.; Femoni, C.; Basile, F. Highly Active Catalysts Based on the $Rh_4(CO)_{12}$ Cluster Supported on $Ce_{0.5}Zr_{0.5}$ and Zr Oxides for Low-Temperature Methane Steam Reforming. *Catalysts* **2019**, *9*, 800. [CrossRef]

13. Lopez, J.S.; Dagle, V.L.; Deshmane, C.A.; Kovarik, L.; Wegeng, R.S.; Dagle, R.A. Methane and Ethane Steam Reforming over $MgAl_2O_4$-Supported Rh and Ir Catalysts: Catalytic Implications for Natural Gas Reforming Application. *Catalysts* **2019**, *9*, 801. [CrossRef]

14. Azara, A.; Benyoussef, E.H.; Mohellebi, F.; Chamoumi, M.; Gitzhofer, F.; Abatzoglou, N. Catalytic Dry Reforming and Cracking of Ethylene for Carbon Nanofilaments and Hydrogen Production Using a Catalyst Derived from a Mining Residue. *Catalysts* **2019**, *9*, 1069. [CrossRef]

Article

Catalytic Performance and Characterization of Ni-Co Bi-Metallic Catalysts in *n*-Decane Steam Reforming: Effects of Co Addition

Qinwei Yu [1], Yi Jiao [1,2], Weiqiang Wang [1], Yongmei Du [1], Chunying Li [1], Jianming Yang [1,*] and Jian Lu [1,*]

[1] State Key Laboratory of Fluorine & Nitrogen Chemicals, Xi'an Modern Chemistry Research Institute, Xi'an 710065, China; qinweiyu204@163.com (Q.Y.); jiaoyiscu@163.com (Y.J.); wqwang07611@163.com (W.W.); dymqw204@sina.com (Y.D.); chunyingli204@163.com (C.L.)

[2] Institute of New Energy and Low-Carbon Technology, Sichuan University, Chengdu 610064, China

* Correspondence: yangjm204@163.com (J.Y.); lujian204@263.net (J.L.); Tel.: +86-029-8829-1367 (J.Y.); +86-029-8829-1213 (J.L.)

Received: 29 September 2018; Accepted: 1 November 2018; Published: 5 November 2018

Abstract: Co-Ni bi-metallic catalysts supported on Ce-Al$_2$O$_3$ (CA) were prepared with different Co ratios, and the catalytic behaviors were assessed in the *n*-decane steam reforming reaction with the purpose of obtaining high H$_2$ yield with lower inactivation by carbon deposition. Physicochemical characteristics studies, involving N$_2$ adsorption-desorption, X-ray diffraction (XRD), H$_2$-temperature-programmed reduction (H$_2$-TPR), NH$_3$-temperature-programmed desorption (NH$_3$-TPD), SEM-energy dispersive spectrometer (EDS), and transmission electron microscope (TEM)/HRTEM, were performed to reveal the textural, structural and morphological properties of the catalysts. Activity test indicated that the addition of moderate Co can improve the hydrogen selectivity and anti-coking ability compared with the mono-Ni/Ce-Al$_2$O$_3$ contrast catalyst. In addition, 12% Co showed the best catalytic activity in the series Co-Ni/Ce-Al$_2$O$_3$ catalysts. The results of catalysts characterizations of XRD and N$_2$ adsorption-desorption manifesting the metal-support interactions were significantly enhanced, and there was obvious synergistic effect between Ni and Co. Moreover, the introduction of 12% Co and 6% Ni did not exceed the monolayer saturation capacity of the Ce-Al$_2$O$_3$ support. Finally, 6 h stability test for the optimal catalyst 12%Co-Ni/Ce-Al$_2$O$_3$ demonstrated that the catalyst has good long-term stability to provide high hydrogen selectivity, as well as good resistance to coke deposition.

Keywords: x%Co-Ni/Ce-Al$_2$O$_3$; steam reforming; regeneration; thermal stability; anti-coking ability

---

1. Introduction

Nowadays, hydrogen is recognized as a clean fuel and energy carrier since its combustion produces only water as product [1,2]. However, how to produce hydrogen from primary energy sources (such as hydrocarbons) in an efficient and economic way should be further researched and developed [3–6]. In the past few decades, the most effective approach was catalytic reforming of hydrocarbons. Currently, over 50% of the world's hydrogen supply is from steam reforming of hydrocarbons [7].

Nowadays, H$_2$ is mainly produced by steam reforming of CH$_4$ and other high-energy density liquid fuels, including ethanol, gasoline, diesel, or jet fuel [8–10]. An interesting option is hydrogen production from diesel steam reforming. *n*-decane, one of the main components of diesel, is considered as an ideal source of hydrogen since its availability, easy handling and storage and, relatively high H/C ratio (produce 31 mol of H$_2$ per mole of reacted *n*-decane) [11,12]. However, the *n*-decane steam reforming reaction is different with CH_4, CH_3OH, C_2H_5OH etc., and always accompanied by other side effects (cracking, isomerization, hydrogen transfer reaction). More so, the catalysts used in *n*-decane steam reforming reaction are easily to lose activity caused by carbon deposition, especially at higher temperatures [13–15]. Therefore, the catalysts use in *n*-decane steam reforming reaction are put forward higher requirements.

$$C_{10}H_{22} + 20H_2O \rightarrow 31H_2 + 10CO_2$$

Hydrocarbons steam reforming reactions have been extensively investigated over noble and transition metals (Pt, Pd, Rh, Ni, Co, etc.) and several oxide supports (Al_2O_3, CeO_2, MgO, ZrO_2, zeolite, etc.) [16–23], so as to develop excellent catalysts to obtain hydrogen as high yield as possible together with high resistance of coke deposition. Transition metals (especially Ni-based) catalysts, which have the high C–C and C–H bonds breaking activity, have been proved to be very effective for hydrocarbons steam reforming reactions as noble metal catalysts [16–18]. Moreover, the lower cost improved its applicability. Therefore, more and more researchers focused on studying hydrocarbon steam reforming over Ni-based catalysts [16–21]. However, coking is easily deposited on the surface of the active phase Ni, which can lower the catalytic activity [24–27]. Therefore, various promoters were introduced into Ni-based catalysts to improve catalytic activity and coking resistance. Lanthanide metals (La, Ce), alkali metals (Na, K), and alkali earth metals (Mg, Ca, Sr, Ba) promoters [28–33], have been found to be effective for improving coking-resistant capacity. However, the addition of these additives influenced Ni dispersion, due to a part of the promoter is in an intimate contact with nickel [34–36].

In order to improve the anti-coking ability of Ni-based catalyst, and have a slight influence on catalytic activity, many scholars introduced another active metal into Ni-based catalyst to form bi-metallic catalysts [37–42]. Wang et al. [37] introduced Pd into Ni-alumina catalysts, the catalytic activity and stability was obviously improved. Vizcaino et al. [43] found that Cu modified Ni-based catalyst showed better anti-coking ability. The addition of Cu is helpful for the process of eliminating the deposited carbon. In our previous work [44], we have added M (Fe, Co, Cu, Zn) as a promoter into the $Ni/Ce-Al_2O_3$ catalyst in order to improve the anti-coking ability. Clearly, Co doped $Ni/Ce-Al_2O_3$ showed an excellent coking-resistant effect. But, the catalytic activity have a slightly reduction at high temperature (650~800 °C). In another study [45], we added Co as another active species into $Ni/Ce-Al_2O_3$ to form Ni-Co bi-metallic catalyst and investigated the catalytic activity, stability and coking inhibition effect during *n*-decane reforming. The results showed that the introduction of Ni and Co synchronously can effectively suppress carbon deposition and obviously improve catalytic activity. There was obvious synergistic effect between Ni and Co. However, the difference among different Co content on the $Co-Ni/Ce-Al_2O_3$ bi-metallic catalyst has not been discussed. Consequently, it would be valuable to investigate the influence of the content of Ni on *n*-decane steam reforming.

In this paper, the steam reforming experiments of *n*-decane over $x\%Co-Ni/Ce-Al_2O_3$ catalysts with different Co loading were carried out. The effect of different Co loading on catalytic activity and the amount of deposited carbon were discussed. The purpose of this work is screening the suitable catalysts for steam reforming process in order to maximize *n*-decane conversion and H_2 yield, and minimize the formation of byproducts and carbon deposition. This work provided some positive suggestions for catalysts preparation and optimization by studying the structure-activity correlations.

2. Results and Discussion

2.1. Catalytic Performance

2.1.1. *n*-Decane Conversion and H_2 Selectivity

n-decane steam reforming is used as the probe reaction. The initial activity tests over the series catalysts were performed from 650 to 800 °C in order to examine the influence of temperature and different promoters on catalytic performance. *n*-decane conversion and H_2 selectivity are considered the main parameters to check the advantages and disadvantages of the catalysts, and the results are shown in Figure 1a–d. The catalytic activities over these catalysts gradually increase with the temperature. Obviously, the presence of Ni or/and Co can effectively promote the rate of the steam reforming reaction, as well as the selectivity of H_2 and *n*-decane conversion. Moreover, the synchronous introduction of Co and Ni further enhanced the catalytic activity compared with the 6%Ni/Ce-Al$_2$O$_3$(NCA) catalyst. This demonstrates that the addition of Co could provide sufficient Ni active sites for the reactants. In addition, the catalytic activity of *x*%Co-Ni/Ce-Al$_2$O$_3$ (CNCA) bi-metallic catalysts with different Co content increases firstly and then decreases with Co addition. The catalytic activity reaches the best when the Co content is 12%. This indicated that moderate Co is favor for promoting the activity. There is a synergistic effect between Co and Ni.

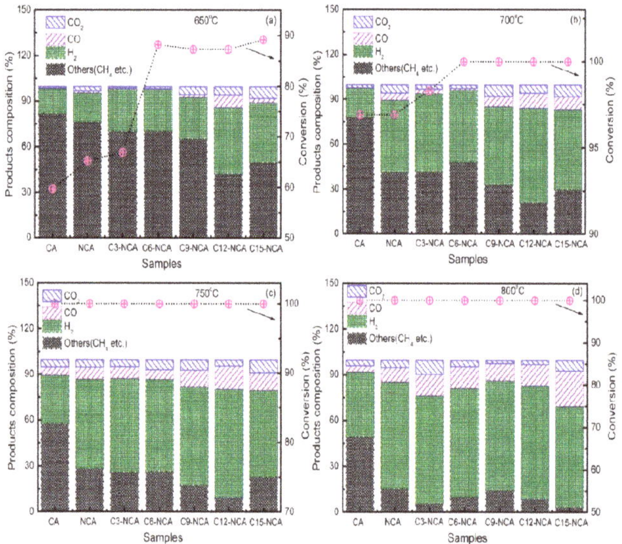

Figure 1. *n*-decane conversions and H_2 selectivity over the series catalysts at 650 °C (**a**), 700 °C (**b**), 750 °C (**c**), and 800 °C (**d**).

2.1.2. Thermal Stability and Regeneration of C12-NCA

To better understand the effect of the ordered co-modification in *n*-decane steam reforming, the 12%Co-Ni/Ce-Al$_2$O$_3$ (C12-NCA) catalyst was screened out with a 6 h stability test at 750 °C and 800 °C, and the results are displayed in Figure 2a. It can be seen in Figure 2a that the H_2 selectivity and *n*-decane conversion have a slightly change within 6 h. The C12-NCA catalyst has a good thermal stability. The C12-NCA catalyst also screened out for regeneration experiment. Carbon deposition on used C12-NCA catalyst was removed by oxygen enriched calcinations at 650 °C. The regenerative C12-NCA catalyst was carried out again in the same reactor as the fresh ones, and the contrast results are presented in Figure 2b. It is found that the *n*-decane conversions and H_2 selectivity over the reused-1(-2) C12-NCA catalyst are approximately equal to the results of fresh one. Therefore, the C12-NCA catalyst is renewable.

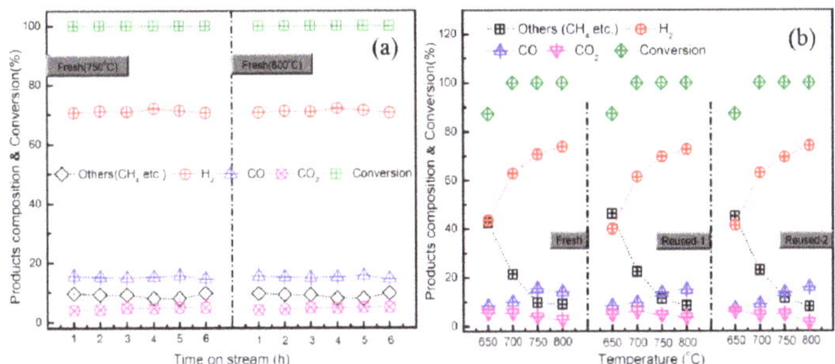

Figure 2. Thermal stability (**a**) and regeneration (**b**) over the C12-NCA catalyst.

2.2. Fresh Catalyst Characterization

2.2.1. N_2 Adsorption-Desorption Measurements

Table 1 shows the results of N_2 adsorption-desorption results of the fresh and used catalysts. The surface areas of different samples in this work are in the range of similar CA support, even if the introduction of Ni and Co species by impregnation method. The value has a slightly decrease with the addition of Co and Ni, and gradually drops with the increase of Co. The loss can be attributed to the fact that the internal surface area of the CA pore system is progressively covered by Ni, Co species forming a layer [45–47].

Table 1. The textural properties of the fresh and used catalysts.

Catalysts	Textural Properties		
	Surface Area *(m^2/g)	Pore Volume (mL/g)	Mean Pore Diameter (nm)
CA	155.9 (69.3) *	0.49	5.42
NCA	150.2 (83.6)	0.47	5.35
C3-NCA	149.3 (89.1)	0.47	5.32
C6-NCA	147.6 (88.6)	0.47	5.34
C9-NCA	143.9 (92.3)	0.45	5.28
C12-NCA	140.5 (96.7)	0.44	5.26
C15-NCA	136.9 (92.4)	0.44	5.26

* The numbers in the parentheses represent the surface area of used catalysts.

On the other hand, the surface areas of all catalysts used decrease with different levels after *n*-decane reforming reactions. CA and NCA catalysts decreased by 56% and 44% respectively compared with the fresh ones. Fortunately, the falling range gradually reduced with the addition of Co. Co as the active species showed a better carbon-resistant ability. The results are consistent with the results of the catalyst characterization.

2.2.2. X-ray Diffraction (XRD) Analysis

Figure 3 depicts the X-ray diffraction analysis of the fresh CA, NCA and bi-metallic CNCA catalysts. All the samples present similar characteristic features of γ-Al_2O_3 at $2\theta = 45.7°, 66.8°$; cubic fluorite structural CeO_2 at 2θ value of 28.5°, 56.3°; and the Ce crystallite at $2\theta = 34.7°, 49.8°$ and 59.2° by Bragg's refections [44,48]. For all the catalysts, there was no CoO_x, $CoAl_2O_4$, NiO, or $NiAl_2O_4$ diffraction peaks detected. This was probably due to the highly dispersion of CoO_x and NiO particles are not easy to be detected by XRD [49]. Moreover, the synchronous addition of Co and Ni did not form Ni-Co alloy phase. This indicated that active species were strongly interacted with CA support, and all of the as-prepared have a good thermal stability. It also can be seen that the degree of crystallization of all the fresh catalysts are smaller, suggesting that these catalysts are stable at high temperatures, which is coincides with the surface area analysis [50,51]. This is in agreement with the XRD analysis previously shown and with the observations made in other studies.

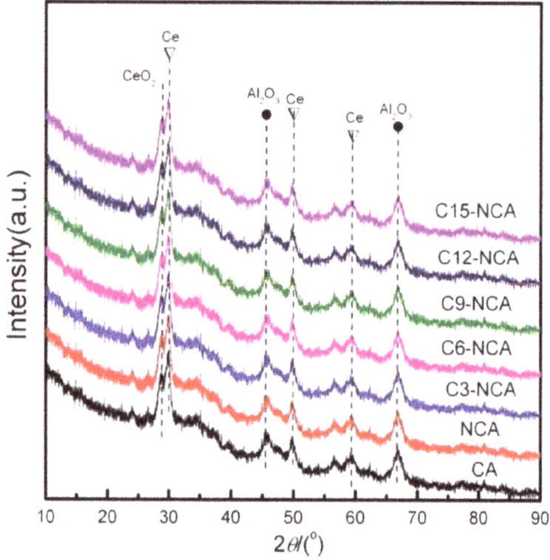

Figure 3. X-ray diffraction (XRD) diffraction spectrum of the series catalysts.

2.2.3. H_2-Temperature-Programmed Reduction (H_2-TPR) and NH_3-Temperature-Programmed Desorption (NH_3-TPD) Analysis

Figure 4 shows the reduction profiles of the CA, and Co, Ni modified CA. It can be seen that there is one or two H_2 consumption peaks for CA support at the region of 250~350 °C, which could be assigned to the reduction of a small amount of CeO_2 to CeO_x [28,29]. For NCA catalyst, reduction peaks of ~460 °C and ~823 °C which are attributed to the reduction of NiO and $NiAl_2O_4$ [40]. The highest reduction temperature is between 780 and 900 °C indicate the existence of species of NiO with strong interaction with Al_2O_3, resulting from the formation of the $NiAl_2O_4$ [42]. For CNCA catalysts with different Co loading, two reduction peaks around ~300 °C and ~610 °C are ascribed to the reduction of Co_2O_3 and CoO, respectively [41]. Significantly, the reduction temperatures differences of NiO and Co_2O_3 between these CNCA catalysts can be attributed to the existence of different interaction between Co and Ni. It is noteworthy that Co addition can obviously promote the reduction of NiO, and reach the optimal effect at 12% Co loading. There was obvious synergistic effect between Ni and Co, which is consistent with the results of the work reported by Jiao et al. [48,52–56].

NH_3-TPD technology was used to investigate the acidities of the series catalysts, such as total amount, nature, and strength distribution, in order to look for the possible interpretation for the above experimental results, and the profiles are shown in Figure 5. The area of desorption peak goes hand in hand with the total amount of surface acid sites, while the peak temperature is closely related to the strength of individual acid site. The peak temperature in the range of 80 to 200 °C is regarded as weak acid sites, and the desorption temperature between 200 to 400 °C is considered the medium acid sites, while the peak temperature locates at 400 to 700 °C corresponds to strong acid sites. Figure 5 shows that CA and NCA catalysts have three desorption peaks at the region of 100~500 °C, which is regarded as the desorption peak of the weak and medium acid. Moreover, all the Co doped NCA catalysts have one strong desorption peak at 100~400 °C. Obviously, the desorption peak temperature moves to a low temperature area by adding Co, suggesting that the amount of acid sites decrease with the introduction of Co. In our previous studies [12,18], we found that the larger acidity and active strong acid centers are easy to give rise to rapid deactivation of the catalyst due to carbon deposition. It is noteworthy that the addition of Co modifier increases the basicity of the NCA catalyst. This process will result in preventing alkenes further reacting into aromatic or heavier products, which is beneficial to reduce the carbon deposition over catalysts and prolong the work life of catalysts.

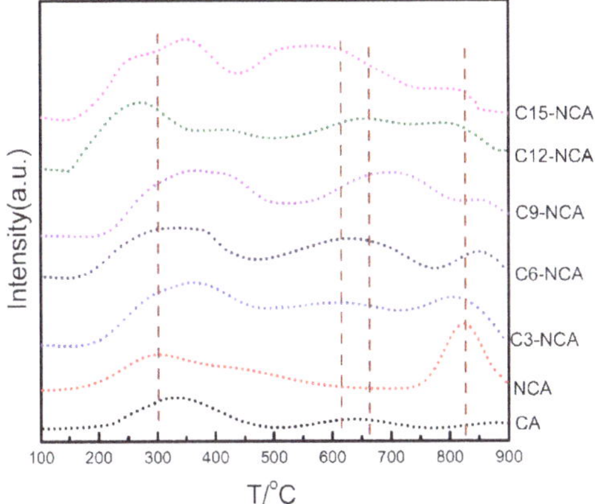

Figure 4. H_2-temperature-programmed reduction (H_2-TPR) results of the series catalysts.

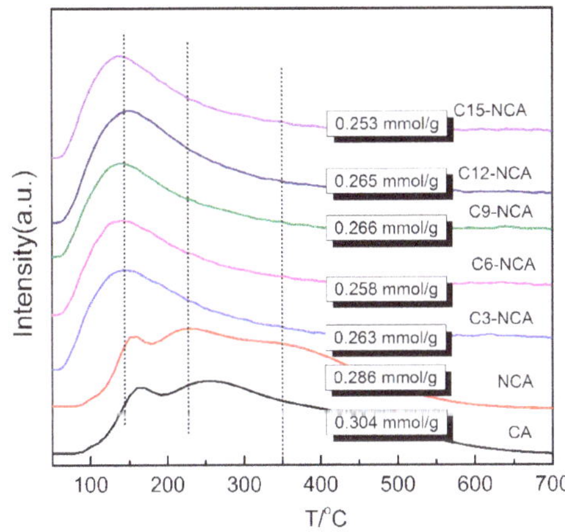

Figure 5. NH_3-temperature-programmed desorption (NH_3-TPD) results of the series catalysts.

2.2.4. Transmission Electron Microscope (TEM) Analysis

In this work, the catalytic activity over the C12-NCA catalyst is better than C15-NCA. However, there is somewhat different texture and structural properties between the two. In order to further study the difference between C12-NCA and C15-NCA, the TEM analysis was used to find the influence of microscopic appearance and dispersion on catalytic activity. Figure 6 shows TEM images and Ni or/and Co particle size distributions of C12-NCA and C15-NCA catalysts. It is found that the Ni or/and Co particle size over C12-NCA is mainly focused on 11–20 nm, while the value for C15-NCA is about 16–25 nm. The average particle size of Ni or/and Co of C15-NCA is significantly larger than C12-NCA. Obviously, poor Ni or/and Co distribution over C15-NCA are observed. It may be the reason of the weaker catalytic activity of C15-NCA catalyst. The results indicate that 6% Ni and 15% Co loading is easy to aggregate after 800 °C calcination, which may be due to redundant Co enriching on the catalyst surface and exceed the monolayer saturation capacity of the CA support.

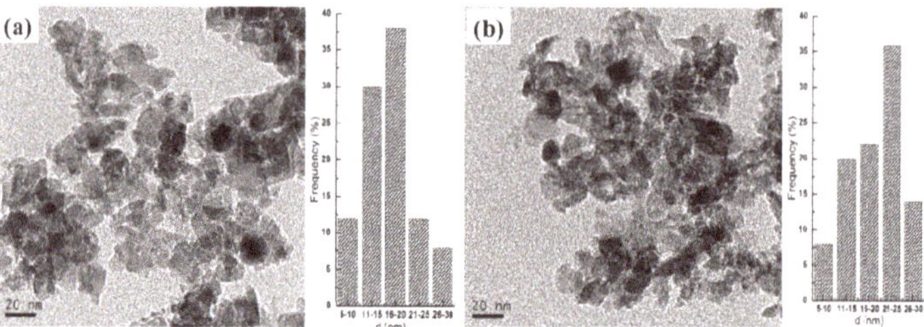

Figure 6. Transmission electron microscope (TEM) micrographs of the C12-NCA (**a**) and C15-NCA (**b**) catalysts.

In order to further understand the Ni dispersion, the energy dispersive spectrometer (EDS) mappings of C12-NCA and C15-NCA catalysts were examined. As shown in Figure 7, the red sections correspond to Co, the blue sections correspond to Ce, the green sections correspond to Ni, while the yellow sections correspond to Al. Obviously, Ni and Co particles evenly dispersed on the surface of the CA support over C12-NCA. However, some regions overlap each other for C15-NCA. This may be caused by the partial sintering of 6% Ni and 15% Co loading, and form bigger crystallite and lower metal dispersion. Poor nickel dispersion not only notably impact the physiochemical properties of the catalyst, but also impact the catalytic performance in steam reforming of reactions. Therefore, the dispersion state of active species is one of the main reasons of the activity difference among these catalysts.

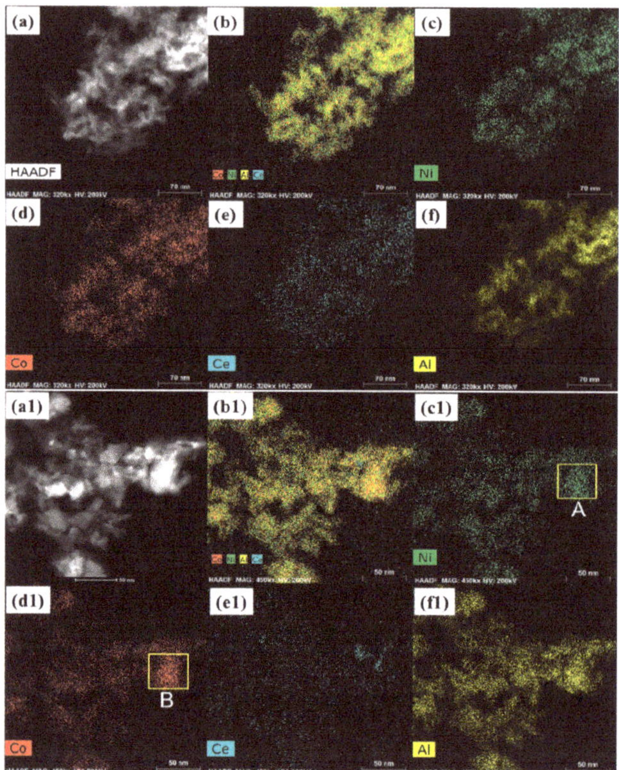

Figure 7. Energy dispersive spectrometer (EDS) mappings of the C12-NCA and C15-NCA catalysts. (**a–f**) belongs for C12-NCA catalyst; (**a1–f1**) belongs for C15-NCA catalyst.

2.3. Carbon Deposition

Table 2 presented the amount of carbon deposition over the series catalysts determined by the temperature programmed oxidation (TPO)-infrared spectrum (IR) analysis. It can be seen that the amount of carbon deposition is varied with the introduction of Ni and Co. Clearly, the amount of carbon deposition over CA catalyst is larger than others, and the values drop off with the Co content increased. For the bi-metallic C12(C15)-NCA catalyst, the amount of carbon deposition (<4.0 mg/g-fuel) is much less than conventional Ni-alumina catalysts. The possible reasons are that the addition of Co reduces the acidity of the CA catalyst and carbon dioxide adsorption, and decreases carbon deposition generation during *n*-decane steam reforming process [48]. According to the literature [57], Co addition can increase the dispersion state of Ni, prevent metal sintering, and inhibit carbon deposition, which is consistent with our results in this work.

Table 2. The amount of carbon deposition over used catalysts.

Catalysts	Carbon Deposition	
	[mg/(g$_{Cat}$·h)]	[mg/gC$_{Feed}$]
CA	258.9	11.6
NCA	194.3	8.55
C3-NCA	178.8	7.87
C6-NCA	143.2	6.30
C9-NCA	114.1	5.02
C12-NCA	88.6	3.90
C15-NCA	75.0	3.30

2.4. Used Catalyst Characterization

The formation of carbon deposition on the used catalyst was also evidenced by SEM observations. In Figure 8a–g, the micrographs of used catalysts are presented. Carbon deposition grows on the surface of catalysts in different range. It is observed from Figure 8h that the Ni species is occupied at the tip of the filamentous carbon and the Ni metal particle size is found to be ca. 30 nm. This can demonstrate that active metal has removed from the surface of catalyst with the generation of filamentous carbon. The results are consistent with the mechanisms of filamentous carbon growth in literature [58]. In addition, the C content over the used CA, NCA, C3-NCA, C6-NCA, C9-NCA, C12-NCA, C15-NCA catalysts are 91.14%, 63.90%, 56.46%, 49.89%, 34.79%, 14.94% and 8.31%, respectively. It clearly indicates that the C content over the used catalysts has a close relationship with the surface area. Obviously, the Brunauer–Emmett–Teller (BET) specific surface area decrease caused by the formation of the carbon deposition is another main reason of the activity difference among these catalysts.

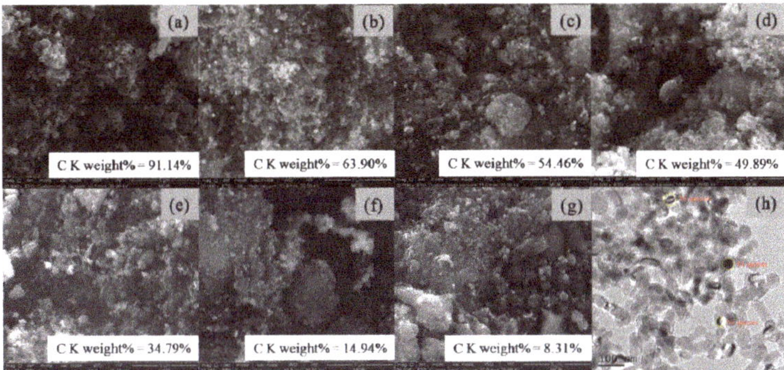

Figure 8. Field emission scanning electron microscopy (FESEM) micrographs of the used catalysts: CA (**a**), NCA (**b**), C3-NCA (**c**), C6-NCA (**d**), C9-NCA (**e**), C12-NCA (**f**), C15-NCA (**g**) and C15-NCA (**h**).

3. Experimental Section

3.1. Catalysts Preparation

Ce-Al$_2$O$_3$ composite oxides (5 wt.% Ce) support was prepared by coprecipitation method, and the complete experimental procedure is detailed in references [8,44]. The catalysts were prepared by incipient wetness impregnation using Ni(NO$_3$)$_2$·6H$_2$O and Co(NO$_3$)$_2$·6H$_2$O as the precursors of Ni and Co, respectively. After impregnation and drying (120 °C, 3 h), the samples were calcined in air at 800 °C for 4 h. Finally, the mono-metallic Ni catalyst (Ni/Ce-Al$_2$O$_3$, 6 wt.% Ni), and Co-Ni bi-metallic catalysts {Co-Ni/Ce-Al$_2$O$_3$, 6 wt.% Ni, x wt.% Co (x = 3, 6, 9, 12, 15)} were made into columnar particles (length of 5 mm, diameter of 5 mm), and marked as NCA, C3-NCA, C6-NCA, C9-NCA, C12-NCA and C15-NCA, respectively.

3.2. Catalysts Characterization

Surface area and pore volume were obtained from N$_2$ adsorption-desorption isotherms at 77 K performed on a Quanta chrome autosorb-1 analyzer. Before analysis, the samples were dried at room temperature and then vacuum degassed at 300 °C for 5 h. X-ray diffraction (XRD) patterns were obtained on a Bruker D8 Advance diffractometer with a copper Kα radiation source operated at 40 kV and 40 mA. During the analysis, the catalysts were scanned from 20° to 80° at a speed of 5 °/min. Furthermore, temperature-programmed reduction (TPR) profiles of fresh catalysts under H$_2$-blanket were carried out using a TP-5076 TPR instrument. The samples (100 mg) were heated to 400 °C at a heating rate of 8 °C/min in a flow of Ar, kept for 45 min, and then cooled to 30 °C. The flow gas was switched to H$_2$ (5%) in Ar (25 mL/min). The reduction was carried out from 30 °C to 920 °C, at a heating rate of 8 °C/min. NH$_3$-temperature-programmed desorption (NH$_3$-TPD) experiments were performed in a TP-5076 TPD instrument (Tian Jin Xian Quan Instrument Co., Ltd., Tianjin, China) equipped with a thermal conductivity detector (TCD) to investigate the surface acidity of catalysts. Prior to analysis, catalysts were pretreated at 400 °C for 1 h in a flow of N$_2$ to clean the catalyst surface and then cooled to room temperature. After the pretreatment, a 2% NH$_3$/N$_2$ gas mixture was passed through the sample at 20 mL/min and the temperature was raised from 50 to 750 °C at a heating rate of 8 °C/min. In addition, morphological characterization was examined by field emission scanning electron microscopy (FESEM) using a Quanta 600FEG Field emission scanning electron microscope (FEI, Hillsboro, OR, USA) and coupled with Oxford-IE-250 energy dispersive spectrometer (EDS) for local elemental composition determination.

Transmission electron microscope (TEM) analysis was obtained using a FEI titan themis 200 transmission electron microscope (FEI, Hillsboro, OR, USA) equipped with Bruker super-X energy-dispersive spectrometer (EDS). Prior to measurement, the samples were dispersed in ethanol at first and then collected on a Cu grid which was covered with carbon film.

3.3. Catalytic Performance Test

n-decane steam reforming reaction was carried out in a fixed-bed stainless steel tubular reactor (i.d. = 12 mm) at atmospheric pressure. 3.0 g catalyst diluted with equi-volume quartz particles was charged for each catalytic assessment. The schematic diagram was shown in Figure 9. In each run, water and *n*-decane were pre-mixed quantitatively at 350 °C and the mixture vapor was fed into the reactor. The volumetric feed flow of water and *n*-decane are both 2.5 mL/min. The reactor was set at 650 °C, 700 °C, 750 °C or 800 °C respectively for evaluating the catalysts' activity.

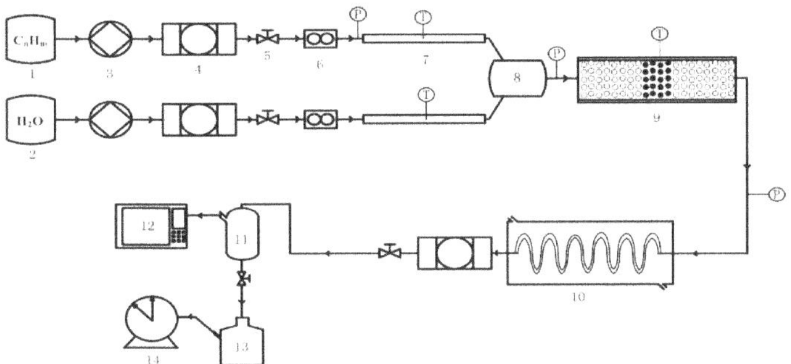

Figure 9. Schematic diagram of apparatus. 1-feed tank; 2-water tank; 3-high pressure metering pump; 4-filter; 5-check valve; 6-mass flow meter; 7-heating system; 8-mixer; 9-reactor; 10-cold trap; 11-gas-liquid separator; 12-gas chromatograph; 13-liquid receiver; 14-wet gas flow meter.

The reaction effluents were on-line analyzed by Gas Chromatography with FID detector (GC-2010, SHIMADZU Co., Ltd., Kyoto, Japan) equipped with HP-Al/S separation capillary column (CH_4, C_2H_4, C_2H_6, C_3H_8, C_3H_6, C_4); and TCD detector with TDX-01 packed column (H_2, CO and CO_2). Solid products deposited covering the catalysts were investigated by temperature programmed oxidation (TPO), and the characteristics signals of CO and CO_2 were tested by IR analyzer.

Activity data were reported as *n*-decane conversion X_A (%), and H_2, CO and CO_2 selectivity (S_i), which are defined as follows:

$$X_A(\%) = \frac{\Delta m_A}{m_{A,in}} = \frac{m_{A,in} - m_{A,out}}{m_{A,in}} \times 100 \tag{1}$$

$$S_i = \frac{moles\,P_i}{\sum_{i=1}^{n} moles\,P_i} \times 100\% \tag{2}$$

where *i* represents the different gas products detected.

4. Conclusions

Series Co-Ni bi-metallic catalysts supported on Ce-Al$_2$O$_3$ (CA) with different Co ratios are prepared for *n*-decane steam reforming towards H_2 preparation, so as to screen appropriate Co content. N_2 adsorption-desorption, XRD, NH_3-TPD, SEM-EDS, and TEM were used to explain the relationship between catalytic activity and textural-structure properties, microscopic appearance and dispersion state of the catalysts. The activity test indicated that the addition of Co resulted in higher selectivity of hydrogen compared with NCA contrast catalyst. The results indicated that 12% Co showed better hydrogen selectivity (74%), *n*-decane conversion (100%), and anti-carbon ability (3.90 mg/g-fuel). Moreover, thermal stability and regeneration performance were also tested, and the results showed that reforming activity over C12-NCA catalyst was stable in 6 h, and the hydrogen selectivity and *n*-decane conversion over regenerated catalysts were almost close to the fresh ones.

The results were mainly interpreted on the basis of characterization studies, which revealed the significant role of textural and structural properties, microscopic appearance and dispersion state on reforming performance. Obviously, the dispersion state and the BET-specific surface area decrease caused by the formation of the carbon deposition are the main reasons for the activity difference among these catalysts.

Author Contributions: Q.Y., Y.J. and J.L. conceived and designed the experiments; Q.Y. and Y.J. performed the experiments; W.W., Y.D. and C.L. contributed with catalyst characterization; J.Y. analyzed the data; J.L. administrated the project; Q.Y. wrote original draft; Q.Y. and Y.J. reviewed and modified the manuscript.

Funding: The research was funded by the special program for Key Basic Research in China (Grant No. 0040202204) and the Key Research& Development Plan in Shaanxi Province (2017ZDXM-GY-042, 2017ZDXM-GY-070).

Conflicts of Interest: The authors declare no conflict of interest.

References

1. Cheekatamarla, P.K.; Finnerty, C.M. Reforming catalysts for hydrogen generation in fuel cell applications. *J. Power Sources* **2006**, *160*, 490–499. [CrossRef]
2. Kaewpanha, M.; Guan, G.; Hao, X.; Wang, Z.; Kasai, Y.; Kakuta, S.; Kusakabe, K.; Abudula, A. Steam reforming of tar derived from the steam pyrolysis of biomass over metal catalyst supported on zeolite. *J. Taiwan Inst. Chem. Eng.* **2013**, *44*, 1022–1026. [CrossRef]
3. Liu, X.; Yang, X.; Liu, C.; Chen, P.; Yue, X.; Zhang, S. Low-temperature catalytic steam reforming of toluene over activated carbon supported nickel catalysts. *J. Taiwan Inst. Chem. Eng.* **2016**, *65*, 233–241. [CrossRef]
4. Yang, R.X.; Chuang, K.H.; Wey, M.Y. Hydrogen production through methanol steam reforming: Effect of synthesis parameters on Ni-Cu/CaO-SiO$_2$ catalysts activity. *Int. J. Hydrog. Energy* **2014**, *39*, 19494–19501. [CrossRef]
5. Oararteta, L.; Remiro, A.; Aguayo, A.T.; Bilbao, J.; Gayubo, A.G. Effect of operating conditions on dimethyl ether steam reforming over a CuFe$_2$O$_4$/γ-Al$_2$O$_3$ bifunctional catalyst. *Ind. Eng. Chem. Res.* **2015**, *54*, 9722–9732. [CrossRef]
6. Dou, B.; Song, Y.; Wang, C.; Chen, H.; Xu, Y. Hydrogen production from catalytic steam reforming of biodiesel byproduct glycerol: Issues and challenges. *Renew. Sust. Energy Rev.* **2014**, *30*, 950–960. [CrossRef]
7. Chen, H.; Ding, Y.; Cong, N.T.; Dou, B.; Dupont, V.; Ghadiri, M.; Williams, P.T. Progress in low temperature hydrogen production with simultaneous CO$_2$ abatement. *Chem. Eng. Res. Des.* **2011**, *89*, 1774–1782. [CrossRef]
8. Jiao, Y.; Sun, D.; Zhang, J.; Du, Y.; Kang, J.; Li, C.; Lu, J.; Wang, J.; Chen, Y. Steam reforming of *n*-decane toward H$_2$ production over Ni/Ce-Al$_2$O$_3$ composite catalysts: Effects of M (M = Fe, Co, Cu, Zn) promoters. *J. Anal. Appl. Pyrol.* **2016**, *120*, 238–246. [CrossRef]
9. Dan, M.; Mihet, M.; Tasnadi-Asztalos, Z.; Imre-Lucaci, A.; Katona, G.; Lazar, M.D. Hydrogen Production by Ethanol Steam Reforming on Nickel Catalysts: Effect of Support Modification by CeO$_2$ and La$_2$O$_3$. *Fuel* **2015**, *147*, 260–268. [CrossRef]
10. Bambal, A.S.; Vecchio, K.S.; Cattolica, R.J. Catalytic effect of Ni and Fe addition to gasifier bed material in the steam reforming of producer gas. *Ind. Eng. Chem. Res.* **2014**, *53*, 13656–13666. [CrossRef]
11. Sangjun, Y.; Youngchan, C.; Jaegoo, L. Hydrogen production from biomass tar by catalytic steam reforming. *Energy Convers. Manag.* **2010**, *51*, 42–47.
12. Jiao, Y.; Du, Y.; Zhang, J.; Li, C.; Xue, Y.; Lu, J.; Wang, J.; Chen, Y. Steam reforming of *n*-decane for H$_2$ production over Ni modified La-Al$_2$O$_3$ catalysts: Effects of the active component Ni content. *J. Anal. Appl. Pyrol.* **2015**, *116*, 58–67. [CrossRef]
13. Dieuzeide, M.L.; Laborde, M.; Amadeo, N.; Cannilla, C.; Bonura, G.; Frusteri, F. Hydrogen production by glycerol steam reforming: How Mg doping affects the catalytic behaviour of Ni/Al$_2$O$_3$ catalysts. *Int. J. Hydrog. Energy* **2016**, *41*, 157–166. [CrossRef]
14. Sun, L.Z.; Tan, Y.S.; Zhang, Q.D.; Xie, H.J.; Song, F.E.; Sun, Y.Z. Effects of Y$_2$O$_3$-modification to Ni/γ-Al$_2$O$_3$ catalysts on auto thermal reforming of methane with CO$_2$ to syngas. *Int. J. Hydrog. Energy* **2013**, *38*, 1892–1900. [CrossRef]
15. Quitete, C.P.B.; Bittencourt, R.C.P.; Souza, M.M.V.M. Coking resistance evaluation of tar removal catalysts. *Catal. Commun.* **2015**, *71*, 79–83. [CrossRef]
16. Rossetti, I.; Biffi, C.; Bianchi, C.L.; Nichele, V.; Signoretto, M.; Menegazzo, F.; Finocchio, E.; Ramis, G.; Di Michele, A. Ni/SiO$_2$ and Ni/ZrO$_2$ catalysts for the steam reforming of ethanol. *Appl. Catal. B Environ.* **2012**, *117*, 384–396. [CrossRef]
17. Mondal, T.; Pant, K.K.; Dalai, A.K. Catalytic oxidative steam reforming of bio-ethanol for hydrogen production over Rh promoted Ni/CeO$_2$-ZrO$_2$ catalyst. *Int. J. Hydrog. Energy* **2015**, *40*, 2529–2544. [CrossRef]
18. Jiao, Y.; Zhang, J.; Du, Y.; Li, F.; Li, C.; Lu, J.; Wang, J.; Chen, Y. Hydrogen production by catalytic steam reforming of hydrocarbon fuels over Ni/Ce-Al$_2$O$_3$, bifunctional catalysts: Effects of SrO addition. *Int. J. Hydrog. Energy* **2016**, *41*, 13436–13447. [CrossRef]
19. Bizkarra, K.; Barrio, V.L.; Yartu, A.; Requies, J.; Arias, P.L.; Cambra, J.F. Hydrogen production from *n*-butanol over alumina and modified alumina nickel catalysts. *Int. J. Hydrog. Energy* **2015**, *40*, 5272–5280. [CrossRef]

20. Souza, G.D.; Marcilio, N.R.; Perezlopez, O.W. Dry reforming of methane at moderate temperatures over modified Co-Al Co-precipitated catalysts. *Mater. Res.* **2014**, *17*, 1047–1055. [CrossRef]
21. Huang, L.; Chong, C.; Chen, L.; Wang, Z.; Zhong, Z.; Campos-Cuerva, C.; Lin, J. Monometallic carbonyl-derived CeO_2-supported Rh and Co bicomponent catalysts for CO-Free, high-yield H_2 generation from low-temperature ethanol steam reforming. *ChemCatChem* **2013**, *5*, 220–234. [CrossRef]
22. Liguras, D.K.; Kondarides, D.I.; Verykios, X.E. Production of hydrogen for fuel cells by steam reforming of ethanol over supported noble metal catalysts. *Appl. Catal. B Environ.* **2003**, *43*, 345–354. [CrossRef]
23. Cavallaro, S. Ethanol steam reforming on Rh/Al_2O_3 catalysts. *Energy Fuels* **2000**, *14*, 1195–1199. [CrossRef]
24. Wang, Z.; Wang, C.; Chen, S.; Liu, Y. Co-Ni bimetal catalyst supported on perovskite-type oxide for steam reforming of ethanol to produce hydrogen. *Int. J. Hydrog. Energy* **2014**, *39*, 5644–5652. [CrossRef]
25. De Lima, S.M.; da Cruz, I.O.; Jacobs, G.; Davis, B.H.; Mattos, L.V.; Noronha, F.B. Study of catalyst deactivation and reaction mechanism of steam reforming, partial oxidation, and oxidative steam reforming of ethanol over Co/CeO_2 catalyst. *J. Catal.* **2009**, *268*, 268–281. [CrossRef]
26. Frusteri, F.; Freni, S.; Spadaro, L.; Chiodo, V.; Bonura, G.; Donato, S.; Cavallaro, S. H_2 production for MC fuel cell by steam reforming of ethanol over MgO supported Pd, Rh, Ni and Co catalysts. *Catal. Commun.* **2004**, *5*, 611–615. [CrossRef]
27. Freni, S.; Cavallaro, S.; Mondello, N.; Spadaro, L.; Frusteri, F. Production of hydrogen for MC fuel cell by steam reforming of ethanol over MgO supported Ni and Co catalysts. *Catal. Commun.* **2003**, *4*, 259–268. [CrossRef]
28. Ashok, J.; Kawi, S. Steam reforming of toluene as a biomass tar model compound over CeO_2 promoted Ni/CaO-Al_2O_3 catalytic system. *Int. J. Hydrog. Energy* **2013**, *38*, 13938–13949. [CrossRef]
29. Tao, J.; Zhao, L.Q.; Dong, C.Q.; Qiang, L.; Du, X.Z.; Dahlquist, E. Catalytic steam reforming of toluene as a model biomass gasification tar compound using Ni-CeO_2/SBA-15 catalysts. *Energies* **2013**, *6*, 3284–3296. [CrossRef]
30. Guan, G.; Chen, G.; Kasai, Y.; Lim, E.W.C.; Hao, X.; Kaewpanha, M.; Abuliti, A.; Fushimi, C.; Tsutsumi, A. Catalytic steam reforming of biomass tar over iron- or nickel-based catalyst supported on calcined scallop shell. *Appl. Catal. B Environ.* **2012**, *115–116*, 159–168. [CrossRef]
31. Blanco, P.H.; Wu, C.; Onwudili, J.A.; Williams, P.T. Characterization and evaluation of Ni/SiO_2 catalysts for hydrogen production and tar reduction from catalytic steam pyrolysis-reforming of refuse-derived fuel. *Appl. Catal. B Environ.* **2013**, *134–135*, 238–250. [CrossRef]
32. Garbarino, G.; Finocchio, E.; Lagazzo, A.; Valsamakis, I.; Riani, P.; Escribano, V.S.; Busca, G. Steam reforming of ethanol-phenol mixture on Ni/Al_2O_3: Effect of magnesium and boron on catalytic activity in the presence and absence of sulphur. *Appl. Catal. B Environ.* **2014**, *147*, 813–826. [CrossRef]
33. Koike, M.; Ishikawa, C.; Li, D.; Wang, L.; Nakagawa, Y.; Tomishige, K. Catalytic performance of manganese-promoted nickel catalysts for the steam reforming of tar from biomass pyrolysis to synthesis gas. *Fuel* **2013**, *103*, 122–129. [CrossRef]
34. Chen, X.; Liu, Y.; Niu, G.; Yang, Z.; Bian, M.; He, A. High temperature thermal stabilization of alumina modified by lanthanum species. *Appl. Catal. A Gen.* **2001**, *205*, 159–172. [CrossRef]
35. Oudet, F.; Courtine, P.; Vejux, A. Thermal stabilization of transition alumina by structural coherence with $LnAlO_3$ (Ln = La, Pr, Nd). *J. Catal.* **1988**, *114*, 112–120. [CrossRef]
36. Borowiecki, T.; Denis, A.; Rawski, M.; Gołębiowski, A.; Stołecki, K.; Dmytrzyk, J.; Kotarba, A. Studies of potassium-promoted nickel catalysts for methane steam reforming: Effect of surface potassium location. *Appl. Surf. Sci.* **2014**, *300*, 191–200. [CrossRef]
37. Wang, Y.H.; Zhang, J.C. Hydrogen production on Ni-Pd-Ce/γ-Al_2O_3 catalyst by partial oxidation and steam reforming of hydrocarbons for potential application in fuel cells. *Fuel* **2005**, *84*, 1926–1932. [CrossRef]
38. Fauteux-Lefebvre, C.; Abatzoglou, N.; Braidy, N.; Achouri, I.E. Diesel steam reforming with a nickel-alumina spinel catalyst for solid oxide fuel cell application. *J. Power Sources* **2011**, *196*, 7673–7680. [CrossRef]
39. Al-Musa, A.A.; Ioakeimidis, Z.S.; Al-Saleh, M.S.; Al-Zahrany, A.; Marnellos, G.E.; Konsolakis, M. Steam reforming of iso-octane toward hydrogen production over mono- and bi-metallic Cu-Co/CeO_2 Catalysts: Structure-activity correlations. *Int. J. Hydrog. Energy* **2014**, *39*, 19541–19554. [CrossRef]
40. Ramasamy, K.K.; Gray, M.; Job, H.; Wang, Y. Direct syngas hydrogenation over a Co-Ni bimetallic catalyst: Process parameter optimization. *Chem. Eng. Sci.* **2015**, *135*, 266–273. [CrossRef]

41. Laosiripojana, N.; Sutthisripok, W.; Charojrochkul, S.; Assabumrungrat, S. Development of Ni-Fe bimetallic based catalysts for biomass tar cracking/reforming: Effects of catalyst support and Co-fed reactants on tar conversion characteristics. *Fuel Process. Technol.* **2014**, *127*, 26–32. [CrossRef]

42. Dave, C.D.; Pant, K.K. Renewable hydrogen generation by steam reforming of glycerol over zirconia promoted ceria supported catalyst. *Renew Energy* **2011**, *36*, 3195–3202. [CrossRef]

43. Vizcaíno, A.J.; Carrero, A.; Calles, J.A. Hydrogen production by ethanol steam reforming over Cu-Ni supported catalysts. *Int. J. Hydrog. Energy* **2007**, *32*, 1450–1461. [CrossRef]

44. Jiao, Y.; Zhang, J.; Du, Y.; Sun, D.; Wang, J.; Chen, Y.; Lu, J. Steam reforming of hydrocarbon fuels over M (Fe, Co, Ni, Cu, Zn)-Ce bimetal catalysts supported on Al_2O_3. *Int. J. Hydrog. Energy* **2016**, *41*, 10473–10482. [CrossRef]

45. Goula, M.A.; Charisiou, N.D.; Papageridis, K.N.; Delimitis, A.; Pachatouridou, E.; Iliopoulou, E.F. Nickel on alumina catalysts for the production of hydrogen rich mixtures via the biogas dry reforming reaction: Influence of the synthesis method. *Int. J. Hydrog. Energy* **2015**, *40*, 9183–9200. [CrossRef]

46. Bereketidou, O.A.; Goula, M.A. Biogas reforming for syngas production over nickel supported on ceria-alumina catalysts. *Catal. Today* **2012**, *195*, 93–100. [CrossRef]

47. Papageridis, K.N.; Siakavelas, G.; Charisiou, N.D.; Avraam, D.G.; Tzounis, L.; Kousi, K.; Goula, M.A. Comparative study of Ni, Co, Cu supported on γ-alumina catalysts for hydrogen production via the glycerol steam reforming reaction. *Fuel Process. Technol.* **2016**, *152*, 156–175. [CrossRef]

48. Jiao, Y.; He, Z.; Wang, J.; Chen, Y. n-decane steam reforming for hydrogen production over mono- and bi-metallic Co-Ni/Ce-Al_2O_3 catalysts: Structure-activity correlations. *Energy Convers. Manag.* **2017**, *148*, 954–962. [CrossRef]

49. Cassinelli, W.H.; Feio, L.S.F.; Araújo, J.C.S.; Hori, C.E.; Noronha, F.B.; Marques, C.M.P.; Bueno, J.M.C. Effect of CeO_2 and La_2O_3 on the activity of CeO_2-La_2O_3/Al_2O_3-supported Pd catalysts for steam reforming of methane. *Catal. Lett.* **2008**, *120*, 86–94. [CrossRef]

50. Peña, J.A.; Herguido, J.; Guimon, C.; Monzón, A.; Santamarí, J. Hydrogenation of acetylene over Ni/$NiAl_2O_4$ catalyst: Characterization, coking, and reaction studies. *J. Catal.* **1996**, *159*, 313–322. [CrossRef]

51. Boukha, Z.; Jiménez-González, C.; Rivas, B.D.; González-Velasco, J.R.; Gutiérrez-Ortiz, J.I.; López-Fonseca, R. Synthesis, characterisation and performance evaluation of spinel-derived Ni/Al_2O_3 catalysts for various methane reforming reactions. *Appl. Catal. B Environ.* **2014**, *158–159*, 190–201. [CrossRef]

52. Nandini, A.; Pant, K.K.; Dhingra, S.C. CeO_2 and Mn-Promoted Ni/Al_2O_3 Catalysts for Stable CO_2 Reforming of Methane. *Appl. Catal. A Gen.* **2005**, *290*, 166–174. [CrossRef]

53. Xu, G.; Shi, K.; Gao, Y.; Xu, H.; Wei, Y. Studies of Reforming Natural Gas with Carbon Dioxide to Produce Synthesis Gas: The Role of CeO_2 and MgO Promoters. *J. Mol. Catal. A Chem.* **1999**, *147*, 47–54. [CrossRef]

54. Siew, K.W.; Lee, H.C.; Gimbun, J.; Cheng, C.K. Production of CO-Rich Hydrogen Gas from Glycerol Gry Reforming over La-promoted Ni/Al_2O_3 Catalyst. *Int. J. Hydrog. Energy* **2014**, *39*, 6927–6936. [CrossRef]

55. Zeng, S.H.; Zhang, L.; Zhang, X.H.; Wang, Y.; Pan, H.; Su, H.Q. Modification Effect of Natural Mixed Rare Earths on Co/g-Al_2O_3 catalysts for CH_4/CO_2 Reforming to Synthesis Gas. *Int. J. Hydrog. Energy* **2012**, *37*, 9994–10001. [CrossRef]

56. Djaidja, A.; Libs, S.; Kiennemann, A.; Barama, A. Characterization and Activity in Dry Reforming of Methane on NiMg/Al and Ni/MgO Catalysts. *Catal. Today* **2006**, *113*, 194–200. [CrossRef]

57. Resini, C.; Delgado, M.C.H.; Presto, S.; Alemany, L.J.; Riani, P.; Marazza, R.; Ramis, G.; Busca, G. Yttria-stabilized zirconia (YSZ) supported Ni-Co alloys (precursor of SOFC anodes) as catalysts for the steam reforming of ethanol. *Int. J. Hydrog. Energy* **2008**, *33*, 3728–3735. [CrossRef]

58. Fenelonov, V.B.; Derevyankin, A.Y.; Okkel, L.G.; Avdeeva, L.B.; Zaikovskii, V.I.; Moroz, E.M.; Salanov, A.N.; Rudina, N.A.; Likholobov, V.A.; Shaikhutdinov, S.K. Structure and texture of filamentous carbons produced by methane decomposition on Ni and Ni-Cu catalysts. *Carbon* **1997**, *35*, 1129–1140. [CrossRef]

Article

In Situ Regeneration of Alumina-Supported Cobalt–Iron Catalysts for Hydrogen Production by Catalytic Methane Decomposition

Anis H. Fakeeha [1], Siham Barama [2], Ahmed A. Ibrahim [1,*], Raja-Lafi Al-Otaibi [3], Akila Barama [2], Ahmed E. Abasaeed [1] and Ahmed S. Al-Fatesh [1,*]

[1] Chemical Engineering Department, College of Engineering, King Saud University, P.O. Box 800, Riyadh 11421, Saudi Arabia; anishf@ksu.edu.sa (A.H.F.); abasaeed@ksu.edu.sa (A.E.A.)

[2] Laboratoire Matériaux Catalytiques et Catalyseen Chimie Organique (LMCCCO), USTHB BP.32 El Alia, Bab Ezzouar, Algiers 16111, Algeria; siham_barama@yahoo.com (S.B.); a_barama@yahoo.fr (A.B.)

[3] King Abdulaziz City for Science and Technology, P.O. Box 6086, Riyadh 11421, Saudi Arabia; raletabi@kacst.edu.sa

* Correspondence: aidid@ksu.edu.sa (A.A.I.); aalfatesh@ksu.edu.sa (A.S.A.-F.); Tel.: +966-11-467-6859 (A.S.A.-F.)

Received: 1 October 2018; Accepted: 16 November 2018; Published: 21 November 2018

Abstract: A novel approach to the in situ regeneration of a spent alumina-supported cobalt–iron catalyst for catalytic methane decomposition is reported in this work. The spent catalyst was obtained after testing fresh catalyst in catalytic methane decomposition reaction during 90 min. The regeneration evaluated the effect of forced periodic cycling; the cycles of regeneration were performed in situ at 700 °C under diluted O_2 gasifying agent (10% O_2/N_2), followed by inert treatment under N_2. The obtained regenerated catalysts at different cycles were tested again in catalytic methane decomposition reaction. Fresh, spent, and spent/regenerated materials were characterized using X-ray powder diffraction (XRD), transmission electron microscopy (TEM), laser Raman spectroscopy (LRS), N_2-physisorption, H_2-temperature programmed reduction (H_2-TPR), thermogravimetric analysis (TGA), and atomic absorption spectroscopy (AAS). The comparison of transmission electron microscope and X-ray powder diffraction characterizations of spent and spent/regenerated catalysts showed the formation of a significant amount of carbon on the surface with a densification of catalyst particles after each catalytic methane decomposition reaction preceded by regeneration. The activity results confirm that the methane decomposition after regeneration cycles leads to a permanent deactivation of catalysts certainly provoked by the coke deposition. Indeed, it is likely that some active iron sites cannot be regenerated totally despite the forced periodic cycling.

Keywords: carbon; combined Co–Fe species; deactivation; hydrogen production; methane decomposition; regeneration

---

1. Introduction

The reduction of catalytic activity over time is an issue of considerable and continuing concern in industrial practices of catalytic processes. Catalyst replacement and process shutdown could cost the industry very large financial resources every year. The catalyst deactivation changes extensively; for example, in the case of catalytic methane decomposition, catalyst deterioration may be on the scale of seconds, whereas in NH_3 synthesis the Fe-catalyst may stay for 5–10 years. Nevertheless, it is unavoidable that all catalysts drop their activities.

Hydrogen is a pure fuel source which can replace fossil fuels. It can be utilized to run numerous devices like fuel cells, engines, vehicles, and electric devices [1]. Subsequently, the attention given to hydrogen production has been progressively growing these past years. H_2 can be obtained from several sources and methods, for example, reforming of hydrocarbons, biomass, electrolysis, and photo-splitting of water, in addition to water-gas shift reaction [2–6]; however, the steam reforming reaction is the main conventional method for hydrogen production. The high H_2/CO ratio of the saturated hydrocarbons, particularly CH_4, makes them the major feedstock for H_2 production by reforming reactions. For example, partial oxidation, if properly controlled, is a more suitable method to produce hydrogen than the dry-reforming process because the H_2/CO ratio is 2 and the reaction is mildly exothermic, whereas dry-reforming being endothermic is energy-intensive with a low ratio: $H_2/CO = 1$ [7]. The graphene hydro/dehydrogenation process might be used as an active and eco-friendly device to yield and store hydrogen from water. Its mechanisms involve water decomposition at the graphene/metal interface at room temperature to hydrogenate the graphene sheet, which is buckled and decoupled from the metal substrate. Likewise, thermal programmed reaction experiments demonstrate that molecular hydrogen can be released upon heating the water-exposed graphene/metal interface above 400 K [8].

The endothermic catalytic methane decomposition (CMD) reaction [9] could provide a promising substitute for the conventional processes, like steam reforming, partial oxidation, and autocatalysis of methane, used for hydrogen production [10–12]. The CMD process produces pure hydrogen and valuable carbon (i.e., free carbon nanotubes) [13–16]. The formation of carbon nanotubes (CNT) can be of high economic interest; it enjoys many applications, such as catalysts, catalyst-support, H_2 storage, electronic components, and polymer additives [17–20]. Transition metals such as Ni, Co, and Fe supported on different oxides, such as MgO, Al_2O_3, and SiO_2, are often used for CMD reaction [21–24]. Furthermore, it has been stated that the properties of a single metal catalyst can be improved by introducing a second metal to form a bimetallic catalyst [24–28]. Numerous articles have been lately published on bimetallic catalytic systems for CMD reaction [29–34].

The utilization of supported bimetallic Ni–Fe, Ni–Co, and Fe–Co catalysts was examined for catalytic methane decomposition by several researchers such as Awadallah et al. [35]. The catalytic data showed that the bimetallic catalyst exhibited remarkably higher activity and stability and a higher yield of multi-walled carbon nanotubes. The Fe and Co play the role of active metal in the catalyst; they are cheap and abundant compared to noble metals and Ni. The catalytic activity of 30 wt % Fe and 15 wt % Co bimetallic catalyst, reduced at 500 °C and operated at 700 °C, showed excellent performance among all the tested catalysts which comprised Fe/Al_2O_3, Co–Fe/Al_2O_3, Ni–Fe/Al_2O_3 with different amounts of Co and Ni loading [13]. The present work underscores the features of catalyst regeneration performance in activity and stability.

Pudukudy et al. [34] investigated the direct decomposition of CH_4 over SBA-15-supported Ni-, Co- and Fe-based bimetallic catalysts. Their results specified that all of the bimetallic catalysts were highly active and stable for the reaction at 700 °C even after 300 min of time on stream. Co–Fe/SBA-15 catalyst revealed high catalytic stability [34]. Bimetallic Ni–Fe, Ni–Co, and Fe–Co supported on MgO catalysts with a total metal content of 50 wt % were examined for CMD by Awadallah et al. [35]. The catalytic data exhibited that the bimetallic 25%Fe–25%Co/MgO catalyst displayed remarkably higher activity and stability up to ~ 10 h of time on stream and a higher yield of multi-walled carbon nanotubes [35]. The deactivation of heterogeneous catalysts due to carbon deposition is a global issue that causes a reduction in catalytic activity with time. Various types of carbon and coke that change in morphology and reactivity are created. The more reactive, amorphous forms of carbon created at low temperatures are transformed into less reactive, graphitic forms at high temperatures over a period of time [36]. Normally, catalyst regeneration is considered to restore catalytic activity by removing carbon, poisons, and site blockage [37]. Hazzim et al.; studied the regeneration for CMD using an activated carbon catalyst [38]. They found that the activity at the start and the final mass gain of the catalyst increased as the reaction temperatures rose from 850 to 950 °C. However, at 850 and 950 °C reaction temperatures, the activity and mass gain declined after each regeneration step. The decrease was slower under severe regenerating conditions [38].

In this work, a cobalt–iron supported on alumina catalyst (noted 15%Co–30%Fe) was synthesized for catalytic methane decomposition (CMD) reaction ($CH_4 \rightarrow C + 2H_2$:$\Delta H° = 75{:}6$ KJ/mol). The effect of the catalyst regeneration was evaluated by using different oxidizing forced periodic cycling. The regenerated catalysts were tested again in CMD reaction for coproduction of hydrogen and carbon nanomaterials. This study focused on identifying the nature of the carbon deposits formed after each CMD testing that precedes the regeneration cycles; also in addition, the obtained yields of hydrogen were quantified to understand the reaction's mechanism.

2. Results and Discussion

Because of the materials' complexity, structural and textural properties and morphology of spent/regenerated catalysts (referred to as SP-180 min, SP-360 min, and SP-720 min) were examined using X-ray diffraction (XRD), transmission electron microscopy (TEM), laser Raman spectroscopy (LRS), N_2-physisorption, hydrogen temperature programmed reduction (H_2-TPR), and thermogravimetric analysis (TGA), and the carbon amount was evaluated by atomic absorption spectrometry (AAS). For each of these three samples, the oxidative regeneration process was performed every 90 min of the test. For comparison, the un-regenerated SP-90 min was used as a reference.

2.1. Structure and Morphology

XRD patterns of fresh catalyst ($15Co–30Fe/Al_2O_3$), SP-90 min, and three spent/regenerated samples are shown in Figure 1. As mentioned previously [39], the diffractogram of the fresh solid showed the presence of γ-Al_2O_3 phase at 2θ ca. 40°, 49°, and 64° in accordance with (ICDD# 004-0875) lines. In this sample, additional peaks related to Fe_3O_4 magnetite were found at 2θ ca. 32°, 35.5°, 45°, 55°, and 60° according to (ICDD# 04-006-6550), whereas Fe_2O_3 hematite was not observed at 2θ: 32°, 35° (most intense) and 25°, 50°, 54° (less intense) in accordance with (ICDD# 089-0598). Peaks observed at 2θ ca. 32°, 37°, 45°, 56°, 59°, and 66° correspond to $CoAl_2O_4$ spinel (ICDD#44-0160). The formation of $CoAl_2O_4$ was ascribed to the strong interactions between cobalt species and γ-Al_2O_3 lattice during the synthesis process. No peak corresponding to Co–Fe mixed oxides like $CoFe_2O_4$ spinel was observed; this was plausibly due to a stronger interaction between cobalt and alumina which have more pronounced acidic properties than that of iron oxides.

The XRD pattern of spent SP-90 min was completely different from that of the fresh catalyst; it was dominated by the reflection of (200) planes observed in 2θ = 20°–30° range, attributed to carbon species with a graphite-like structure. No reflections corresponding to $CoAl_2O_4$ spinel, Fe_3O_4 magnetite, or Fe_2O_3 hematite phases were detected. The presence of γ-Al_2O_3 phase was not clearly ascertained. However, a close inspection of this diffractogram revealed the presence of the common reflection of cobalt and iron chemically in the oxidation state zero, indicating the reduction under the reaction mixture. Part of $CoAl_2O_4$ and Fe_3O_4 directly forms Co and Fe metallic species or probably Co–Fe alloy. The authors supposed that the reduction of $CoAl_2O_4$ and Fe_3O_4 was incomplete under CH_4 atmosphere. The presence of unreduced species $CoAl_2O_4$ spinel and Fe_3O_4 could not be clearly checked because their lines are probably overlapped by those of the Co and Fe metallic species.

In the case of the SP-180 min sample, the pattern performed after the oxidative regeneration process was similar to that of SP-90 min, but the intensity of the peaks was different. The SP-180 min pattern revealed the presence of graphitic carbon with very low-intensity peaks due to the removal of surface carbon in the form of CO_2. Cobalt and iron were always observed in the metallic form, that is, no cobalt and iron oxides were observed after oxidative treatment. From these results, it might be inferred that filamentous carbon stabilized the cobalt and iron in oxidation state zero.

In the SP-360 min and SP-720 min samples, the intensity of the carbon lines increased with the time of CMD reaction due to an additional deposit of carbon in well-structured forms that is difficult to eliminate during the regeneration process. The presence of Fe^0 metallic species was always ascertained, but that of Co^0 is unlikely. However, the diffractogram revealed the presence of a new phase attributed to CoC_2 cobalt carbide and identified after Rietveld refinement; this identification is observed by a slight peak splitting situated at $2\theta \approx 45°$ (ICDD#44-0962). No reflections corresponding to iron carbide species (like Fe_5C_2 or Fe_3C) could be observed in these samples.

Nevertheless, with a refinement scan, some traces of Fe_3O_4 magnetite and Fe_2O_3 hematite were identified indicating the re-oxidation of iron during the regeneration process. Unlike the spent/regenerated sample SP-180 min, the SP-360 min and SP-720 min samples revealed the presence of a new phase recorded as θ-Al_2O_3 observed at 2θ positions of 29.8°, 41.0°, 44.2°, 50.81°, 61.8°, 67.9°, and 79.1° in accordance of (ICDD#086-1410). From these results, it could be inferred that the θ-Al_2O_3 would probably help to make and stabilize more iron species at the surface of SP-360 min and SP-720 min samples [40,41].

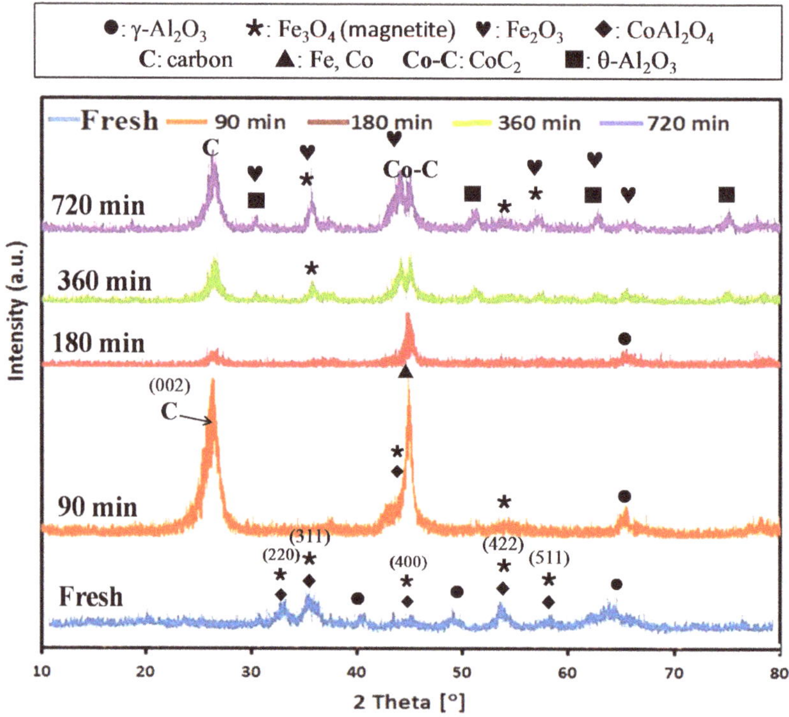

Figure 1. X-ray diffraction (XRD) patterns of fresh, spent, and spent/regenerated catalysts.

Transmission electron microscopy (TEM) was performed in order to get more information on the carbon morphology detected by XRD. TEM micrographs of the spent and spent/regenerated catalysts are shown in Figure 2. In all cases, the formation of filamentous-type carbon was confirmed. It is known that the fibers of carbon nanotubes (CNTs) are composed of two kinds: (1) single-walled carbon nanotubes (SWCNTs) which consist of a single tube of graphite and (2) multi-walled carbon nanotubes (MWCNTs). Generally, in TEM analysis, it is difficult to establish the length of nanotubes to compare them. However, it is well known that nanotube diameters can range from just a few nanometers for SWCNTs to several tens of nanometers for MWCNTs. In their study, Jorio et al. [42] demonstrated by TEM that diverse SWCNT diameters varied between 0.7 and 3.0 nm. On the other hand, Hou et al. [43], using the same technique, found that MWCNT were generally in the diameter range from 10 to 200 nm. In this work, only MWCNT fibers with 27–53 nm diameters were identified (Figure 2). Similarly, in an earlier report [39] of CMD on SP-90 min, the lengths of the produced carbon (CNT) at 90 min of CMD (observed by TEM) varied between 14.09 and 72.0 nm, which were the characteristics of MWCNTs.

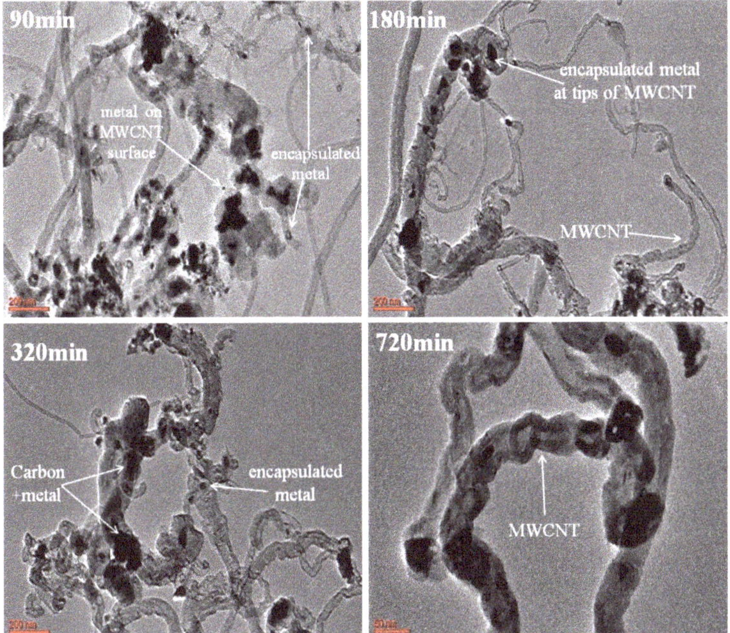

Figure 2. Top surface transmission electron microscopy (TEM) micrographs of spent and spent/regenerated 15Co–30Fe/Al$_2$O$_3$ catalysts obtained after different times (90, 180, 320, and 720 min).

For all spent and spent/regenerated samples, TEM micrographs showed large amounts of graphitic carbon layers around the Co metal particles. In fact, these Co particles were mostly encapsulated within CNTs; on the other hand, the Fe particles clung to the CNT surface in accordance with XRD analysis, which showed, after oxidative regeneration, a re-oxidation of iron metallic species in Fe$_3$O$_4$ magnetite or in Fe$_2$O$_3$ hematite (according to XRD observations); it however showed no oxidized form of cobalt which was encapsulated and stabilized with filamentous carbon. According to the literature [44], a metal encapsulated into the CNTs was not very accessible to reactants and therefore resistive to any oxidative regeneration process. Besides metal particles encapsulated (Co) and metallic particles located on top of nanocarbons (Fe), all TEM images (Figure 2) showed agglomerate black particles clearly observed in the tips of MWCNTs and dispersed on the CNT surface. These black particles are more numerous on the unregenerate sample (SP-90 min) and are attributed to condensation of carbon nanoparticles. These black particles could be easily removed in CO$_2$ form (C $\rightarrow$ CO$_2$) during the regeneration process.

Table 1 summarizes the average size of carbon crystallites determined by XRD and obtained by TEM. The crystallite size (XRD) was estimated using the Debye–Scherrer formula [45] for the most intense (002) peak of carbon and using the equation $\tau = \frac{0.94 \times \lambda}{\text{FWHM} \times \text{Cos}(\frac{2\theta}{2})}$, where $\lambda_{cu} = 1.5406$ Å, τ (Å) is the carbon crystallite size, 2θ (radian) is the diffraction peak position, and Full width at half maximum (FWHM) (radian) is full width at half maximum describing the width measurement of carbon peak. In TEM, the calculation of sizes was realized using the ImageJ software package (1.52 h, National Institutes of Health, Madison, WI, USA, 2018). Except for SP-720 min (with instrumental broadening $\times 50$ against $\times 200$), the TEM investigation displays homogeneous distribution of carbon species with a particle size 52.31, 49.28, and 53.23 nm for 90, 180, and 360 min samples, respectively. Similar trends were observed by XRD for these three samples (49.74 and 42.63 Å). In the case of the SP-720 min, the crystallite sizes (XRD or TEM) were the smallest as compared to those of other catalysts because of the removal of amorphous carbon species from the MWCNT surface (loss of carbon C $\rightarrow$ CO$_2$) [46].

Table 1. Estimation of carbon crystallite size using XRD and TEM techniques.

Samples (SP)	Crystallite Size [*] of Carbon Graphite (Å)	Average Size [**] of Carbon Particles (nm)
90 min	49.7	52.3
180 min	42.6	49.3
360 min	59.6	53.2
720 min	39.7	27.5

[*]: XRD calculations with Debye–Scherrer formula on the (002) most intense reflection; [**]: TEM calculations from 05–08 zones of each micrograph.

As a conclusion, XRD and TEM measurements are complementary techniques, but not comparable: XRD involves some heterogeneity properties such as crystalline structures, types (carbon, carbide, etc.), and amorphous regions, whereas TEM gives information on the particle organization of carbon (homogeneous distribution), dispersion, and distribution of metals.

To confirm the carbon nature and structure observed by XRD, laser Raman spectroscopy (LRS) was applied to provide useful information about material crystallinity, phase transition, and structural disorder. The spectra of the spent and spent/regenerated samples are shown in Figure 3. The single band observed between 1000 and 2000 cm^{-1} could be associated with different carbon bonds; it revealed the presence of a carbon graphite structure on the molecular level. According to previous reports [47], this band with a maximum at 1363.6 cm^{-1} was attributed to the D-band of sp^2 carbon material. Deconvolution of this band revealed two components (Figure 3); the second component, centered at 1580.54 cm^{-1} is associated with the G-band. According to Ferrari [48], the Raman G-band was a characteristic of graphite with high crystallinity and the D-band was attributed to defects and lattice distortions in the carbon structures; an increase in the intensity of the D-band reflected an increase in the disorder of the carbon atoms and thus a restructuring of the nanotubes. Therefore, the ratio of D-band intensity on G-band intensity (I_D/I_G) gave an indication of defects or graphitic order. In Figure 3, the ratio I_D/I_G was calculated after deconvolution. The I_D/I_G ratios, which varied in the order of 1.84 (360 min) > 1.75 (180 min) >1.66 (720 min) > 1.51 (90 min), indicated the formation of new phases detected by XRD in spent/regenerated samples, which could be directly responsible for the higher level of lattice distortion in the carbon graphite structures. However, the values (I_D/I_G = 1.66–1.84), obtained after regeneration, had slightly increased compared to that of SP-90 min ($I_D/I_G \sim 1.51$), which means that with the oxidative treatment, the structure of the nanotubes was disturbed but not totally damaged.

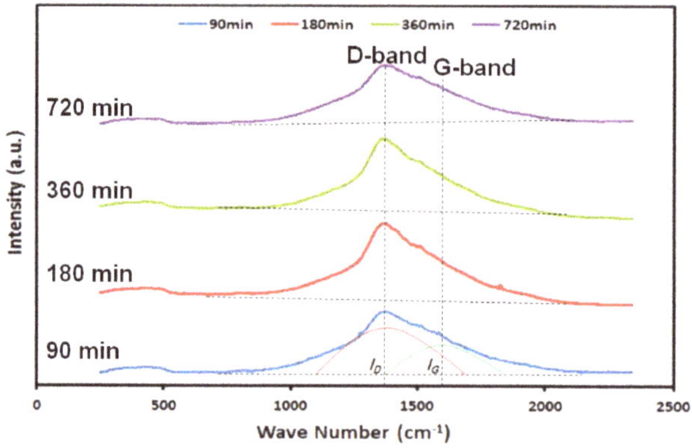

Figure 3. Laser Raman spectra of SP samples (90, 180, 360, and 720 min).

2.2. Surface Characterization

The Brunauer–Emmett–Teller BET surface areas were determined by nitrogen adsorption-desorption at 77 K. All nitrogen adsorption–desorption isotherm curves and variations of the specific surface areas are depicted in Figure 4. According to the International Union of Pure and Applied Chemistry (IUPAC) standard, the catalysts exhibited the IV type isotherm showing that the materials were essentially mesoporous. The fresh sample exhibited H4 hysteresis loop which corresponded to the narrow slit-like pores, whereas the hysteresis loops of spent (90 min) and spent/regenerated (SP-180, 360, and 720 min) samples had the characteristic of a type H3 loop. Compared to fresh and not regenerated samples, the spent/regenerated samples showed a lower amount of nitrogen adsorbed at the high relative pressure ($P/P_0 = 1$), which suggests a strong decrease in the mesoporosity because of the aggregated pores between the CNTs. In addition, as is shown in Figure 4, the displacement of the hysteresis loops toward the lower partial pressures is an indication of the porosity's evolution.

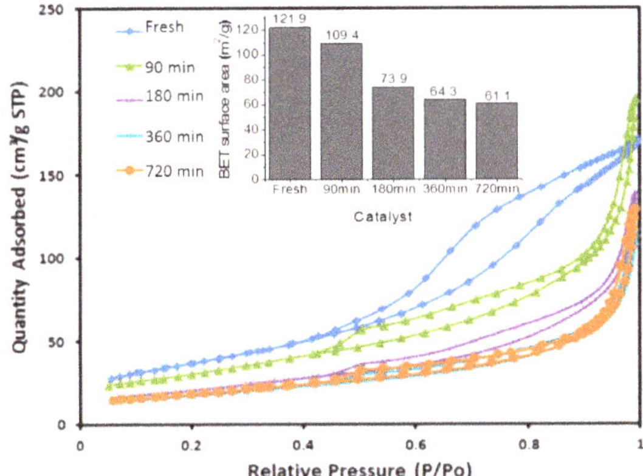

Figure 4. N_2 adsorption-desorption isotherms and variation of specific surface areas for fresh, spent, and spent/regenerated catalysts at different reaction times.

BET surface areas of spent and spent/regenerated samples were relatively small compared to that of the fresh sample. They decreased linearly with the time of CMD reaction in the following order: fresh-sample (122 m^2/g) > SP-90 min (109 m^2/g) > SP-180 min (74 m^2/g) > SP-360 min (64 m^2/g) > SP-720 min (61 m^2/g). During the CMD reaction or oxidative regeneration process, the formation of new phases (such as Co and Fe metallic species, CoCx, carbon, and θ-Al$_2$O$_3$, etc.) caused the blocking of the pores with the subsequent loss of surface area as evidenced by XRD analysis. This result was consistent with AAS findings, which showed an increase in the mass of carbon deposited with the number of reaction cycles. The decline in catalyst activity might be related to the reduction in porosity which occurred during the decomposition process.

2.3. H_2-TPR and TG Analysis

Hydrogen temperature programmed reduction experiments were carried out in 100 to 1000 °C temperature range under a hydrogen atmosphere. H_2-TPR profiles of fresh, spent, and spent/regenerated 15Co–30Fe/Al$_2$O$_3$ samples are shown in Figure 5a,b. Figure 5a is for fresh catalyst, whereas Figure 5b is for the spent catalyst. The reduction profiles can be classified into three regions. In region I (200–450 °C), the reduction of hematite (Fe$_2$O$_3$) to Fe$_3$O$_4$ and the reduction of Co$_3$O$_4$ to CoO take place. In region II (450–710 °C), the reduction of Fe$_3$O$_4$ to FeO and the reduction of CoO to Co occur. In region III (710–900 °C), the reduction FeO to Fe happens. Similar results were observed by other investigators [49–51].

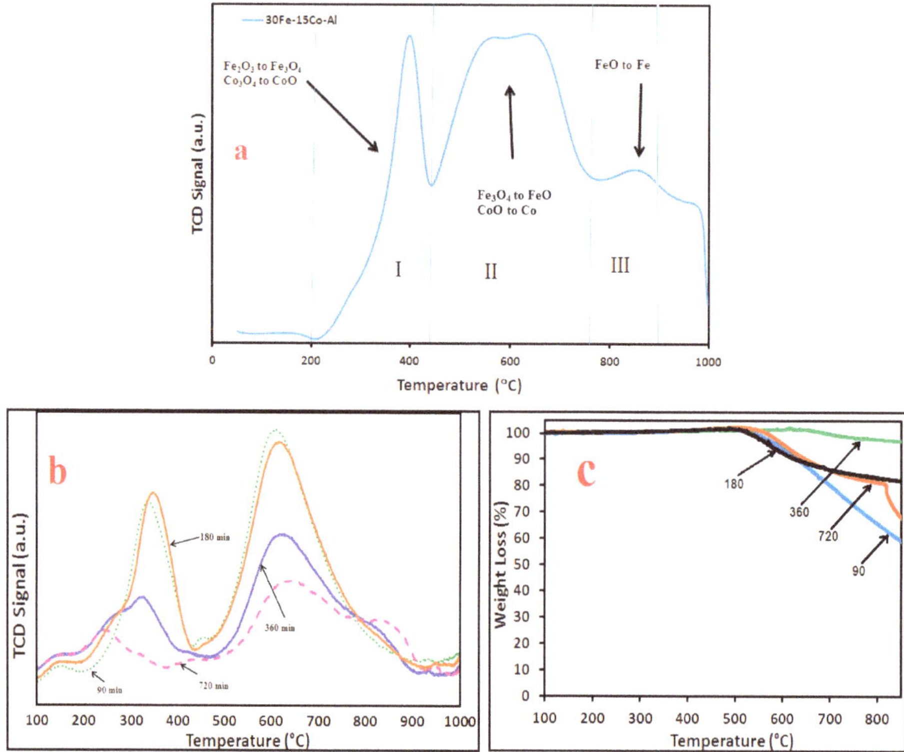

Figure 5. (**a**) H$_2$-TPR for fresh catalyst, (**b**) H$_2$-TPR for used catalyst, and (**c**) thermogravimetric (TG) profiles, of 30Fe-15Co/Al$_2$O$_3$ catalysts at different regeneration temperatures.

From the XRD analysis, the possible reducible phases in fresh sample were the following: Fe$_3$O$_4$, CoAl$_2$O$_4$ spinel, and probably some amorphous (or highly dispersed) Fe^{2+} and Co^{2+} species which could not be detected easily by XRD. Al$_2$O$_3$ was non-reducible under this work's conditions due to the high binding energy between Al and O. Large differences in H$_2$-TPR profiles were observed that depended on the time of reaction and the number of decomposition-oxidation cycles. However, for all samples, the reduction peaks observed below 450 °C were ascribed to the reduction of Co^{2+} species to nanoparticles of metallic cobalt, whereas the peaks observed between 450 and 900 °C were ascribed to the reduction of Fe^{3+} species of metallic iron. In the case of the SP-360 min and SP-720 min samples compared to other samples, several reduction peaks were observed at low and high temperatures; they could be ascribed to successive steps of the reduction to cobalt and iron species and/or to the presence of various types of Co and Fe species of different interactions with the support or with different particle size. Indeed, it is well known that the increase in the degree of the agglomeration and consequently the size of metal oxide particles could cause a decrease in the metal-support interaction and make reduction easier. From H$_2$-TPR results, it should be pointed out that the H$_2$ consumption decreased with the increasing reaction time (or the number of decomposition-regeneration cycles), probably because of the increase in the encapsulation rate of metallic particles in a zero-valence state that makes their oxidative regeneration difficult due to their inaccessibility to hydrogen molecules. This could justify the lower reducibility of both SP-360 min and SP-720 min samples. During the H$_2$-TPR process, a hydrogasification of carbon deposits could also happen at temperatures above 700 °C by releasing methane (C$_{(s)}$ + 2H$_2$ → CH$_4$). An analysis of gaseous effluents (using, for example, mass spectrometry) is however necessary to assess this hypothesis.

TGA results of spent and spent/regenerated catalysts are shown in Figure 5c that shows a weight change in the temperature range 100–850 °C. All samples displayed weight loss (WL) from 500 to 850 °C which was related to the combustion of different kinds of carbon (cobalt carbide amorphous carbon and carbon nanotubes). The TGA curves showed that the weight loss decreased in the order WL(SP-90 min) > WL(SP-720 min) > WL(SP-360 min) > WL(SP-180 min). As expected, the highest weight loss was observed in the spent and not regenerated sample (SP-90 min). Comparing the regenerated samples, it was found that SP-180 min exhibited the lowest weight loss, indicating that the accumulation of encapsulating carbon deposition (difficult to remove) increased with the increasing number of decomposition-oxidation cycles. This result corresponded to the XRD finding which exhibited a more intense peak of carbon for samples having undergone several decomposition–regeneration cycles. In the case of SP-720 min, a change was noticed in the slope of the TGA curve, which suggested two kinds of coke deposit may be related to the more abundant presence of cobalt carbide over this catalyst.

2.4. Catalysts Behavior under CMD-Regeneration Cycles

In most cases, during CMD reaction, carbon encapsulates the active sites and this carbon accumulation is responsible for catalyst deactivation. To increase the catalyst lifetime, regeneration cycles by gasification using oxygen as an oxidizing agent was introduced as has been described before. The regeneration step obviously required the interruption of the production process for 10 min. The catalytic activity of catalysts was determined by evaluating methane conversion and hydrogen yield obtained during the CMD process. In the present work, Weight hourly space velocity (WHSV) is calculated from the feed gas consisting of 15 mL/min of CH_4 and 10 mL/min of N_2 and weight of catalyst 0.3 g, as follows:

$$\text{WHSV} = \frac{(15 + 10)\ \text{mL/min} \times 60\ \text{min/h}}{0.3\ \text{g-cat}} = \frac{25 \times 60\ \text{mL/ (h)}}{0.3\ (\text{g-cat})} = 5000\ \text{mL/(h·g-cat.)}$$

The methane conversion obtained here for regenerated catalysts is above 60%. Zhou et al. conducted an Fe catalyst for CMD and investigated the effect of space velocity on the methane conversion. When a space velocity (3750 mL/g-cat/h), close to the one used in the present work (5000 mL/g-cat/h), was considered, they found a methane conversion of about 38% [52].

Figure 6 compared the evolution, with time of stream (TOS), of the methane conversion obtained over the fresh catalyst and the spent catalysts subjected to successive cycles of CMD/regeneration. As can be seen, the results were not the same after the regeneration cycles. Only the points on the curve at 90, 180, 360, and 720 min corresponding to the samples SP-(90, 180, 360, 720 min) were considered. As is shown, the methane conversion decreased, as the reaction time was increased, in the order of 68% (90 min, zero cycle) > 63% (180 min, after two cycles) > 62% (360 min, after four cycles) > 56% (720 min, after eight cycles). This degradation of catalytic activity and loss of stability of the catalysts (about Δ% = 12% from 90 to 720 min of TOS) was due probably to irreversible deactivation caused by covering and encapsulating of active sites by deposited carbon. The same trend was observed in the BET surface area which decreased when the number of decomposition/regeneration cycles was increased, as is shown in Figure 3. On the other hand, it was shown that the reducibility of catalysts decreased when the number of decomposition/oxidation cycles was increased; this was attributed to the blocking of metallic particles (in the oxidation state zero) by the formation of encapsulating carbon species, consequently leading to the loss of active sites and catalyst deactivation. From these results, it could be concluded that the degradation of physicochemical and catalytic properties was essentially related to carbon formation that was not completely removed after repetitive oxidative regeneration steps. Therefore, the higher the amount of carbon deposited, the lower the catalytic activity.

Through this study's experiments, the authors were able to verify that significant catalyst deactivation occurred toward higher regeneration times (involving a succession of cycles) despite a very interesting catalytic formulation. The formation of cobalt carbide CoC_2 may then complicate the regeneration of metal catalysts (Figure 1), as reported in the literature [52].

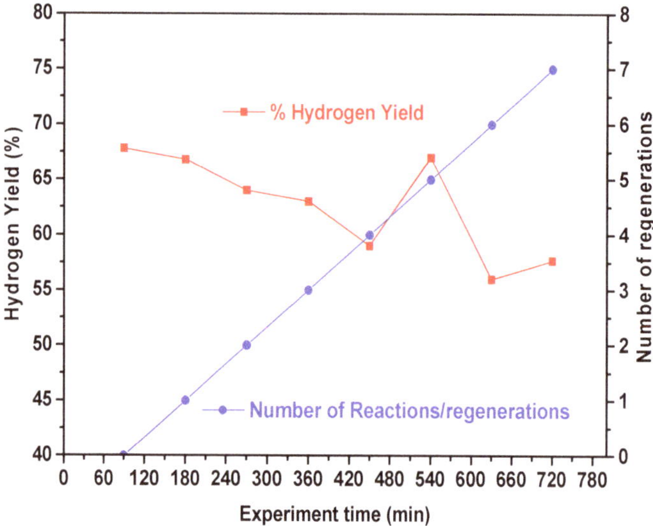

Figure 6. Time on stream (TOS) yield of hydrogen versus different regeneration times with 15%Co–30%Fe/Al$_2$O$_3$ spent-catalyst. TOS conditions: T$_{regeneration-activation}$ = 500 °C; T$_{reaction}$ = 700 °C; (WHSV = 5000 mL/H·g-cat), total flow rate = 25 mL/min and CH$_4$/N$_2$ = 1.5.

Figures 6–8 present, respectively, the evolution with time on stream (TOS) of H$_2$ yield and the coproduced carbon on SP-(90, 180, 360, 720 min) samples. The hydrogen production by 67.8% (90 min) > 66.8% (180 min) > 63.7% (320 min) > 57.7% (720 min) followed the same evolution as the methane conversion. Similar to the conversion of methane, the hydrogen yield was affected by the number of regeneration cycles. It decreased with the increase in the regeneration number, but to a lesser extent. On the other hand, in the case of the carbon deposit, an increasing trend of carbon with the increase in the regeneration number was observed: 1.2 (180 min) < 1.5 (320 min) < 3.2 (720 min).

As reported by some authors [53], the catalyst deactivation mechanism rests on progressive pore blocking by carbon deposition. The carbon formation involved in the first step is the dissociative adsorption of methane on the metal surface (iron and cobalt) producing hydrogen and carbon atoms. The adsorbed carbon atoms could dissolve and diffuse through the metal. Those on the surface form, thereby encapsulating carbon, blocking the access of methane to the active sites, and causing a reduction of activity, as proposed by the following steps. During CMD reaction, the carbon may chemisorb firmly as a monolayer or physically adsorb in multilayers while releasing hydrogen [54]. In either case, this blocks the reactants from reaching the metal surface. The behavior is complex, because the carbon threads may grow from the top surface of the metal particles or the carbon may spread into the metal and form bulk carbides (e.g., formation of CoC$_2$), as observed in the study's XRD. According the H$_2$-TPR conclusions, it can be stated that methane molecules may be formed through the reaction between carbon encapsulating particles and chemisorbed hydrogen gases.

The formation of encapsulating carbon on the metallic surface could be limited by C$_{ads}$ + H$_2$ → CH$_4$ reverse reaction (with H$_2$ produced by reaction or added to the feed). This step was favored in the presence of a high hydrogen concentration in the feed which depended on the methane conversion. Besides coke deposition, the sintering of metallic active phase (Co, Fe) occurred during CMD reaction and oxidative regeneration process and was the second factor responsible for catalyst deactivation. Similar to carbon deposition, sintering brings about the loss of surface area and porosity. Despite the coke deposits and many oxidative treatments, the present catalysts remained active with a methane conversion and hydrogen yield exceeding 56% and 58%, respectively (after eight reaction cycles).

Further tests were needed to optimize the regeneration step to eliminate quantitatively the carbon deposits and improve the catalytic performance.

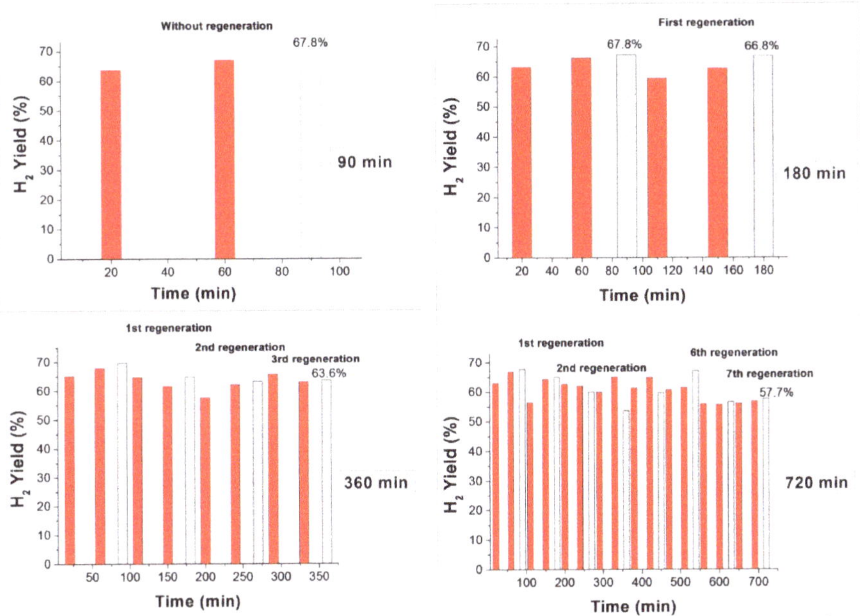

Figure 7. Comparison of different H_2 yields (%) obtained from 90, 180, 360, and 720 min of reaction (the white bar is a limit between each regeneration cycle).

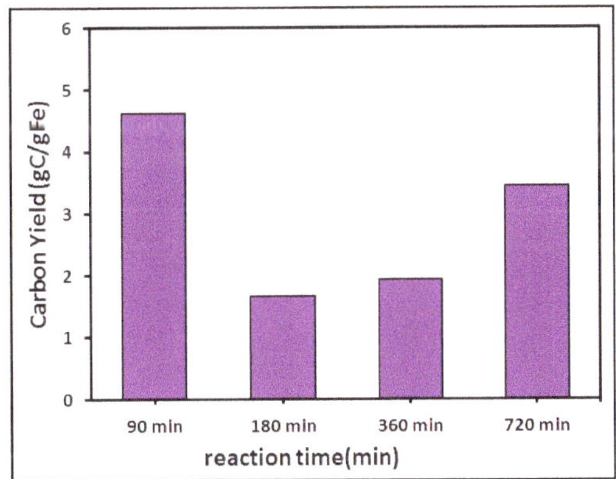

Figure 8. Carbon yields obtained after 90, 180, 360, and 720 min of reaction.

3. Experimental Section

3.1. Preparation of Fresh Catalyst

A wet-impregnation technique was used to prepare a combined 15%Co–30%Fe supported on alumina catalyst. The synthesis steps are described as follows:

Alumina (γ-Al$_2$O$_3$; SA6175) were obtained from Norton Chemical Company (Short Hills, NJ, USA). The alumina were disintegrated into tiny particles before being employed as a ultimate support material. A first solution was prepared from a certain volume of double-distilled water and stoichiometric ferric nitrate amount ($Fe(NO_3)_3 \cdot 9H_2O$, Ass. $\geq$ 99%, Sigma-Aldrich, Taufkirchen, Germany). Dissolving a certain mass of cobalt nitrate ($Co(NO_3)_3 \cdot 9H_2O$, Ass. $\geq$ 99%, Sigma-Aldrich, Taufkirchen, Germany) in double-distilled water, a second solution was obtained. As in a typical wet-impregnation procedure, the two solutions of Co^{2+} and Fe^{3+} were simultaneously impregnated on the γ-Al_2O_3 support. The mixture was stirred at 80 °C for 3 h. After the stirring/heating step, more than 50% of the water contained in the mixture–solution evaporated. The final mixture was dried at 120 °C for 12 h (without washing) and followed by calcination at 450 °C for 3 h under air atmosphere. The mixed catalyst was designated as 15Co–30Fe/Al_2O_3.

3.2. Characterization Methods

XRD: X-ray powder diffraction (XRD) was carried out using a Rigaku (Miniflex) diffractometer (Rigaku, Bahrain, Saudi Arabia). The unit was set, using a Cu Kα radiation run at 40 kV and 40 mA, to investigate the diffraction peaks of the catalysts before and after the reaction. The 2θ step and the scanning range were set at 0.02° and 10–85°, respectively. The instrumental raw data was analyzed via X'pert HighScore Plus software (3.0.5, Malvern Panalytical Ltd, Malvern, UK, 2017). An ASCII file of the peak intensity was produced at granularity 8, bending factor 5, minimum, peak significance 1, minimum peak width 0.40, maximum tip width 1, and peak base width 2 by minimum second derivatives. Additional dissimilar phases with their marks were accorded using the JCPDS data bank (N.B.S*AIDS83, International Center for Diffraction Data, Newtown Square, PA, USA, 1980) and X'pert HighScore Plus software.

TEM: Samples were arranged for transmission electron microscopy (TEM) investigation by crushing the powders between clean glass slides and then scattering them onto a lacey carbon film held on a Cu mesh grid. Bright-field transmission electron microscopy (BF-TEM) and selected area electron diffraction (SAED) experiments were performed using a JEOL 2000FX microscope (JEOL, Peabody, MA, USA) fitted with a thermionic LaB6 source working at 200 kV.

LRS: Laser Raman spectroscopy apparatus was represented by a highly sensitive CCD Q15 spectrograph (Thorlabs, Munich, Germany) having two 215 excitation lasers of 532 and 785 nm with CCD cooling temperature reaching up to 60 °C and high-throughput laboratory fiber optic probes. Moreover, the scanning was fixed between 250 and 2350 cm^{-1}.

N_2-physisorption: The distribution of the pore size and the specific surface area of the catalysts were computed from N_2 adsorption–desorption data. Measurements of the BET surface area were conducted by nitrogen adsorption at -196 °C using a Micromeritics Tristar II 3020 (Micromeritics, Riyadh, Saudi Arabia) surface area and porosity analyzer. In each test, 300 mg of catalyst was degassed at 300 °C for 3 h to remove the wetness and adsorbed gases from the catalyst surface. Pore size distribution was computed via the Barrett-Joyner-Halenda (BJH) technique.

H_2-TPR: Micromeritics Auto Chem II apparatus (Micromeritics, Riyadh, Saudi Arabia) was employed in the temperature programmed reduction (TPR) to examine the reducibility by taking 0.07 g for each test. High purity argon flow was first passed thought the samples at 150 °C for 30 min. Then, the samples were brought to 25 °C. Lastly, the furnace temperature was set to 1000 °C at a 10 °C/min rate while running 40 mL/min flow rate of H_2/Ar mixture that had 10 vol % of H_2. A thermal conductivity detector (TCD) checked the H_2 consumption signals.

TGA: The quantitative investigation of coke formation on the used catalysts after the duration of 90, 180, 360, and 720 min reaction at 700 °C was conducted using thermogravimetric analysis (TGA) in the presence of air via a Shimadzu TGA (Thermo-gravimetric/Differential) analyzer (Shimadzu, Jebel Ali Free Zone, Dubai). The temperature of the spent catalysts weighing 10–15 mg was increased from 25 °C to 1000 °C at a heating rate of 20 °C/min, and the mass reduction was recorded.

AAS: Atomic absorption spectrometry was employed to examine samples of Fe and Co components in the catalysts. The AAS analyses were conducted using a Model 951 AA/AE (Berkeley Nucleonics corporation, San Rafael, CA, USA) spectrophotometer with graphite furnace and Model

254 Auto sampler. The composition of Co and Fe in catalysts were theoretically determined and compared to those experimentally obtained by AAS.

3.3. Activity Test (Regeneration Procedure)

CMD activity tests were done in a fixed-bed quartz reactor (id = 9 mm). Before the CMD reaction, 0.3 g of the catalyst was reduced in situ under hydrogen flow (40 mL/min) at 500 °C for 60 min. Then 20 min of N_2 flushing was carried out. Subsequently, the reactor temperature was raised to 700 °C under N_2. For all runs, a fixed feed gas composed of 15 mL of CH_4 and 10 mL of N_2 was used.

The regeneration cycles were performed in situ at 700 °C under diluted O_2 gasifying agent ($10\%O_2/N_2$), followed by inert treatment under N_2. The obtained regenerated catalyst was tested again in CMD reaction. First, the CMD reaction was carried out for 90 min in CMD reaction. Then, the spent catalyst (labeled SP-90 min) was removed for characterization as a reference. Then, the reactor was charged with fresh catalyst and the above steps were repeated for each of the following experiments:

In the first regeneration experiment: the CMD reaction was performed for 90 min. At this point, 10 mL/min of O_2 was added to the system for 10 min. Then, the reactor was fed with N_2 at a rate of 20 mL/min for 20 min and the CMD reaction was continued for another 90 min to finally obtain 180 min (= 2 × 90 min). Then, the spent/regenerated catalyst (noted as SP-180 min) was removed for characterization.

In the last regeneration experiment: the reactor was again recharged with a fresh catalyst. The CMD reaction was allowed to run for 90 min. Then, 10 mL/min of O_2 was used to regenerate the catalyst for 10 min. Then, 20 mL/min of N_2 was used to flush the reactor for 20 min and the reaction was continued for another 90 min. After that, the above addition of O_2 and N_2 was periodically repeated for eight cycles to give a total time of reaction/regeneration of 720 min (= 8 × 90 min). Then, the spent/regenerated catalyst (labeled as SP-720 min) was removed for characterization.

Finally, in this work, three regenerations of periodic cycling (with 2×, 4×, and 8× of 90 min) were chosen and the spent samples at 180, 360, and 720 min were removed for characterizations.

The CH_4 reactant and H_2 product were evaluated using an online GC (Alpha MOS PR 2100, Alpha MOS, Toulouse, France,) fitted with a sampling valve and two thermal conductivity detectors for examining heavier and lighter gases. CH_4 conversion and H_2 yield were computed using Equations (1) and (2):

$$CH_4 \text{ conversion}(\%) = \frac{CH_{4in} - CH_{4out}}{CH_{4in}} \times 100, \tag{1}$$

$$H_2 \text{Yield}(\%) = \frac{H_{2out}}{2 \times CH_{4 \text{ converted}}} \times 100. \tag{2}$$

4. Conclusions

In this work, an alumina-supported cobalt–iron sample ($15Co–30Fe/Al_2O_3$) was prepared and used in the catalytic methane decomposition and regeneration process. In situ regeneration of spent catalysts was performed with different oxidizing forced periodic cycling at 180, 360, and 720 min. The X-ray powder diffraction results showed the presence of γ-Al_2O_3 phase, Fe_3O_4 magnetite and $CoFe_2O_4$ spinel in fresh catalyst, whereas SP-90 min spent catalyst provided only graphite, Fe^0 and CoC_2 reflections. The X-ray powder diffraction profile of SP-720 min spent/regenerated catalyst exhibited a new allotropic phase identified as θ-Al_2O_3. Transmission electron microscopy of spent and spent/regenerated samples indicated the formation of filamentous multi-walled carbon nanotubes; these nanotubes were formed by large amounts of encapsulating carbon on the surface which played an important role in the explanation of the reaction mechanism. Furthermore, different carbon bonds, detected by laser Raman spectroscopy, revealed the presence of graphite carbon structure on the molecular level. The reduction peaks were ascribed to the reduction of Co^{2+} species to nanoparticles of Co^0 (below 450 °C) and $Fe^{3+} \rightarrow Fe^{2+} \rightarrow Fe^0$ (between 450 and 900 °C). The catalytic activities of spent/regenerated catalysts exhibited lower activity at a higher cycle number. Hydrogen yield was affected by the oxidizing regeneration number (same trend as methane conversion); on the other hand, carbon deposit (calculated by atomic absorption spectroscopy) followed the reverse trend. Indeed, the formation of carbon encapsulating on the metallic surface could be limited by the $C_{ads} + H_2 \rightarrow CH_4$ reverse reaction. This observation has been confirmed by the hydrogasification of carbon deposit (in H_2-TPR toward 700 °C) by releasing methane ($C_{(s)} + 2H_2 \rightarrow CH_4$). The authors believe that the main factors responsible for the catalyst deactivation are coke deposition and weak sintering of the metallic active phase (Co–Fe) which occurred during the catalytic methane decomposition reaction and regeneration process.

Author Contributions: A.S.A.-F., A.H.F., A.A.I., and A.E.A. carried out all experiments and characterization tests as well as shared in the analysis of the data and shared in the writing of the manuscript. Both S.B. and A.B. wrote the paper and shared data analysis. R.-L.A.-O. performed TGA measurements as well as sharing in the writing of the manuscript.

Funding: This research was funded by the Deanship of Scientific Research at King Saud University, Project No. RG-1435-078.

Acknowledgments: The authors would like to express their sincere appreciation to the Deanship of Scientific Research at King Saud University for its funding for this research group Project No. RG-1435-078.

Conflicts of Interest: The authors declare no conflict of interest.

References

1. Dodds, P.E.; Staffell, I.; Hawkes, A.D.; Li, F.; Grunewald, P.; McDowall, W.; Ekins, P.D. Hydrogen and fuel cell technologies for heating: A review. *Int. J. Hydrog. Energy* **2015**, *40*, 2065–2083. [CrossRef]

2. Deng, J.; Lv, X.; Gao, J.; Pu, A.; Li, M.; Sun, X.; Zhong, J. Facile synthesis of carbon-coated hematite nanostructures for solar water splitting. *Energy Environ. Sci.* **2013**, *6*, 1965–1970. [CrossRef]

3. Lima, S.M.; Cruz, I.O.; Jacobs, G.; Davis, B.H.; Mattos, L.V.; Noronha, F.B. Steam reforming, partial oxidation, and oxidative steam reforming of ethanol over Pt/CeZrO$_2$ catalyst. *J. Catal.* **2008**, *257*, 356–368.

4. Siriwardane, R.; Tian, H.; Fisher, J. Production of pure hydrogen and synthesis gas with Cu-Fe oxygen carriers using combined processes of chemical looping combustion and methane decomposition/reforming. *Int. J. Hydrog. Energy* **2015**, *40*, 1698–1708. [CrossRef]

5. Al-Fatesh, S.A.; Ibrahim, A.A.; Abu-Dahrieh, J.K.; Al-Awadi, A.S.; El-Toni, A.M.; Fakeeha, A.H.; Abasaeed, A.E. Gallium-Promoted Ni Catalyst Supported on MCM-41 for Dry Reforming of Methane. *Catalysts* **2018**, *8*, 229. [CrossRef]

6. Osman, A.I.; Abu-Dahrieh, J.K.; Cherkasov, N.; Fernandez-Garcia, J.; Walker, D.; Walton, R.I.; Rooney, D.W.; Rebrov, E. A highly active and synergistic Pt/Mo$_2$C/Al$_2$O$_3$ catalyst for water-gas shift reaction. *Mol. Catal.* **2018**, *455*, 38–47. [CrossRef]

7. Osman, A.I.; Meudal, J.; Laffir, F.; Thompson, J.; Rooney, D. Enhanced catalytic activity of Ni on η-Al$_2$O$_3$ and ZSM-5 on addition of ceria zirconia for the partial oxidation of methane. *App. Catal. B Environ.* **2017**, *212*, 68–79. [CrossRef]

8. Politano, A.; Cattelan, M.; Boukhvalov, D.W.; Campi, D.; Cupolillo, A.; Gnoli, S.; Apostol, N.G.; Lacovig, P.; Lizzit, S.; Far, D.; et al. Unveiling the Mechanisms Leading to H$_2$ Production Promoted by Water Decomposition on Epitaxial Graphene at Room Temperature. *ACS Nano* **2016**, *10*, 4543–4549. [CrossRef] [PubMed]

9. Abbas, H.F.; Daud, W.M.A. Hydrogen production by methane decomposition: A review. *Int. J. Hydrog. Energy* **2010**, *35*, 1160–1190. [CrossRef]

10. Diehm, C.; Deutschmann, O. Hydrogen production by catalytic partial oxidation of methane over staged Pd/Rh coated monoliths: Spatially resolved concentration and temperature profiles. *Int. J. Hydrog. Energy* **2014**, *39*, 17998–18004. [CrossRef]

11. Vigneault, A.; Grace, J.R. Hydrogen production in multi-Channel membrane reactor via steam methane reforming and methane catalytic combustion. *Int. J. Hydrog. Energy* **2015**, *40*, 233–243. [CrossRef]

12. Wang, Y.; Peng, J.; Zhou, C.; Lim, Z.-Y.; Wu, C.; Ye, S.; Wang, G. Effect of Pr addition on the properties of Ni/Al$_2$O$_3$ catalysts with an application in the autothermal reforming of methane. *Int. J. Hydrog. Energy* **2014**, *39*, 778–787. [CrossRef]

13. Fakeeha, A.H.; Khan, W.U.; Al-Fatesh, A.S.; Abasaeed, A.E.; Naeem, M.A. Production of hydrogen and carbon nanofibers from methane over Ni-Co-Al catalysts. *Int. J. Hydrog. Energy* **2015**, *40*, 1774–1781. [CrossRef]

14. Al-Hassani, A.A.; Abbas, H.F.; Wan Daud, W.M.A. Production of CO_x-free hydrogen by the thermal decomposition of methane over activated carbon: Catalyst deactivation. *Int. J. Hydrog. Energy* **2014**, *39*, 14783–14791. [CrossRef]

15. Anjaneyulu, C.; Kumar, S.N.; Kumar, V.V.; Naresh, G.; Bhargava, S.K.; Chary, K.V.R.; Venugopal, A. Influence of La on reduction behavior and Ni metal surface area of $Ni-Al_2O_3$ catalysts for CO_x free H_2 by catalytic decomposition of methane. *Int. J. Hydrog. Energy* **2015**, *40*, 3633–3641. [CrossRef]

16. Al-Hassani, A.A.; Abbas, H.F.; Wan Daud, W.M.A. Hydrogen production via decomposition of methane over activated carbons as catalysts: Full factorial design. *Int. J. Hydrog. Energy* **2014**, *39*, 7004–7014. [CrossRef]

17. Pereira, M.F.R.; Figueiredo, J.L.; Órfão, J.J.M.; Serp, P.; Kalck, P.; Kihn, Y. Catalytic activity of carbon nanotubes in the oxidative dehydrogenation of ethylbenzene. *Carbon* **2004**, *42*, 2807–2813. [CrossRef]

18. Serp, P.; Corrias, M.; Kalck, P. Carbon nanotubes and nanofibers in catalysis. *Appl. Catal. A Gen.* **2003**, *253*, 337–358. [CrossRef]

19. Iijima, S. Helical microtubes of graphitic carbon. *Nature* **1991**, *354*, 56–58. [CrossRef]

20. Lee, M.S.; Lee, S.Y.; Park, S.J. Preparation and characterization of multi-walled carbon nanotubes impregnated with polyethyleneimine for carbon dioxide capture. *Int. J. Hydrog. Energy* **2015**, *40*, 3415–3421. [CrossRef]

21. Takenaka, S.; Ishida, M.; Serizawa, M.; Tanabe, E.; Otsuka, K. Formation of carbon nanofibers and carbon nanotubes through methane decomposition over supported cobalt catalysts. *J. Phys. Chem. B* **2004**, *108*, 11464–11472. [CrossRef]

22. Li, D.; Chen, J.; Li, Y. Evidence of composition deviation of metal particles of a $NieCu/Al_2O_3$ catalyst during methane decomposition to CO_x-free hydrogen. *Int. J. Hydrog. Energy* **2009**, *34*, 299–307. [CrossRef]

23. Jana, P.; de la Pena, V.A.; Shea, O.; Coronado, J.M.; Serrano, D.P. Cobalt based catalysts prepared by pechini method for CO_2-free hydrogen production by methane decomposition. *Int. J. Hydrog. Energy* **2010**, *35*, 10285–10294. [CrossRef]

24. Pinilla, J.L.; Utrilla, R.; Lazaro, M.J.; Moliner, R.; Suelves, I.; García, A.B. Ni- and Fe-based catalysts for hydrogen and carbon nanofilament production by catalytic decomposition of methane in a rotary bed reactor. *Fuel Process. Technol.* **2011**, *92*, 1480–1488. [CrossRef]

25. Politano, A.; Chiarello, G. Unveiling the Oxidation Processes of $Pt_3Ni(111)$ by Real-Time Surface Core-Level Spectroscopy. *ChemCatChem* **2016**, *8*, 713–718. [CrossRef]

26. Ren, H.; Humbert, M.P.; Carl, A.; Menning, C.A.; Chen, J.G.; Shu, Y.; Singh, U.G.; Cheng, W.C. Inhibition of coking and CO poisoning of Pt catalysts by the formation of Au/Pt bimetallic surfaces. *Appl. Catal. A* **2010**, *375*, 303–309. [CrossRef]

27. Baraldi, A.; Bianchettin, L.; Gironcoli, S.D.; Vesselli, E.; Lizzit, S.; Petaccia, L.; Comelli, G.; Rosei, R. Enhanced chemical reactivity of under-coordinated atoms at Pt-Rh bimetallic surfaces: A spectroscopic characterization. *J. Phys. Chem. C* **2011**, *115*, 3378–3384. [CrossRef]

28. Nilekar, A.U.; Xu, Y.; Zhang, J.; Vukmirovic, M.B.; Sasaki, K.; Adzic, R.; Mavrikakis, M. Bimetallic and Ternary Alloys for Improved Oxygen Reduction Catalysis. *Top. Catal.* **2007**, *46*, 276–284. [CrossRef]

29. Reshetenko, T.V.; Avdeeva, L.B.; Ushakov, V.A.; Moroz, E.M.; Shmakov, A.N.; Kriventsov, V.V.; Kochubey, D.I.; Pavlyukhin, Y.T.; Chuvilin, A.L.; Ismagilov, Z.R. Coprecipitated iron-containing catalysts ($Fe-Al_2O_3$, $Fe-Co-Al_2O_3$, $Fe-Ni-Al_2O_3$) for methane decomposition at moderate temperatures. Part II: Evolution of the catalysts in reaction. *Appl. Catal. A* **2004**, *270*, 87–99. [CrossRef]

30. Shen, W.; Huggins, F.E.; Shah, N.; Jacobs, G.; Wang, Y.; Shi, X.; Huffman, G.P. Novel Fe–Ni nanoparticle catalyst for the production of CO- and CO_2-free H_2 and carbon nanotubes by dehydrogenation of methane. *Appl. Catal. A* **2008**, *351*, 102–110. [CrossRef]

31. Avdeeva, L.B.; Reshetenko, T.V.; Ismagilov, Z.R.; Likholobov, V.A. Iron-containing catalysts of methane decomposition: Accumulation of filamentous carbon. *Appl. Catal. A* **2002**, *228*, 53–63. [CrossRef]

32. Latorre, N.; Cazana, F.; Martínez-Hansen, V.M.; Royo, C.; Romeo, E.; Monzón, A. Ni-Co-Mg-Al catalysts for hydrogen and carbonaceous nanomaterials production by CCVD of methane. *Catal. Today* **2011**, *172*, 143–151. [CrossRef]

33. Shah, N.; Panjala, D.; Huffman, G.P. Hydrogen production by catalytic decomposition of methane. *Energy Fuels* **2001**, *15*, 1528–1534. [CrossRef]

34. Pudukudy, M.; Yaakob, Z.; Zubair, Z.; Akmal, S. Direct decomposition of methane over Pd promoted Ni/SBA-15 catalysts. *Appl. Surf. Sci.* **2015**, *353*, 127–136. [CrossRef]

35. Awadallah, A.E.; Aboul-Enein, A.A.; El-Desouki, D.S.; Aboul-Gheit, A.K. Catalytic thermal decomposition of methane to CO_x-free hydrogen and carbon nanotubes over MgO supported bimetallic group VIII catalysts. *Appl. Surf. Sci.* **2014**, *296*, 100–107. [CrossRef]

36. Bartholomew, C.H. Carbon deposition in steam reforming and methanation. *Catal. Rev. Sci. Eng.* **1982**, *24*, 67–112. [CrossRef]

37. Hartenstein, H.U.; Hoffman, T. Method of Regeneration of SCR Catalyst. U.S. Patent 7723251B2, 25 May 2010.

38. Hazzim, F.; Abbas, W.M.A.; Wan, D. Thermocatalytic decomposition of methane for hydrogen production using activated carbon catalyst: Regeneration and characterization studies. *Int. J. Hydrog. Energy* **2009**, *34*, 8034–8045.

39. Al-Fatesh, A.S.; Fakeeha, A.H.; Khan, W.U.; Ibrahim, A.A.; He, S.; Seshan, K. Production of hydrogen by catalytic methane decomposition over alumina supported mono-, bi- and tri-metallic catalysts. *Int. J. Hydrog. Energy* **2016**, *41*, 22932–22940. [CrossRef]

40. Osman, A.I.; Abu-Dahrieh, J.K.; McLaren, M.; Laffir, F.; Rooney, D.W. Characterization of Robust Combustion Catalyst from Aluminum Foil Waste. *ChemistrySelect* **2018**, *3*, 1545–1550. [CrossRef]

41. Zárate, J.; Rosas, G.; Pérez, R. Structural Transformations of the Pseudo boehmite to a-Alumina. *Adv. Technol. Mater. Mater. Process. J.* **2005**, *1*. [CrossRef]

42. Jorio, A.; Saito, R.; Hafner, J.H.; Lieber, C.M.; Hunter, M.; McClure, T.; Dresselhaus, G.; Dresselhaus, M.S. Structural (*n, m*) determination of isolated single-wall carbon nanotubes by resonant Raman scattering. *Phys. Rev. Lett.* **2001**, *86*, 1118–1121. [CrossRef] [PubMed]

43. Hou, P.X.; Xu, S.T.; Ying, Z.; Yang, Q.H.; Liu, C.; Cheng, H.M. Hydrogen adsorption/desorption behavior of multi-walled carbon nanotubes with different diameters. *Carbon* **2003**, *41*, 2471–2476. [CrossRef]

44. Miners, S.A.; Rance, G.A.; Khlobystov, A.N. Chemical reactions confined within carbon nanotubes. *Chem. Soc. Rev.* **2016**, *45*, 4727–4746. [CrossRef] [PubMed]

45. Cullity, B.D. *Elements of X-ray Diffractions*; Addison-Wesley: New York, NY, USA, 1956.

46. Al-Fatesh., A.S.; Barama, S.; Ibrahim, A.A.; Barama, A.; Khan, W.U.; Fakeeha, A. Study of Methane Decomposition on Fe/MgO-Based Catalyst Modified by Ni, Co, and Mn Additives. *Chem. Eng. Commun.* **2017**, *204*, 739–749. [CrossRef]

47. Al-Fatesh, A.S.; Amin, A.; Ibrahim, A.A.; Khan, W.U.; Soliman, M.A.; AL-Otaibi, R.L.; Fakeeha, A.H. Effect of Ce and Co Addition to Fe/Al$_2$O$_3$ for Catalytic Methane Decomposition. *Catalysts* **2016**, *6*, 40. [CrossRef]

48. Ferrari, A.C. Raman spectroscopy of graphene and graphite: Disorder, electron–phonon coupling, doping and non-adiabatic effects. *Solid State Commun.* **2007**, *143*, 47–57. [CrossRef]

49. Shi, B.; Zhang, Z.; Zha, B.; Liu, D. Structure evolution of spinel Fe-MII (M=Mn, Fe, Co, Ni) ferrite in CO hydrogeneration. *Mol. Catal.* **2018**, *456*, 31–37. [CrossRef]

50. Khan, A.; Smirniotis, P.G. Relationship between temperature-programmed reduction profile and activity of modified ferrite-based catalysts for WGS reaction. *J. Mol. Catal. A Chem.* **2008**, *280*, 43–51. [CrossRef]

51. Jo, S.B.; Chae, H.J.; Kim, T.Y.; Lee, C.H.; Oh, J.U.; Kang, S.-H.; Kim, J.W.; Jeong, M.; Lee, S.C.; Kim, J.C. Selective CO hydrogenation over bimetallic Co-Fe catalysts for the production of light paraffin hydrocarbons (C2-C4): Effect of H$_2$/CO ratio and reaction temperature. *Catal. Commun.* **2018**, *117*, 74–78. [CrossRef]

52. Zhou, L.; Enakonda, L.R.; Harb, M.; Saih, Y.; A-Tapiaa, A.; Ould-Chikh, S.; Hazemann, J.L.; Li, J.; Wei, N.; Gary, D.; et al. Fe catalysts for methane decomposition to produce hydrogen and carbon nano materials. *Appl. Catal. B Environ.* **2017**, *208*, 44–59. [CrossRef]

53. Muradov, N.; Chen, Z.; Smith, F. Fossil hydrogen with reduced CO$_2$ emission: Modeling thermocatalytic decomposition of methane in a fluidized bed of carbon particles. *Int. J. Hydrog. Energy* **2005**, *30*, 1149–1158. [CrossRef]

54. Moliner, R.; Suelves, I.; Lázaro, M.J.; Moreno, O. Thermo-catalytic decomposition of methane over activated carbons: Influence of textural properties and surface chemistry. *Int. J. Hydrog. Energy* **2005**, *30*, 293–300. [CrossRef]

Article

A Facile Fabrication of Supported Ni/SiO$_2$ Catalysts for Dry Reforming of Methane with Remarkably Enhanced Catalytic Performance

Yan Xu [1,2], Qiang Lin [1], Bing Liu [1], Feng Jiang [1], Yuebing Xu [1] and Xiaohao Liu [1,*]

[1] Department of Chemical Engineering, School of Chemical and Material Engineering, Jiangnan University, Wuxi 214122, China; xuyan8787@163.com (Y.X.); lq15355615253@163.com (Q.L.); liubing@jiangnan.edu.cn (B.L.); jiangfeng@jiangnan.edu.cn (F.J.); xuyuebing@jiangnan.edu.cn (Y.X.)

[2] School of Chemistry and Chemical Engineering, Xuzhou University of Technology, Xuzhou 221018, China

* Correspondence: liuxh@jiangnan.edu.cn

Received: 12 January 2019; Accepted: 13 February 2019; Published: 15 February 2019

Abstract: Ni catalysts supported on SiO$_2$ are prepared via a facile combustion method. Both glycine fuel and ammonium nitrate combustion improver facilitate the formation of much smaller Ni nanoparticles, which give excellent activity and stability, as well as a syngas with a molar ratio of H$_2$/CO of about 1:1 due to the minimal side reaction toward revserse water gas shift (RWGS) in CH$_4$ dry reforming.

Keywords: Ni catalysts; combustion method; dry reforming of methane; RWGS reaction; improved stability

---

1. Introduction

The availability of natural gas (or shale gas) in large reserves makes CH$_4$ serve as a suitable feedstock used in C1 chemistry to produce desired fuels and chemicals [1]. Unfortunately, the chemical inertness of CH$_4$ results in direct conversion, which constitutes a great challenge for highly efficient utilization [2]. Ideally, the best use of CH$_4$ occurs when it is converted into syngas, which can facilitate further downstream conversion [3] by means of the methanol route [4] and Fischer–Tropsch synthesis (FTS) [5–12] due to good reactivity, unlike the CH$_4$ which has a high dissociation energy C–H bond [1]. Among the most widely investigated technologies, there are comparable advantages associated with the dry reforming of CH$_4$ (DRM) with CO$_2$ for producing syngas [13]. On the one hand, compared to the other reforming processes, there is a 20% lower operating cost for DRM [14]; on the other hand, the reforming of CH$_4$ using CO$_2$ not only produces high purity syngas [15,16] but also reduces the emissions of two abundantly available greenhouse gases to alleviate global climate change [17–20].

In spite of the above-mentioned merits, DRM suffers from serious carbon deposits on the surface of Ni nanoparticles, which leads to a remarkable loss of active sites [21–25]. Recently, DRM research efforts have resulted in strategies to improve the stability of the catalyst [26]. Based on the fact that smaller Ni nanoparticles efficiently improve catalytic performance by avoiding carbon accumulation [27–32], the general concept is to develop the catalyst preparation protocol to obtain small Ni nanoparticles encapsulated in the support or confined by the stable porous oxide layer to prevent sintering [33,34]. For example, Tomishige et al. reported that the solid solution catalyst of nickel–magnesia, which was prepared by the co-precipitation method, showed high and stable activity without carbon deposits for 100 days [35,36]. Kawi et al. synthesized a Ni-yolk@Ni@SiO$_2$ nanocomposite with a yolk-satellite shell structure to efficiently inhibit the sintering of Ni, which resulted in negligible carbon deposition, and the CH$_4$ conversion was 10% after the first 2 hours of reaction under the conditions of 800 °C, a gas hourly space velocity (GHSV) of 1440 L·g^{-1}cat·h^{-1}, a Wcat of 0.01 g, and a CO$_2$:CH$_4$:N$_2$ ratio of 1:1:1 [37].

Similarly, Wang et al. pointed out that the Ni nanoparticle cores encapsulated by the mesoporous Al_2O_3 shells show superior coke resistance because of the confinement effects which prevent the Ni nanoparticles from agglomeration at high temperatures, and the CH_4 and CO_2 conversions under the reaction conditions of 800 °C, CO_2/CH_4 of 1/1, and a weight hourly space velocity (WHSV) of 36 $L\cdot h^{-1}\cdot gcat^{-1}$ were about 88% and 92%, respectively [38].

Herein, different from the above-mentioned encapsulated Ni catalysts with relatively complicated preparation procedures, we propose a facile one-step strategy to prepare the SiO_2 supported Ni catalysts toward the controlled formation of nanoparticle size and Ni-support interaction, which could lead to high activity and stability. Following the conventional impregnation method, glycine ($C_2H_5NO_2$) and ammonium nitrate (NH_4NO_3) were introduced into the impregnated solution of nickel precursor ($Ni(NO_3)_2\cdot 6H_2O$), as shown in Scheme 1. It was expected that the mixed materials with $C_2H_5NO_2$ as fuel and NH_4NO_3 as combustion improver reacted exothermically after ignition which finished within a short time-frame with a very high temperature and release of a large quantity of gases, such as CO_2, water, and N_2. We thought this process might facilitate the formation of smaller crystalline materials and regulate the metal-support interaction, resulting in improved catalytic performance in the DRM reaction. To demonstrate the effects of the above combustion process on the catalytic performance, several characterizations, such as Brunauer-Emmett-Teller (BET), transmission electron microscopy (TEM), X-ray diffraction (XRD), H_2 temperature-programmed reduction (H_2-TPR) and thermogravimetric (TG), were employed to characterize the catalyst.

2. Results and Discussion

2.1. Characterization of the Catalyst Sample

As shown in Figure S1, all the fresh Ni/SiO_2 catalysts exhibit apparent diffraction peaks at 2θ values of 37.3°, 43.2°, 63.0°, 75.4°, and 79.4° assigned to the NiO (JCPDS 22-1189). For the reduced catalysts (Figure 1a), Ni/SiO_2-0/0 prepared by the conventional wetness impregnation method displayed the most intensive diffraction peaks at 2θ values of 44.5°, 52.2°, and 77.0°, which are the characteristic peaks of metallic Ni (JCPDS 1-1206). According to Figure S1, the peak at 37.3° should be assigned to NiO. As the NH_4NO_3 was introduced into the impregnated solution with nickel nitrate, the resulting catalyst (Ni/SiO_2-0/1) exhibited almost the same diffraction peak intensity at 44.5°. However, for the case of $C_2H_5NO_2$, Ni/SiO_2-2/0 displays a much weaker diffraction peak. Interestingly, the addition of both $C_2H_5NO_2$ and NH_4NO_3 results in almost no detectable diffraction peaks for Ni nanoparticles (Ni/SiO_2-2/1), suggesting that smaller Ni nanoparticles can be obtained by synergistic effects of fuel and combustion improver in the combustion process, as presented in Scheme 1.

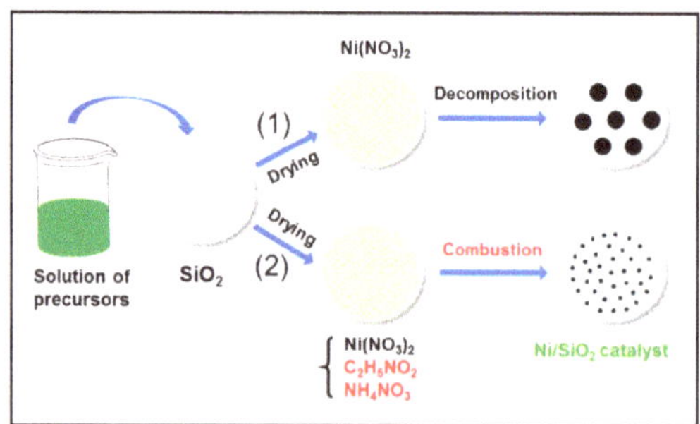

Scheme 1. One-step facile synthesis of Ni catalysts supported on silica (SiO_2) prepared by the combustion of $Ni(NO_3)_2$–$C_2H_5NO_2$–NH_4NO_3 impregnated in the porous SiO_2.

TEM images of the reduced catalysts are depicted in Figure 1b,c. The Ni/SiO$_2$-2/1 displays an average Ni nanoparticle size of only 6.1 ± 2.7 nm which is significantly smaller than that for Ni/SiO$_2$-0/0 (31.3 ± 13.5 nm). The significant difference in the Ni nanoparticle size further confirms the synergistic effects of C$_2$H$_5$NO$_2$ and NH$_4$NO$_3$ in reducing the Ni nanoparticle size. The combustion process between N$_2$O and NH$_3$ is highly exothermic. The decomposition of nickel nitrate produces N$_2$O gas at 250 °C, while the decomposition of C$_2$H$_5$NO$_2$ gives NH$_3$ along with CO$_2$ and H$_2$O. The combustion process is triggered by the reaction between N$_2$O and NH$_3$ to form N$_2$ and H$_2$O [39]. When NH$_4$NO$_3$ is further added, NH$_3$ and N$_2$O can be formed via its decomposition at a low temperature of about 200 °C, thereby promoting combustion. The high-temperature stage in a short-duration favors the formation of ultra-small nanoparticles in a short time which may be in the order of seconds [40].

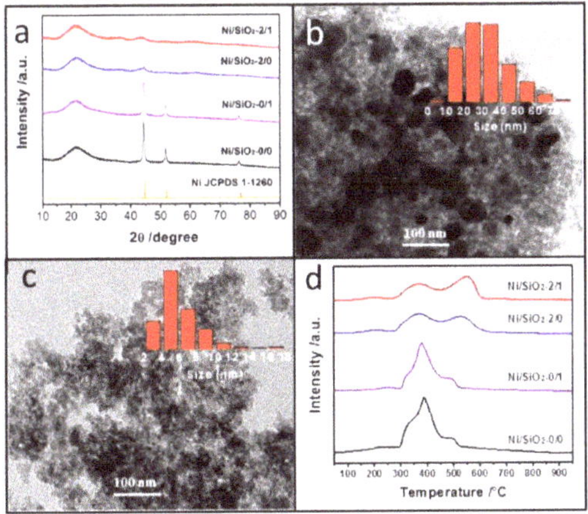

Figure 1. (**a**) XRD patterns of reduced Ni/SiO$_2$ catalysts prepared with the combustion method by using different ratios of C$_2$H$_5$NO$_2$ to NH$_4$NO$_3$. (**b,c**) TEM images and Ni size distribution of the reduced Ni/SiO$_2$-0/0 and Ni/SiO$_2$-2/1 catalysts, respectively. (**d**) H$_2$-TPR profiles of the fresh Ni/SiO$_2$ catalysts prepared with the combustion method.

Figure 1d exhibits the reduction behavior of fresh Ni/SiO$_2$ catalysts with different molar ratios of C$_2$H$_5$NO$_2$ to NH$_4$NO$_3$. As expected, NH$_4$NO$_3$ does not obviously change the H$_2$-TPR profile compared to the case of Ni/SiO$_2$-0/0, as both catalysts show a strong reduction peak at 300–450 °C with a small right shoulder peak at 450–510 °C. However, C$_2$H$_5$NO$_2$ only (Ni/SiO$_2$-2/0) notably weakens the peak at lower temperatures, accompanied by a shift in the right shoulder peak to the higher reduction temperature with enhanced intensity. For Ni/SiO$_2$-2/1, the high temperature reduction peak is further intensified and shifts to a higher reduction temperature range. This result suggests that the smaller Ni nanoparticle size results in a more difficult reduction owing to a stronger metal-support interaction [41]. The reduction profiles correspond to the XRD and TEM results.

2.2. Activity Evaluation

Figure 2 shows the catalytic performance in the CH$_4$ dry reforming reaction with CO$_2$ over the as-prepared catalysts. In the case of catalytic activity, the CH$_4$ conversion over Ni/SiO$_2$-0/0 exhibits a rapid drop from 78.3% to 53.0% in the early ten hours and then gradually becomes stable. In contrast, Ni/SiO$_2$-0/1 gives a milder and continuous decrease in CH$_4$ conversion until the end of the reaction. Surprisingly, Ni/SiO$_2$-2/0 exhibits a stable and higher CH$_4$ conversion over the whole reaction period of 50 hours. Furthermore, Ni/SiO$_2$-2/1 displays a more stable and even higher CH$_4$ conversion. The CO$_2$ and CH$_4$ conversions are similar for all Ni/SiO$_2$ catalysts. However, in the corresponding reaction period, the CO_2 conversion is always slightly higher compared to the CH_4 conversion. When the DRM reaction over Ni/SiO$_2$-2/1 is stable, the conversion rates of CH_4 and CO_2 are 83.6% and 90.6%, respectively, which are slightly lower than their equilibrium conversion rates at 91% and 95% calculated by HSC chemistry 6.0 (Table S1). Also, in the case of the H_2/CO molar ratio, it follows the same trend as that for CH_4 conversion over all the Ni/SiO$_2$ catalysts. Specifically, for the Ni/SiO$_2$-2/1 catalyst, the desired H_2/CO molar ratio at the value of 1/1 is obtained, which results from the efficiently suppressed reverse water gas shift (RWGS) reaction. How the Ni/SiO$_2$ morphology affects the catalytic performance is discussed briefly in the following part.

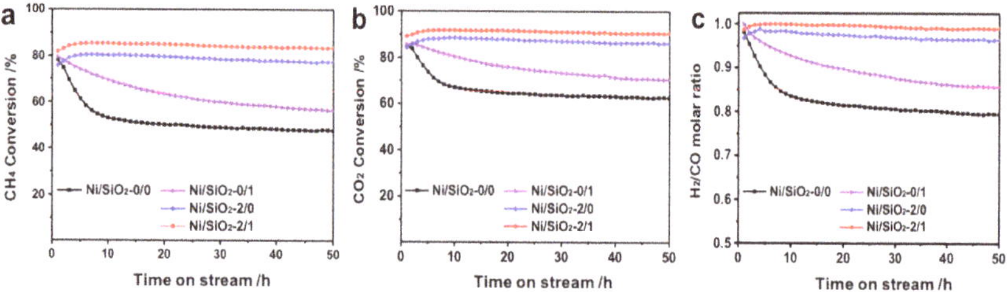

Figure 2. CH_4 conversion (**a**), CO_2 conversion (**b**), and H_2/CO molar ratio (**c**) as a function of time on stream for Ni/SiO$_2$ catalysts prepared with the combustion method at different molar ratios of $C_2H_5NO_2$ to NH_4NO_3 (black line: conventional wetness impregnation method; purple line: NH_4NO_3 only; blue line: $C_2H_5NO_2$ only; red line: 2/1 ratio of $C_2H_5NO_2$ to NH_4NO_3). The reaction was carried at 800 °C with 200 mg of catalyst and a molar ratio of CH_4/CO_2/N_2 = 9/9/2 with 160 mL/min.

The DRM reaction is extremely endothermic. Equation (1) shows that the DRM process can produce a syngas with an H_2/CO ratio of 1:1. During the DRM process, several reactions simultaneously occur, like CH_4 dissociation (Equation (2)), reduction of CO_2 to CO (Equation (3)), and the RWGS reaction (Equation (4)).

$$CH_4 + CO_2 = 2CO + 2H_2 \ (\Delta H_{298K} = +247 \text{ kJ mol}^{-1}) \tag{1}$$

$$CH_4 = C(s) + 2H_2 \ (\Delta H_{298K} = +75 \text{ kJ mol}^{-1}) \tag{2}$$

$$C(s) + CO_2 = 2CO \ (\Delta H_{298K} = +171 \text{ kJ mol}^{-1}) \tag{3}$$

$$CO_2 + H_2 = CO + H_2O \ (\Delta H_{298K} = +41.2 \text{ kJ mol}^{-1}) \tag{4}$$

The driving force for Equations (2)–(4) strongly depends on the temperature, reactant partial pressure and catalyst structures. In the investigated Ni/SiO$_2$ catalysts, both activation of CH_4 and CO_2 can occur on the active Ni surface since SiO$_2$ support is inert material. It is believed that CH_4 activation tends to form an intermediate, like CH_x or a formyl group, but dissociates directly to C species and H_2 at high temperature. Essentially, the DRM reaction of Ni catalysts might follow a dynamic redox type mechanism as the CO_2 oxidizes Ni^0 to $Ni^{+\delta}$ to give CO, and the oxidative state $Ni^{+\delta}$ is reduced to Ni^0 by C species as a result of CH_4 dissociation. As seen from the above reaction cycle, it is clear that the presence of O from CO_2 helps the dissociation of CH_4. To avoid the catalyst deactivation resulting from carbon accumulation, the C species from CH_4 dissociation must react timely with CO_2 to give CO. The reaction rate of this step is closely related to the Ni nanoparticle size, as the larger Ni surface favors the formation of multicarbon C_n species, which are potential precursors of carbon deposits such as coke. The smaller Ni nanoparticles allow a smaller amount of carbon species on the Ni nanoparticle surface. Thus, it is easier to keep the monoatomic C species isolated, and in time, they are oxidized by CO_2 to CO. By minimizing the rate of C species combination, the carbon accumulation could be effectively suppressed. Indeed, as shown in Figure 3, Ni/SiO$_2$-0/0, with an average nanoparticle size of 31.3 ± 13.5 nm, gives the highest amount of carbon deposits with 2.7 mg carbon deposits gCH$_4^{-1}$

as the BET surface area is decreased to the largest extent (Table S2). In contrast, the Ni/SiO$_2$-2/1 with a smaller nanoparticle size of 6.1 $\pm$ 2.7 nm is significantly coke-resistant, as the amount of carbon deposits decreases to 0.9 mg carbon deposits gCH$_4{}^{-1}$. The above experimental results reflect that the smaller Ni nanoparticle size is favorable to lower carbon deposits and thereby improve the catalyst stability, as shown in Figure S2. It should be noted that, in spite of the significant decrease in carbon deposits over Ni/SiO$_2$-2/1 catalyst, a considerable amount of coke is still formed during the DRM reaction of 50 hours. It can be deduced that most of the carbon deposits might not locate on the Ni nanoparticle surface but are located on the SiO$_2$ support since the catalytic activity is quite stable. It is reasonable for us to imagine that the Ni nanoparticles are lying on the SiO$_2$ support and not confined by porous layer material, which provides a chance for the carbon species to grow continuously along the SiO$_2$ support surface initiated by the Ni nanoparticle and finally form strips of nanofiber.

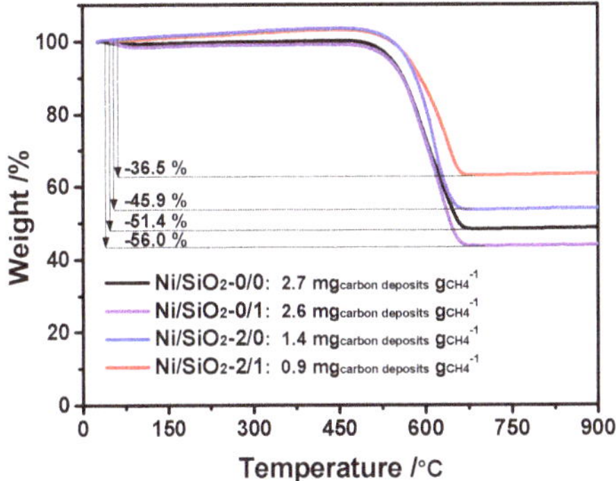

Figure 3. TG patterns of spent Ni/SiO$_2$ catalysts after the dry reforming (DRM) reaction of 50 hours. The catalytic results are shown in Figure 2.

As seen from Figure 2, the H$_2$/CO molar ratio is highly dependent on the CO$_2$ conversion. A lower CO$_2$ conversion can cause a decrease in the molar ratio of H$_2$/CO to a large extent as a result of the RWGS reaction, as the higher concentration of CO$_2$ drives the reaction to the right side (Equation 4). At 800 °C, the standard free energy for the RWGS reaction ($\Delta G^0 = 8545 + 7.84T$) and the reduction of CO$_2$ to CO ($\Delta G^0 = 39810 - 40.87T$) [13] is -132.68 kJ mol^{-1} and -4043.51 kJ mol^{-1}, respectively. It can be speculated that the reduction of CO$_2$ to CO, C(s) + CO$_2$ = 2CO, occurs more easily as a result of the lower ΔG. Comparing the value of ΔG in the RWGS reaction, the CO$_2$ that oxidizes the C species to CO is more thermodynamically favored than its RWGS reaction. As the lower CO$_2$ conversion corresponds to lower CH$_4$ conversion, the C(s) species dissociated from CH$_4$ is not sufficient for its reaction with CO$_2$. Therefore, the CO$_2$ reacting with H$_2$ toward the RWGS reaction is promoted. In order to minimize the side reaction toward the RWGS, it is necessary to operate the DRM reaction with a high CO$_2$ conversion rate.

3. Materials and Methods

3.1. Catalyst Preparation

The supported Ni catalysts were prepared with SiO$_2$ support (Tosoh Kabushiki-gaisha, Tokyo, Japan) by the combustion method, and the combustible materials contained hydrate glycine (C$_2$H$_5$NO$_2$), ammonium nitrate (NH$_4$NO$_3$) and nickel nitrate (Ni(NO$_3$)$_2$·6H$_2$O) with different C$_2$H$_5$NO$_2$/Ni(NO$_3$)$_2$·6H$_2$O and NH$_4$NO$_3$/Ni(NO$_3$)$_2$·6H$_2$O molar ratios. Briefly, the aqueous solution of the desired amounts of C$_2$H$_5$NO$_2$, NH$_4$NO$_3$ and Ni(NO$_3$)$_2$·6H$_2$O was added into the SiO$_2$ support at room temperature by incipient wetness impregnation, followed by drying with a rotary evaporator for 2 hours at 80 °C, and then overnight at 120 °C. Afterwards, the dried solid materials were calcined in air for 1 hour at 300 °C with a heating rate of 1 °C/min and another 3 hours at 550 °C with a heating rate of 2 °C/min. The calcined samples were denoted as Ni/SiO_2-x/y, where x and y indicate the molar ratio of $C_2H_5NO_2/Ni(NO_3)_2 \cdot 6H_2O$ and $NH_4NO_3/Ni(NO_3)_2 \cdot 6H_2O$, respectively. The metallic Ni loading was 10 wt%. The samples were then crushed and sieved into a 40–60 mesh size for subsequent catalytic tests.

3.2. Catalyst Characterization

Fresh, reduced and spent samples were characterized by several techniques to identify and infer the effects of combustible materials such as $C_2H_5NO_2$ and NH_4NO_3 on the catalyst morphology and the resulting catalytic performance. N_2 adsorption-desorption isotherms for each sample were collected on a Micromeritics ASAP 2020 system (Micromeritics, Norcross, GA, USA). The surface area, pore size and pore volume were calculated with the N_2 adsorption-desorption isotherms via the conventional Barrett–Joyner–Halenda (BJH) and Brunauer–Emmett–Teller (BET) methods. Prior to the measurements, the samples were outgassed under vacuum for 5 hours at 200 °C. The X-ray diffraction (XRD) patterns of each reduced sample were obtained with a Bruker AXS D8 Advance diffractometer (Bruker AXS, Karlsruhe, Germany) with Cu Kα radiation (λ=1.5406 Å) at a scanning rate of 6°/min with the 2θ range of 10–90°. The reducibility of the catalyst was studied by the H_2 temperature-programmed reduction (H_2-TPR) in an auto-controlled flow reactor system of TP-5076, which is equipped with a thermal conductivity detector (TCD, Tianjin Xianquan Co., China). The sample of 50 mg was pretreated in N_2 stream at 200 °C for 1 hour. Additionally, when the temperature cooled down to 30 °C, the sample was heated to 950 °C at a heating rate of 10 °C/min in the H_2/N_2 flow (5 vol.% H_2 in N_2) of 30 mL/min. The H_2-TPR spectra were obtained at the temperature range of 50–950 °C. The carbon accumulation in spent samples after reaction for 50 hours was determined by thermogravimetric (TG) analysis on a Mettler–Toledo TGA-1100SF thermogravimetric analyzer (Mettler-Toledo, Greifensee, Switzerland).

3.3. Catalytic Test

The dry reforming of CH_4 with CO_2 was performed at atmospheric pressure in a continuous-flow fixed bed quartz tube reactor with an inner diameter of 9 mm. For the typical experiment, 200 mg of shaped catalyst was filled into the center of the reactor. Before starting the reforming reaction, the catalyst was pre-reduced to 750 °C and atmospheric pressure for 2 hours in an H_2 flow of 60 mL/min with a heating rate of 10 °C/min. After that, the reactor temperature was elevated to 800 °C, and then a flow of gas mixture with a molar ratio of $CH_4/CO_2/N_2 = 9/9/2$ was fed with a flow rate of 160 mL/min. The products were analyzed by online gas chromatography (Agilent GC 7820A, Agilent, USA). CH_4, CO_2, H_2, N_2 and CO were measured by a TCD detector with a 5A molecular sieve column and a Porapak Q column. Additionally, 10% of N_2 was employed as an internal standard. The conversions of CH_4 and CO_2 were calculated with the following formulas:

$$X_{CH4} = (F_{CH4\text{-}in} - F_{CH4\text{-}out})/F_{CH4\text{-}in} \times 100\% \tag{5}$$

$$X_{CO2} = (F_{CO2\text{-}in} - F_{CO2\text{-}out})/F_{CO2\text{-}in} \times 100\% \tag{6}$$

where X and F indicate the conversion and flow rate of i gas in the feed or the effluent, respectively.

4. Conclusions

In summary, the combustion method was applied to prepare SiO_2 supported Ni catalysts which showed remarkably smaller Ni nanoparticle sizes due to the synergistic effects of $C_2H_5NO_2$ and NH_4NO_3 in the combustion process. This kind of Ni/SiO_2 catalyst exhibits excellent coke-resistance performance and effectively suppresses the side reaction toward RWGS compared to that prepared with the conventional wetness impregnation method. As a result, there is almost no loss of activity with the H_2/CO molar ratio close to the theoretical value at 1/1 after a 50-hour stability test over the Ni/SiO_2-2/1 catalyst.

Supplementary Materials: The following are available online at http://www.mdpi.com/2073-4344/9/2/183/s1, Figure S1: XRD patterns of fresh Ni/SiO_2 catalysts prepared with the combustion method by using different ratios of $C_2H_5NO_2$ to NH_4NO_3, Figure S2: TEM images of spent Ni/SiO_2-0/0 catalyst ((a) and (b)) and Ni/SiO_2-2/1 catalyst ((c) and (d)) after 50-hours of reaction, Table S1: The equilibrium conversions of CH_4 and CO_2, H_2/CO molar ratio, and selectivity to H_2O calculated by HSC chemistry 6.0, Table S2: BET surface area of as-prepared Ni/SiO_2 catalysts.

Author Contributions: Conceptualization, X.L. and Y.X. (Yan Xu); methodology, X.L., Y.X. (Yan Xu) and Q.L.; formal analysis, Y.X. (Yan Xu) and Q.L.; investigation, Y.X. (Yan Xu) and Q.L.; writing—original draft preparation, Y.X. (Yan Xu) and X.L.; writing—review and editing, B.L., F.J. and Y.X. (Yuebing Xu); funding acquisition, X.L.

Funding: This work is supported by the National Natural Science Foundation of China (21576119, 21878127), the Fundamental Research Funds for the Central Universities (JUSRP51720B, JUSRP11813), and the Program of Introducing Talents of Discipline to Universities (111 Project B13025).

Conflicts of Interest: The authors declare no conflict of interest.

References

1. Caballero, A.; Pérez, P.J. Methane as Raw Material in Synthetic Chemistry: The Final Frontier. *Chem. Soc. Rev.* **2013**, *42*, 8809–8820. [CrossRef] [PubMed]

2. Schwach, P.; Pan, X.L.; Bao, X.H. Direct Conversion of Methane to Value-Added Chemicals over Heterogeneous Catalysts: Challenges and Prospects. *Chem. Rev.* **2017**, *117*, 8497–8520. [CrossRef] [PubMed]

3. Pruett, R.L. Synthesis Gas: A Raw Material for Industrial Chemicals. *Science* **1981**, *211*, 11–16. [CrossRef] [PubMed]

4. Reubroycharoen, P.; Yamagami, T.; Vitidsant, T.; Yoneyama, Y.; Ito, M.; Tsubaki, N. Continuous Low-Temperature Methanol Synthesis from Syngas Using Alcohol Promoters. *Energy Fuels* **2003**, *17*, 817–821. [CrossRef]

5. Xiong, H.F.; Jewell, L.L.; Coville, N.J. Shaped Carbons as Supports for the Catalytic Conversion of Syngas to Clean Fuels. *ACS Catal.* **2015**, *5*, 2640–2658. [CrossRef]

6. Zhang, Q.H.; Kang, J.C.; Wang, Y. Development of Novel Catalysts for Fischer-Tropsch Synthesis: Tuning the Product Selectivity. *ChemCatChem* **2010**, *2*, 1030–1058. [CrossRef]

7. Jiang, F.; Liu, B.; Li, W.P.; Zhang, M.; Li, Z.J.; Liu, X.H. Two-Dimensional Graphene-Directed Formation of Cylindrical Iron Carbide Nanocapsules for Fischer–Tropsch Synthesis. *Catal. Sci. Technol.* **2017**, *7*, 4609–4621. [CrossRef]

8. Zheng, J.; Cai, J.; Jiang, F.; Xu, Y.B.; Liu, X.H. Investigation of the Highly Tunable Selectivity to Linear α-Olefins in Fischer-Tropsch Synthesis over Silica-Supported Co and CoMn Catalysts by Carburization Reduction Pretreatment. *Catal. Sci. Technol.* **2017**, *7*, 4736–4755. [CrossRef]

9. Jiang, F.; Zhang, M.; Liu, B.; Xu, Y.B.; Liu, X.H. Insights into the Influence of Support and Potassium or Sulfur Promoter on Iron-based Fischer-Tropsch Synthesis: Understanding the Control of Catalytic Activity, Selectivity to Lower Olefins, and Catalyst Deactivation. *Catal. Sci. Technol.* **2017**, *7*, 1245–1265. [CrossRef]

10. Cai, J.; Jiang, F.; Liu, X.H. Exploring Pretreatment Effects in Co/SiO_2 Fischer-Tropsch Catalysts: Different Oxidizing Gases Applied to Oxidation-Reduction Process. *Appl. Catal. B Environ.* **2017**, *210*, 1–13. [CrossRef]

11. Liu, X.H.; Tokunaga, M. Controllable Fischer-Tropsch Synthesis by in Situ-Produced 1-Olefins. *ChemCatChem* **2010**, *2*, 1569–1572. [CrossRef]

12. Liu, X.H.; Linghu, W.S.; Li, X.H.; Asami, K.; Fujimoto, K. Effects of Solvent on Fischer-Tropsch Synthesis. *Appl. Catal. A Gen.* **2006**, *303*, 251–257. [CrossRef]

13. Pakhare, D.; Spivey, J. A review of Dry (CO_2) Reforming of Methane over Noble Metal Catalysts. *Chem. Soc. Rev.* **2014**, *43*, 7813–7837. [CrossRef] [PubMed]

14. Ross, J.R.H. Natural Gas Reforming and CO_2 Mitigation. *Catal. Today* **2005**, *100*, 151–158. [CrossRef]

15. Bitter, J.H.; Seshan, K.; Lercher, J.A. On the Contribution of X-ray Absorption Spectroscopy to Explore Structure and Activity Relations of Pt/ZrO_2 Catalysts for CO_2/CH_4 Reforming. *Top. Catal.* **2000**, *10*, 295–305. [CrossRef]

16. Galuszka, J.; Pandey, R.N.; Ahmed, S. Methane Conversion to Syngas in a Palladium Membrane Reactor. *Catal. Today* **1998**, *46*, 83–89. [CrossRef]

17. Vooradi, R.; Bertran, M.O.; Frauzem, R.; Anne, S.B.; Gani, R. Sustainable Chemical Processing and Energy-Carbon Dioxide Management: Review of Challenges and Opportunities. *Chem. Eng. Res. Des.* **2018**, *131*, 440–464. [CrossRef]

18. Peter, S.C. Reduction of CO_2 to Chemicals and Fuels: A Solution to Global Warming and Energy Crisis. *ACS Energy Lett.* **2018**, *3*, 1557–1561. [CrossRef]

19. Ha, K.S.; Bae, J.W.; Woo, K.J.; Jun, K.W. Efficient Utilization of Greenhouse Gas in a Gas-to-Liquids Process Combined with Carbon Dioxide Reforming of Methane. *Environ. Sci. Technol.* **2010**, *44*, 1412–1417. [CrossRef]

20. De Vasconcelos, B.R.; Minh, D.P.; Lyczko, N.T.; Phan, S.; Sharrock, P.; Nzihou, A. Upgrading Greenhouse Gases (Methane and Carbon Dioxide) into Syngas using Nickel-Based Catalysts. *Fuel* **2018**, *226*, 195–203. [CrossRef]

21. Ginsburg, J.M.; Pina, J.; El Solh, T.; De Lasa, H.I. Coke Formation over a Nickel Catalyst under Methane Dry Reforming Conditions: Thermodynamic and Kinetic Models. *Ind. Eng. Chem. Res.* **2005**, *44*, 4846–4854. [CrossRef]

22. Schulz, L.A.; Kahle, L.C.S.; Delgado, K.H.; Schunk, S.A.; Jentys, A.; Deutschmann, O.; Lercher, J.A. On the Coke Deposition in Dry Reforming of Methane at Elevated Pressures. *Appl. Catal. A Gen.* **2015**, *504*, 599–607. [CrossRef]

23. Bitter, J.H.; Seshan, K.; Lercher, J.A. Deactivation and Coke Accumulation during CO_2/CH_4 Reforming over Pt Catalysts. *J. Catal.* **1999**, *183*, 336–343. [CrossRef]

24. Mette, K.; Kühl, S.; Tarasov, A.; Willinger, M.G.; Kröhnert, J.; Wrabetz, S.; Trunschke, A.; Scherzer, M.; Girgsdies, F.; Düdder, H.; et al. High-Temperature Stable Ni Nanoparticles for the Dry Reforming of Methane. *ACS Catal.* **2016**, *6*, 7238–7248. [CrossRef]

25. Pawar, V.; Ray, D.; Subrahmanyam, C.; Janardhanan, V.M. Study of Short-Term Catalyst Deactivation due to Carbon Deposition during Biogas Dry Reforming on Supported Ni Catalyst. *Energy Fuels* **2015**, *29*, 8047–8052. [CrossRef]

26. Li, S.R.; Gong, J.L. Strategies for Improving the Performance and Stability of Ni-Based Catalysts for Reforming Reactions. *Chem. Soc. Rev.* **2014**, *43*, 7245–7256. [CrossRef] [PubMed]

27. Aleksandrov, H.A.; Pegios, N.; Palkovits, R.; Simeonov, K.; Vayssilov, G.N. Elucidation of the Higher Coking Resistance of Small Versus Large Nickel Nanoparticles in Methane Dry Reforming via Computational Modeling. *Catal. Sci. Technol.* **2017**, *7*, 3339–3347. [CrossRef]

28. Christensen, K.O.; Chen, D.; Lødeng, R.; Holmen, A. Effect of Supports and Ni Crystal Size on Carbon Formation and Sintering during Steam Methane Reforming. *Appl. Catal. A Gen.* **2006**, *314*, 9–22. [CrossRef]

29. Gonzalez-Delacruz, V.M.; Pereniguez, R.; Ternero, F.; Holgado, J.P.; Caballero, A. Modifying the Size of Nickel Metallic Particles by H_2/CO Treatment in Ni/ZrO_2 Methane Dry Reforming Catalysts. *ACS Catal.* **2011**, *1*, 82–88. [CrossRef]

30. Baudouin, D.; Rodemerck, U.; Krumeich, F.; de Mallmann, A.; Szeto, K.C.; Ménard, H.; Veyre, L.; Candy, J.P.; Webb, P.B.; Thieuleux, C.; et al. Particle size Effect in the Low Temperature Reforming of Methane by Carbon Dioxide on Silica-Supported Ni Nanoparticles. *J. Catal.* **2013**, *297*, 27–34. [CrossRef]

31. Margossian, T.; Larmier, K.; Kim, S.M.; Krumeich, F.; Fedorov, A.; Chen, P.; Müller, C.R.; Copéret, C. Molecularly-Tailored Nickel Precursor and Support Yield a Stable Methane Dry Reforming Catalyst with Superior Metal Utilization. *J. Am. Chem. Soc.* **2017**, *139*, 6919–6927. [CrossRef] [PubMed]

32. Abba, M.O.; Gonzalez-DelaCruz, V.M.; Colón, G.; Sebti, S.; Caballero, A. In Situ XAS Study of an Improved Natural Phosphate Catalyst for Hydrogen Production by Reforming of Methane. *Appl. Catal. B Environ.* **2014**, *150–151*, 459–465. [CrossRef]

33. Tian, H.; Li, X.Y.; Zeng, L.; Gong, J.L. Recent Advances on the Design of Group VIII Base-Metal Catalysts with Encapsulated Structures. *ACS Catal.* **2015**, *5*, 4959–4977. [CrossRef]

34. Gould, T.D.; Izar, A.; Weimer, A.W.; Falconer, J.L.; Medlin, J.W. Stabilizing Ni Catalysts by Molecular Layer Deposition for Harsh, Dry Reforming Conditions. *ACS Catal.* **2014**, *4*, 2714–2717. [CrossRef]

35. Tomishige, K.; Yamazaki, O.; Chen, Y.G.; Yokoyama, K.; Li, X.H.; Fujimoto, K. Development of Ultra-Stable Ni Catalysts for CO_2 Reforming of Methane. *Catal. Today* **1998**, *45*, 35–39. [CrossRef]

36. Chen, Y.G.; Tomishige, K.; Yokoyama, K.; Fujimoto, K. Promoting Effect of Pt, Pd and Rh Noble Metals to the $Ni_{0.03}Mg_{0.97}O$ Solid Solution Catalysts for the Reforming of CH_4 with CO_2. *Appl. Catal. A Gen.* **1997**, *165*, 335–347. [CrossRef]

37. Li, Z.W.; Mo, L.Y.; Kathiraser, Y.; Kawi, S. Yolk–Satellite–Shell Structured Ni–Yolk@Ni@SiO_2 Nanocomposite: Superb Catalyst toward Methane CO_2 Reforming Reaction. *ACS Catal.* **2014**, *4*, 1526–1536. [CrossRef]

38. Huang, Q.; Fang, X.Z.; Cheng, Q.Z.; Li, Q.; Xu, L.J.; Xu, X.L.; Liu, W.M.; Gao, Z.X.; Zhou, W.F.; Wang, X. Synthesis of Highly Active and Stable Ni@Al$_2$O$_3$ Embedded Catalyst for Methane Dry Reforming: On the Confinement Effects of Al$_2$O$_3$ Shells for Ni Nanoparticles. *ChemCatChem* **2017**, *9*, 3563–3571. [CrossRef]

39. Varma, A.; Mukasyan, A.S.; Rogachev, A.S.; Manukyan, K.V. Solution Combustion Synthesis of Nanoscale Materials. *Chem. Rev.* **2016**, *116*, 14493–14586. [CrossRef]

40. Manukyan, K.V.; Chen, Y.S.; Rouvimov, S.; Li, P.; Li, X.; Dong, S.; Liu, X.Y.; Furdyna, J.K.; Orlov, A.; Bernstein, G.H.; et al. Ultrasmall α-Fe$_2$O$_3$ Superparamagnetic Nanoparticles with High Magnetization Prepared by Template-Assisted Combustion Process. *J. Phys. Chem. C* **2014**, *118*, 16264–16271. [CrossRef]

41. Gao, X.Y.; Hidajat, K.; Sawi, S. Facile synthesis of Ni/SiO$_2$ catalyst by sequential hydrogen/air treatment: A superior anti-coking catalyst for dry reforming of methane. *J. CO$_2$ Util.* **2016**, *15*, 146–153. [CrossRef]

 *catalysts*

Article

Combined Magnesia, Ceria and Nickel catalyst supported over γ-Alumina Doped with Titania for Dry Reforming of Methane

Ahmed Sadeq Al-Fatesh [1],*, **Samsudeen Olajide Kasim** [1], **Ahmed Aidid Ibrahim** [1], **Anis Hamza Fakeeha** [1], **Ahmed Elhag Abasaeed** [1], **Rasheed Alrasheed** [2], **Rawan Ashamari** [2] and **Abdulaziz Bagabas** [2],*

[1] Chemical Engineering Department, College of Engineering, King Saud University, P.O. Box 800, Riyadh 11421, Saudi Arabia; sofkolajide2@gmail.com (S.O.K.); aidid@ksu.edu.sa (A.A.I.); anishf@ksu.edu.sa (A.H.F.); abasaeed@ksu.edu.sa (A.E.A.)

[2] National Petrochemical Technology Center (NPTC), Materials Science Research Institute (MSRI), King Abdulaziz City for Science and Technology, P.O. Box 6086, Riyadh 11442, Saudi Arabia; rrisheed@kacst.edu.sa (R.A.); ralshammary@kacst.edu.sa (R.A.)

* Correspondence: aalfatesh@ksu.edu.sa (A.S.A.-F.), abagabas@hotmail.com (A.B.); Tel.: +966-11-467-6859 (A.S.A.-F.); Tel.: +966-11-481-3790 (A.B.)

Received: 21 January 2019; Accepted: 13 February 2019; Published: 18 February 2019

Abstract: This study investigated dry reforming of methane (DRM) over combined catalysts supported on γ-Al_2O_3 support doped with 3.0 wt. % TiO_2. Physicochemical properties of all catalysts were determined by inductively coupled plasma/mass spectrometry (ICP-MS), nitrogen physisorption, X-ray diffraction, temperature programmed reduction/oxidation/desorption/pulse hydrogen chemisorption, thermogravimetric analysis, and scanning electron microscopy. Addition of CeO_2 and MgO to Ni strengthened the interaction between the Ni and the support. The catalytic activity results indicate that the addition of CeO_2 and MgO to Ni did not reduce carbon deposition, but improved the activity of the catalysts. Temperature programmed oxidation (TPO) revealed the formation of carbon that is mainly amorphous and small amount of graphite. The highest CH_4 and CO_2 conversion was found for the catalyst composed of 5.0 wt. % NiO-10.0 wt. % CeO_2/3.0 wt. %TiO_2-γ-Al_2O_3 (Ti-CAT-II), resulting in H_2/CO mole ratio close to unity. The optimum reaction conditions in terms of reactant conversion and H_2/CO mole ratio were achieved by varying space velocity and CO_2/CH_4 mole ratio.

Keywords: CH_4; CeO_2; dry reforming; MgO; Ni; TiO_2

---

1. Introduction

Global warming has become an alarming issue. Emissions of greenhouse gases including carbon dioxide (CO_2) and methane (CH_4) actively contribute to global warming. Methods of transforming these CO_2(g) and CH_4(g) into useful products are an important area of study to generate industrially important fuels and chemicals [1–3]. In this context, numerous reforming reactions of CH_4 have been employed using several oxidants (e.g., H_2O, CO_2, O_2, etc.) to produce H_2(g) or synthesis gas (syngas, a mixture of H_2(g) and CO(g)) with an equimolar ratio of H_2(g)/CO(g). Methane reforming processes include steam reforming, auto thermal reforming, tri-reforming, etc. [4–11]. Methane reforming using CO_2, known as dry reforming (DRM), is attractive because it mitigates the emission of CH_4 and CO_2, produces syngas, the starting material in the Fischer-Tropsch process to generate hydrocarbons and oxygenates, and generates clean energy through the combustion of hydrogen [12]. CH_4 is a cost-effective feedstock for syngas production. The primary reaction that governs the process is as follows:

$$CH_4 + CO_2 \rightarrow 2H_2 + 2CO \, \Delta H_{298} = +247 \text{ kJ mol}^{-1} \tag{1}$$

The reaction is energetically unfavorable, thus requiring high temperatures to achieve acceptable conversion. Both noble metals (i.e., Ru, Rh, or Pt) and first-row transition metals (i.e., Ni, Fe, Co) are common active elements in that catalyze CO_2 reforming of CH_4. Although noble metals display high activity and stability, their limited availability and high price have rendered them inappropriate for industrial use [13,14]. On the other hand, the first-row transition metals are cheaper and possess similar activity, but their stability is hampered by carbon deposition and particle sintering [15–18]. Therefore, development of Ni-based catalysts with high activity and resistance to deactivation due to carbon formation and metal sintering is essential for DRM. Catalytic performance can be influenced by many factors such as the active metal, support type, and texture. The support can enhance the catalyst selectivity, activity, and stability by increasing the surface area and dispersion of the active metal [19]. For example, Ni deposited on alumina supports result in high catalytic activity, but rapidly deactivates due to sintering, coke deposition, and formation of surface nickel aluminate phase. To increase the catalytic performance of $Ni/\gamma\text{-}Al_2O_3$, various parameters can be incorporated in the catalyst.

Titania (TiO_2) is characterized by low specific surface area and poor mechanical strength, and undergoes a phase transformation from anatase to rutile at high temperatures, making it unsuitable for high temperature reactions [20]. Previous studies have shown enhanced thermal stabilization of TiO_2 by introducing a thermally stable second metal oxide (i.e., SiO_2, Al_2O_3, etc.) [21,22]. Incorporation of TiO_2 in Al_2O_3 supports can improve metal dispersion, reduce particle sintering, increase thermal stability, and enhance oxygen storage capacity to assists in gasifying carbon produced in reforming reaction [23]. Tauster et al. investigated the effects of support modification on the oxidation state of Ru and the catalytic performance of Ru/TiO_2 catalysts under conditions of partial oxidation of methane. It was found that doping of TiO_2 with small amounts of WO_3 favored oxygen adsorption on Ru under reaction conditions, resulting in a stabilization of a fraction of the catalyst in its oxidized form [24]. Addition of metal oxide promoters has been used to improve Ni metal catalysts. For instance, Shamskar et al. investigated the addition of CeO_2, La_2O_3, and ZrO_2 to Ni/Al_2O_3 catalyst used for DRM and found that ceria-promoted catalyst reduced the carbon formation [25]. Ni-MgO-Al_2O_3 catalysts were used for steam reforming of methane by Jang et al. [26]. Al-Fatesh et al. studied the promotional effect of ceria in the catalytic DRM and found that the Ni doping with ceria resulted in an excellent activity and lowered coke formation [27]. MgO promoters enhance CH_4 conversion and mitigated the effect of the potassium poisoning of the Ni-based catalyst. The MgO promoter is beneficial in suppressing carbon formation.

In the present work, supported combinations of MgO, CeO_2 and NiO catalysts were developed to retain high activity and stability while reducing the formation of coke during DRM. The effect of using MgO and CeO_2 as separate and combined promoters, for 5.0 wt. % NiO supported over $\gamma\text{-}Al_2O_3$ doped with 3.0 wt. % TiO_2 was studied. We determine the impact of each modifier on observed catalytic performance.

2. Results and Discussion

2.1. X-ray Powder Diffraction (XRD)

The XRD patterns of all the fresh catalysts are displayed in Figure S1 in the electronic supplementary information (ESI). All the patterns consisted of various metal oxides, where the presence metal oxide phases depended on the added components used to prepare the catalysts. Three metal oxides existed in all catalysts, where these metal oxides were the component of the support: cubic gamma-aluminum oxide, $\gamma\text{-}(Al_2O_3)_{1.333}$ (PDF 01-075-0921), cubic synthesized honguiite titanium oxide, $(TiO_{0.8})_{0.913}$ (PDF 01-085-1380), and aluminum silicate, $Al_{0.5}Si_{0.75}O_{2.25}$ (PDF 00-037-1460). Rhombohedral nickel oxide, NiO (PDF 00-044-1159) was found in Ti-CAT-I, Ti-CAT-II, Ti-CAT-III, and

Ti-CAT-V (these notations are defined in Section 3.2). When magnesium was added, cubic magnesium nickel oxide, $MgNiO_2$ (PDF 00-024-0712) formed. Cubic synthesized cerianite (Ce) (ceria), CeO_2 (PDF 00-034-0394), was detected in Ti-CAT-I, Ti-CAT-II, Ti-CAT-IV, and Ti-CAT-VI. Addition of magnesium strongly influenced the interaction of cerium with the other components of the catalyst. Monoclinic magnesium cerium oxide, $MgCeO_3$ (PDF 00-004-0641), and cubic magnesium cerium titanium oxide, Mg_2CeTiO_6 (PDF 00-058-0550) were present in Ti-CAT-I and Ti-CAT-IV. Cubic periclase magnesium oxide, MgO (PDF 01-071-1176) was detected in Ti-CAT-I, Ti-CAT-III, and Ti-CAT-IV.

2.2. Inductively Coupled Plasma Mass Spectroscopy (ICP-MS)

ICP-MS analysis was carried out to quantify the metallic components as metal oxides for the best two catalysts. The results are shown in Table 1a,b.

Table 1 summarizes the results of ICP analysis of the metallic components in the prepared catalysts and compares it with the theoretical values. The experimental results were found to be in excellent agreement with the nominal values.

Table 1. ICP metal oxide microanalysis of Ti-CAT.

(a)

Catalyst Component	Ti-CAT-I					
	NiO	CeO_2	MgO	TiO_2	SiO_2	Al_2O_3
Theoretical, wt/wt. %	5.00	10.00	1.00	3.00	2.00	79.00
Experimental, wt/wt. %	5.21	9.91	1.02	2.86	1.93	78.05

(b)

Catalyst Component	Ti-CAT-II					
	NiO	CeO_2	MgO	TiO_2	SiO_2	Al_2O_3
Theoretical, wt/wt. %	5.00	10.00	0.00	3.00	2.00	79.00
Experimental, wt/wt. %	4.98	10.11	0.00	2.92	2.07	81.03

2.3. Temperature Programmed Desorption (CO_2-TPD)

The CO_2-TPD experiment was performed to study the basicity of the catalysts. The results obtained are shown in Figure 1. The basicity of the catalyst has a paramount influence on the catalytic performance in DRM due to the acidic nature of CO_2. Thus, strong basic sites can enhance catalytic activity and increase the chemisorption and reaction of reacting gases [28]. Distribution of basic sites on the catalyst (i.e., weak, intermediate, strong, and very strong) correspond to the different desorption peaks in the temperature ranges of 20–150, 150–300, 300–450, and >450 °C, respectively, in the CO_2-TPD profile [29,30].

All catalysts, except Ti-CAT-V and Ti-CAT-VI, showed the same basic site classification, because the CO_2 desorption peaks appeared at almost the same different temperature ranges (Figure 1). Both Ti-CAT-V and Ti-CAT-VI have basic sites correspoding to site of high and strong basicity centered at a temperature around 310 °C.

For the peaks appearing at different temperature ranges, peaks in the temperature range of 50–125 °C correspond to weak basic sites, peaks at 160–185 °C fall under the category of intermediate strength basic sites, while the peaks at 260 °C correspond to strong basicity sites. An elbow peak was observed for all of the samples, except for Ti-CAT-V and Ti-CAT-VI, at temperature centered around 500 °C. This peak had no significant CO_2 uptake.

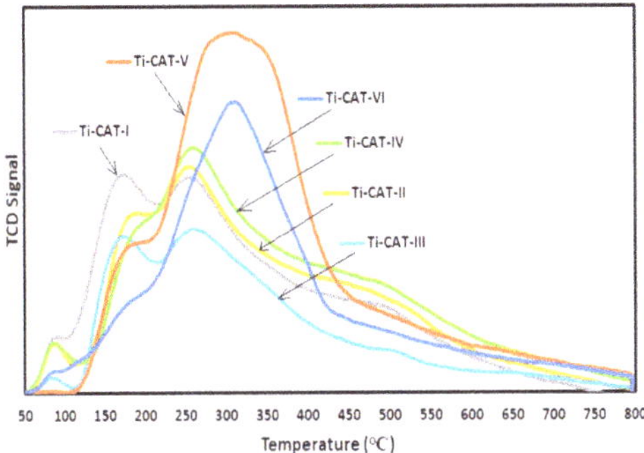

Figure 1. CO_2-TPD profiles of the synthesized catalysts.

2.4. Surface Characterization

The textural properties of the fresh catalysts were studied using nitrogen adsorption-desorption isotherms. The results obtained from the N_2 physisorption are shown in Table 2 and that of the isotherms are presented in Figure 2. The results give an insight into the variations in the activities of the catalysts. In accordance to IUPAC classifications of isotherms, the isotherms in Figure 2 fall under the category of type II, with an H3-type hysteresis loop, which results from capillary condensation and evaporation at high relative pressures [31].

Table 2. N_2 physisorption results for the different catalysts.

Catalyst	BET Surface Area (m^2/g)	Av. Pore Diameter (nm)	Pore Volume (cm^3/g)
Ti-CAT-I	284	11.5	0.40
Ti-CAT-II	283	12.4	0.43
Ti-CAT-III	326	11.8	0.43
Ti-CAT-IV	256	12.3	0.39
Ti-CAT-V	334	12.4	0.43
Ti-CAT-VI	299	12.5	0.40

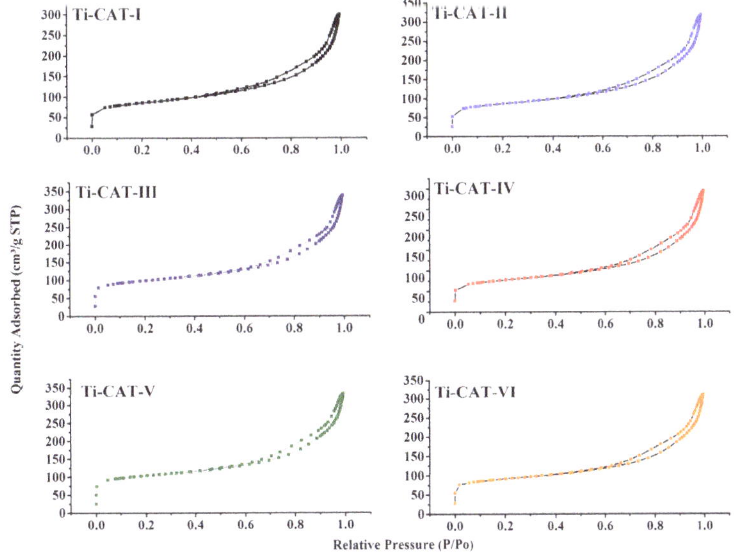

Figure 2. *Cont.*

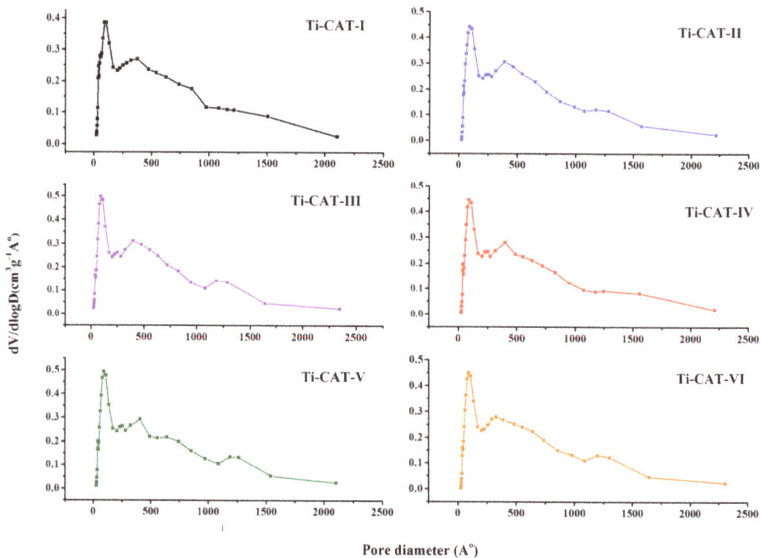

Figure 2. N_2 adsorption-desorption isotherms and BJH desorption pore size distribution curves for Ti-CAT samples.

Type II isotherms constitute macroporous adsorbents, and for detailed study, the BJH pore size distribution is represented in Figure 2. All the Ti-CAT samples displayed a bimodal mesoporous/macroporous distribution curve with average pore size in the range 11.5–12.5 nm, typical for macroporous adsorbents with large surface area. For example, Jiang et al. [32] and Zhao et al. [33] synthesized macro-mesoporous bimodal titania with high surface areas. Here, the effect of surface area variation is observed when Mg, Ce, and Ni were combined. Table 2 shows that surface areas of the combined metal catalysts are reduced in relation to single-metal component catalysts. This observation is due to the combined metal deposition on the porous structure of the support and filling pores [34].

2.5. H_2-TPR

The reduction behavior of the different catalyst samples was investigated using H_2-TPR and the profiles are presented in Figure 3. The nickel reduction peaks for Ti-CAT-x (x = I, II, III) samples containing Ni combined with other metals, are characterized by three reduction regions at low, medium and high temperature ranges. Their ranges are dependent on the degree of dispersion and interaction of the active metal with the support. The nickel phase reducibility was influenced by the combination of the metal oxides. The reduction peak in the temperature range of 280–380 °C is assigned to the reduction of NiO having weak interaction with the support. Higher temperature peaks (600–700 °C) are likely due to the reduction of NiO species having strong interactions with the support. The reduction peak of Ni^{2+} derived from spinel is found at around 810 °C [35].

For Ti-CAT-V (the catalyst with only Ni), the NiO reduction peaks appeared narrower and more intense in temperature ranges lower than those of combined metal counterparts.

Only two reduction peaks are observed for Ti-CAT-VI at temperature ranges centered at 260 and 325 °C. Similar reduction peaks are expected for CeO_2 promoted samples, but appear to have merged with the peaks for NiO that appeared around that temperature range.

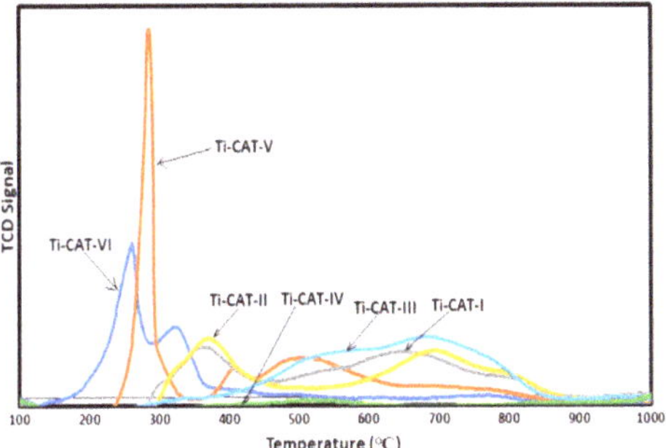

Figure 3. TPR profiles of the promoted and un-promoted catalysts.

2.6. Effect of MgO and CeO₂ Combination on the Catalytic Performance

The effects of combining CO_2 and MgO on Ti-CAT-V and their catalytic performance were studied by comparing the activities of Ti-CAT-V catalyst with that of Ti-CAT-I, Ti-CAT-II, and Ti-CAT-III. CH_4, CO_2 conversions, H_2/CO mole ratio, and their selectivity at 700 °C, for 7.0 h time-on-stream for DRM were calculated and plotted as shown in Figure 4. All the promoted catalysts have CH_4 and CO_2 conversions higher than that of the Ti-CAT-V catalyst except for Ti-CAT-VI and Ti-CAT-IV, which showed no sign of reaction during the DRM. Ti-CAT-II had the highest CH_4 conversion at the start of the reaction (~55%) and maintained stability at around 52%. The high specific surface area of the catalyst (283 m^2/g) enhanced the adsorption, diffusion, and contact of the reactant gases. The high average pore diameter and pore volume of Ti-CAT-II is a likely factor for the best-in-class performance. The Ce and Mg promoted catalysts enhanced the activity. The improvement of the activity is accompanied by the formation of graphitic carbon in comparison with the unpromoted catalysts, as depicted in the TG analysis.

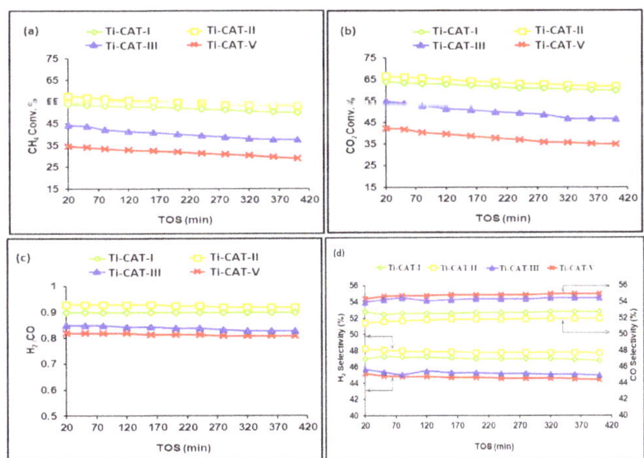

Figure 4. Catalytic performance of Ti-CAT-I, Ti-CAT-II, Ti-CAT-III, and Ti-CAT-V (**a**) CH_4 conversion (**b**) CO_2 conversion and (**c**) H_2/CO ratio and (**d**) H_2, CO selectivity.

The same trend was observed for CO_2 conversion, with the Ti-Cat-V catalyst showing the least conversion. For all the catalysts under investigation, CO_2 conversion was observed to be higher than CH_4 conversion, which is suggestive of the occurrence of reverse water gas shift (RWGS) reaction. Wang et al. gave the same observation in their study on catalytic hydrogenation of carbon dioxide [36].

$$H_2 + CO_2 \rightarrow CO + H_2O \qquad \Delta H_{298} = +41.2 \text{ kj/mol}$$

In addition, the H_2/CO mole ratio showed values less than 1 for all the catalysts. The deviation from the stoichiometric ratio is also suggestive of the occurrence of RWGS reaction. Ti-CAT-II appeared to be the most selective towards H_2 (~48%) and least selective towards CO (~52%), while Ti-CAT-V has the least H_2 selectivity (~45%) but the highest CO selectivity (~55%). In all cases, the as-prepared catalysts showed higher CO selectivity than H_2.

Ti-CAT-II catalyst resulted in a H_2/CO mole ratio value closest to 1, compared to the tested catalysts. The desirable value of the syngas ratio suitable for downstream Fischer-Tropsch synthesis is unity [37], thus making it the best option for the dry reforming.

2.7. H_2-Pulse Chemisorption

To understand the effect of the active Ni component on the catalytic performance of the best two catalysts, Ti-CAT-I and Ti-CAT-II, we carried out H_2 pulse chemisorption to determine the degree of Ni dispersion on the surface of the support and Ni metallic surface area. The results of H_2-pulse chemisorption are displayed in Table 3. We found that both catalysts had high Ni metallic surface areas of ~90% and good dispersion of ~13%, which is responsible for the good catalytic performance of the two catalysts. The small relative higher catalytic performance of Ti-CAT-II than that of Ti-CAT-I could be attributed to the slightly higher Ni metallic surface area and dispersion of Ti-CAT-II.

Table 3. Ni metallic surface area and dispersion obtained by H_2 chemisorption.

Catalyst	Ni Metallic Surface Area, m^2/g	Ni Dispersion, %
Ti-CAT-I	89	13.3
Ti-CAT-II	91	13.6

2.8. Temperature Programmed Oxidation (TPO) of the Spent Catalysts

TPO is a useful technique that can be employed to determine the nature of the carbon deposited onto the surface of the catalysts. Several forms of carbon deposition have been reported in dry reforming reactions—ranging from atomic carbon, to graphitic, and amorphous carbon. The carbon can undergo gasification to form CO_2 under oxidative atmosphere and at different temperature ranges. The atomic carbon, amorphous, and graphitic carbon can be gasified at temperatures less than 250, 250–600, and >600 °C, respectively [38]. The TPO profiles of Ti-CAT-I and Ti-CAT-II spent catalysts are shown in Figure 5. Each of the catalysts exhibited a broad peak near 600 °C and a low-intensity shoulder at 100–250 °C. According to the TPO results, the carbon deposited on both Ti-CAT-I and Ti-CAT-II spent catalysts revealed the formation of carbon atoms, mostly amorphous carbon, and a small amount of extent graphitic carbon.

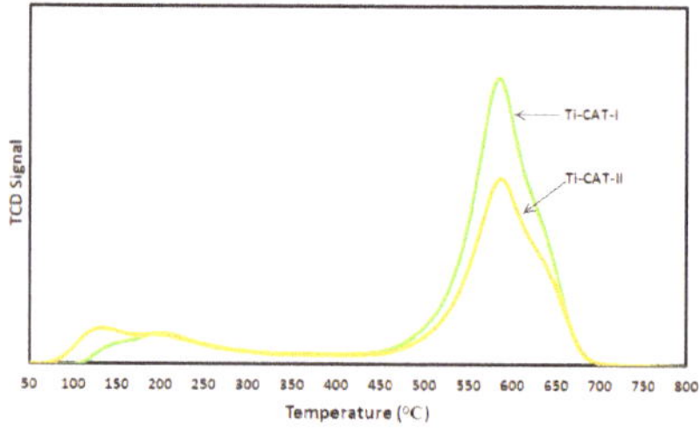

Figure 5. TPO profiles for both Ti-CAT-I and Ti-CAT-II.

2.9. SEM and TG Analysis

We used scanning electron microscopy (SEM) to determine the change in morphology of the spent catalysts. Figure 6 shows the SEM micrographs for the best two catalysts: Ti-CAT-I and Ti-CAT-II. Similar morphology, based on agglomerated, spherical nanoparticles, was observed for both fresh catalysts (Figure 6A,B). Such an observation was expected, because both catalysts were synthesized using an identical preparation procedure and had similar components.

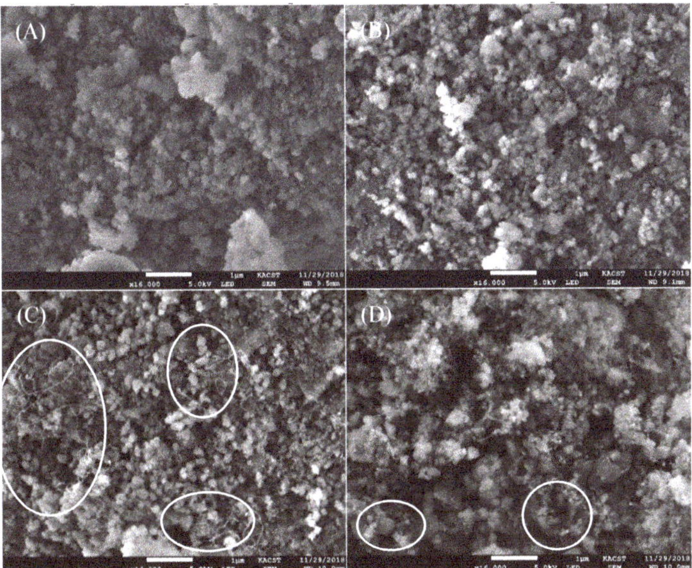

Figure 6. SEM micrographs for fresh catalysts (**A**) Ti-CAT-I, (**B**) Ti-CAT-II, and spent catalysts (**C**) Ti-CAT-I, (**D**) Ti-CAT-II. White circles are for some areas where CNTs are present.

The morphology of the spent catalysts was similar to that one of the fresh samples, except for the presence of carbon nanotubes (CNTs) on the surface of the spent catalysts (Figure 6C,D). Detection of CNTs on the surface of the spent catalyst is in agreement with the TPO results (c.f. Figure 5) and confirms the results of TGA of spent catalysts (Figure 7). The presence of CNTs on the surface of the spent catalysts could be attributed to Boudouard reaction, which in turn would be responsible for reducing the catalytic performance.

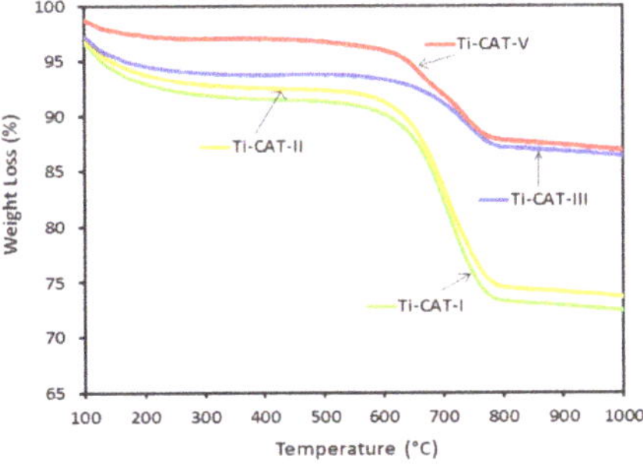

Figure 7. TGA profiles for the spent catalysts.

After the 7 h reaction, we analyzed the used catalysts by thermal gravemetric analysis (TGA), a quantitative analysis that determines the amount of carbon deposition. Figure 7 shows the result of the analysis. Catalysts that showed no sign of reaction were not reported. The weight loss (%) for virtually all the catalysts began at around 620 °C. The TGA profiles revealed that both Ti-CAT-V and Ti-CAT-III catalysts had the lowest weight loss, ~15.0%, while the two most reactive catalysts (Ti-CAT-II and Ti-CAT-I) had the highest amount of carbon deposition, corresponding to a weight loss of 25.0%.

From these TGA results of spent catalysts, it can be inferred that the combined metal catalysts, namely Ti-CAT-II and Ti-CAT-I, enhance the feed conversion capacity of the catalysts and gasify the carbon deposited over the surface to a considerable extent.

2.10. Effect of Space Velocity

The effect of gas hourly space velocity (GHSV) was studied on the catalyst that showed the best performance in the previous section (i.e., Ti-CAT-II catalyst). GHSV of 19,500 and 78,000 $\frac{\text{feed flow rate}}{\text{mass of cat.}}$ $\left(\frac{\text{mL}}{\text{g} \cdot \text{h}}\right)$ were considered at 700 °C and time-on-stream over 7.0 h for DRM, while keeping the mass of the catalyst constant. These GHSV values are half and twice as much as the initial GHSV of 39,000 mL g^{-1} h^{-1}, respectively. The results, in terms of CO_2 and CH_4 conversions, as well as H_2/CO mole ratio, were calculated and plotted in Figure 8A,B. As the GHSV increased, the CH_4 and CO_2 conversions decreased, with the highest conversions for both CH_4 and CO_2 being obtained at a GHSV of 19,500 $\frac{\text{feed flow rate}}{\text{mass of cat.}}$ $\left(\frac{\text{mL}}{\text{g} \cdot \text{h}}\right)$. The decrease in conversions can be attributed to the feed having less residence time at higher GHSV [39]. A similar trend was observed with H_2/CO mole ratio, where it decreased from a ratio of 1 to ~0.8. However, the results at GHSV of 39,000 were the most stable in comparison to those obtained at other GHSV values.

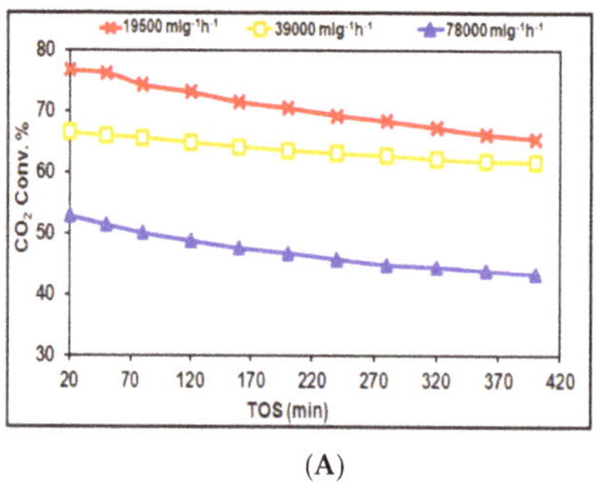

(A)

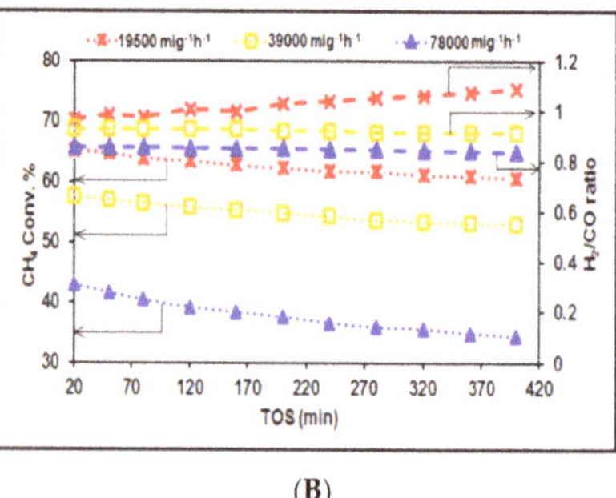

(B)

Figure 8. (**A**) CO_2 conversion for Ti-CAT-II at different gas hourly space velocity (**B**) CH_4 and H_2/CO ratio for Ti-CAT-II catalyst at different space velocities.

2.11. Effect of GHSV on Carbon Deposition

Quantitative analysis of carbon deposition was performed on the catalyst Ti-CAT-II used in methane dry reforming at 3 different space velocities 19,500, 39,000 and 78,000 mL g^{-1} h^{-1}.

The results obtained after the completion of the reactions are shown in Figure 9. The analysis for the reaction performed at 19,500 mL g^{-1} h^{-1} showed the least amount of carbon deposition of about 18%, which shows that, relatively, more contact time between the catalyst and the feed stream was allowed at this space velocity, giving room for gasification of the coke that was deposited during the reaction. The reactions carried out at 39,000 and 78,000 mL g^{-1} h^{-1} showed higher carbon deposition of about 26 and 25%, respectively. This is an indication that at the higher space velocities, the residence time was not enough for the gasification of the carbon deposit, which variably continued to pile up. Lalit et al. reported similar findings in their study of the effect of GHSV on the conversion of CH_4 and CO_2 [39].

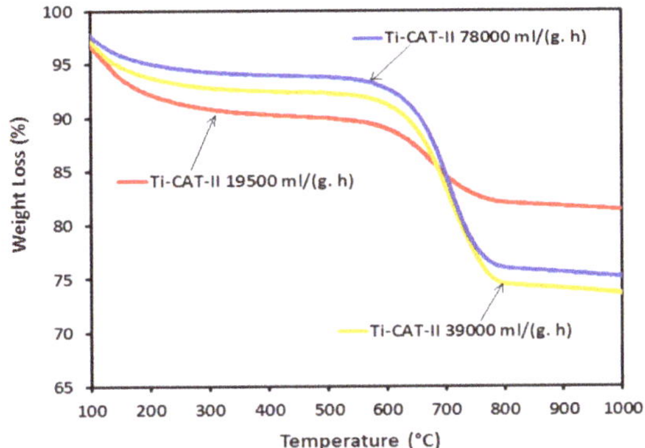

Figure 9. TGA curves for Ti-Cat-II at 19,500, 39,000 and 78,000 mL/(g·h) GHSV values.

From the results of the investigation, it can be inferred that deactivation is expected at all the space velocity investigated, since carbon deposit was evident; however, at 19,500 mL/(g·h) space velocity, the catalyst will stay active for a longer time than at 39,000 and 78,000 mL/(g·h).

2.12. Effect of Different CO_2/CH_4 Ratios

The mole ratio of CO_2 to CH_4 was changed at a fixed total flow rate to study the performance of Ti-CAT-II catalyst when CH_4 was supposed to act as the limiting reagent in excess of CO_2 at 700 °C and 39,000 mL g^{-1} h^{-1} GHSV. The results are shown in Figure 10a–c. The highest CH_4 conversion of about 78% was obtained when CO_2 was 20% in excess of CH_4, while the least conversion of CH_4 (~43%) resulted when the amount of CO_2 was 50% of the required stoichiometric amount in the feed. This observation was expected, as CH_4 would have enough CO_2 to undergo dry reforming. On the other hand, the highest CO_2 conversion of (~90%) was observed when CO_2 was the limiting reagent. This observation could be due to excess CH_4 present in the feed. The CO_2 conversion was reduced with the reaction time-on-stream. Such observation could be ascribed to the disproportionation of carbon monoxide into CO_2 and graphite, a transformation known as the Boudouard reaction:

$$2CO_{(g)} = CO_{2(g)} + C_{(s)}$$

Comparing the different CO_2/CH_4 ratios, it was observed that CH_4 conversion increased with the ratio up to 1.2 (i.e., 0.5 < 1.0 < 1.2) and then declined slightly at 1.5. However, the conversion for CO_2 was observed to decrease as the ratio increased (i.e., 1.5 < 1.2 < 1.0 < 0.5).

Figure 10c displays the H_2/CO mole ratio results. It was observed that at the lowest CO_2/CH_4 mole ratio, the H_2/CO mole ratio was greater than one. This observation could be owing to the insufficient amount of CO_2 for complete dry reforming of the available CH_4 and to the thermal decomposition of unreformed CH_4, giving more H_2 than the stoichiometric amount. Moreover, the Boudouard reaction might contribute to the increase of hydrogen production, because the formed CO_2 from Boudouard reaction would shift the DRM equilibrium to the product side.

On the other hand, H_2/CO mole ratio was close to one of the cases where CO_2 was in excess of CH_4, where it was noticed that the H_2/CO mole ratio increased with the reaction time-on-stream. Once again, the Boudouard reaction might be responsible for such observation.

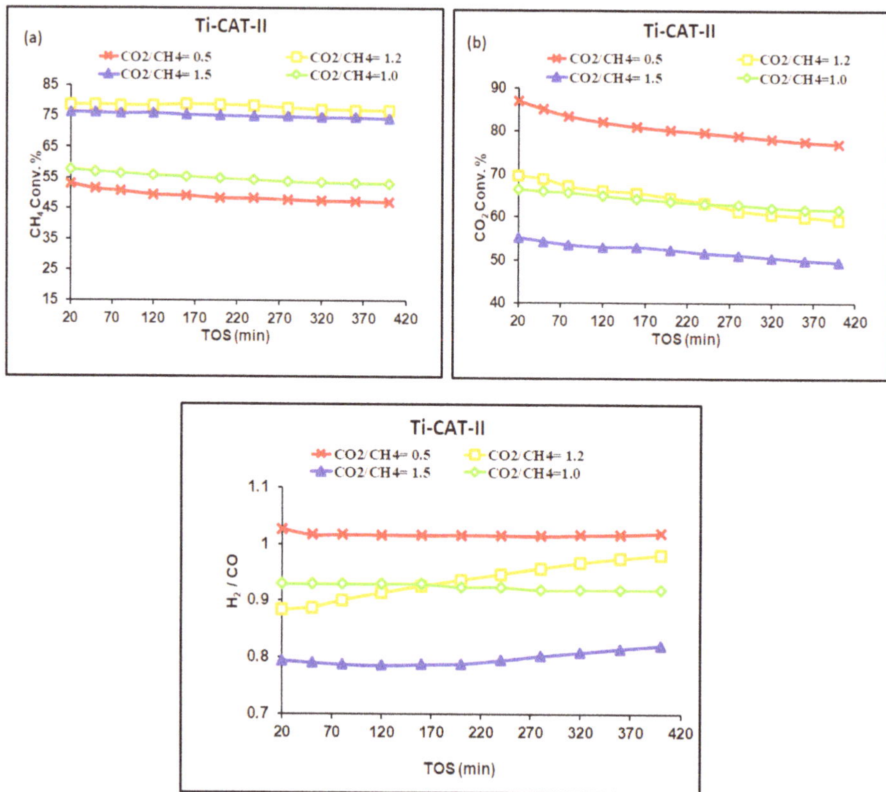

Figure 10. (a) CH_4 (b) CO_2 conversions, and (c) H_2/CO ratio for different CO_2/CH_4 ratio over Ti-CAT-II.

3. Experimental Section

3.1. Materials

Nickel nitrate hexahydrate [$Ni(NO_3)_2.6H_2O$, 98%, Alfa Aesar], cerium nitrate hexahydrate [$Ce(NO_3)_3.6H_2O$, 99.0% assay on Ce basis, general purpose reagent, BDH], magnesium acetate tetra-hydrate [$Mg(O_2CCH_3)_2.4H_2O$, 99.5-102.0%, Merck & Co., Inc., Kenilworth, NJ, USA] were commercially available and were used without further purification. γ-Alumina doped with titania (3.0 wt. % TiO_2/γ-Al_2O_3) in the shape of pellets, was a gift from Tiancun Xiao, Senior Research Fellow, Inorganic Chemistry Laboratory, Oxford University. Ultrapure deionized water (18.2 MΩ.cm) was obtained from a Milli-Q water purification system (Millipore, Burlington, MA, USA).

3.2. Catalyst Preparation

The required amounts of $Ni(NO_3)_2.6H_2O$, Ce $(NO_3)_3.6H_2O$, Mg $(O_2CCH_3)_2.4H_2O$, and support were mixed and were ground together to fine powder by pestle and mortar. A small amount of ultrapure water was used to convert the solid mixture into a paste, which was spun mechanically until dryness. The paste and spinning process was repeated three times. The final solid was calcined in a digital, programmed muffle furnace at 600 °C for three hours by ramping temperature from room temperature by a rate of 3.0 °C/min. The notation of the prepared catalyst samples and their wt. % loadings of nickel oxide, ceria, and magnesia at 600 °C calcination are given below in Table 4.

Table 4. Prepared catalyst samples and the wt. % of their composition.

Catalyst	Concentration, wt. %		
	NiO	CeO$_2$	MgO
Ti-CAT-I	5.0	10.0	1.0
Ti-CAT-II	5.0	10.0	0.0
Ti-CAT-III	5.0	0.0	1.0
Ti-CAT-IV	0.0	10.0	1.0
Ti-CAT-V	5.0	0.0	0.0
Ti-CAT-VI	0.0	10.0	0.0

3.3. Catalyst Characterization

The metallic component composition of all catalysts was determined by an Agilent 7800 inductively coupled plasma mass spectrometry (ICP) at the laboratory of IDAC Merieux NutriSciences, Riyadh, Saudi Arabia. Carbon deposition on the used catalysts was measured by thermogravimetric analysis (TGA) under air by using a Shimadzu TGA-51(Shimadzu Corp., Kyoto, Japan). A certain amount from the spent catalyst (10 mg) was subjected to heat treatment within the temperature range 25 °C–1000 °C. Ramping temperature was maintained at 20 °C/min. Temperature programmed oxidation (TPO) was performed in an oxidative atmosphere to determine the kind of carbon deposited over the surface of the catalyst using Micromeritics AutoChem II over a temperature range of 50–800 °C under a flow of 10% O$_2$/He mixture at 40 mL/min. The spent catalyst was first pretreated in the presence of high purity Argon at 150 °C for 30 min and subsequently cooled to room temperature. The Brunauer-Emmet-Teller technique was adopted in calculating the surface area per unit mass of the samples using a device that analyses surface area and porosity, i.e., Micromeritics Tristar II 3020 (Micromeritics Instrument Corporation, Norcross, GA, USA). For nitrogen physisorption measurements, an amount of 0.20–0.30 g weighed from the catalyst was subjected to degassing at 300 °C for three hours prior to analysis. The reducibility of the fresh catalysts was determined by the Micromeritics AutoChem II (Micromeritics Instrument Corporation, Norcross, GA, USA). A sample weight of 75.0 mg was analyzed. Samples were first heated under argon (99.9%) at 150 °C for 30 min, thereafter cooled to 25 °C. Afterwards, samples were heated to 1000 °C at 10 °C/min by allowing the flow of 10% H$_2$/Ar gas at 40 mL/min. A thermal conductivity detector (TCD) was used to follow the H$_2$ consumption. Temperature programmed desorption of carbon dioxide (CO$_2$-TPD) and CO pulse chemisorption measurements were obtained from an automatic chemisorption equipment (Micromeritics AutoChem II 2920) with a TCD. At the start, a 70 mg sample was heated at 200 °C for 1 h under helium (He) flow to remove adsorbed components. Then, CO$_2$ adsorption was carried out at 50 °C for 60 min in the flow of He/CO$_2$ gas mixture (90/10 L/L) with a flow rate of 30.0 mL/min. Afterwards, a linear temperature rise at a rate of 10 °C/min until 800 °C was registered by the TCD of CO$_2$ desorption signal. The nickel metallic surface area and dispersion were determined by H$_2$ pulse chemisorption by using Micromeritics AutoChem II. A sample of 50.0 mg was heated to 150 °C under vacuum for sixteen hours. The sample was then transferred to the sample tube and was heated at temperature rate of 10.0 °C/min to 400 °C under flow rate of 10.0 mL/min of 10%H$_2$/Ar for one hour. The sample was then flushed with highly pure Ar for one hour at 400 °C. The temperature was then reduced to 70.0 °C and pulses of H$_2$ gas were introduced for one hour for determining the H$_2$ uptake. X-ray powder diffraction patterns for the samples were recorded on a Bruker D8 Advance (Bruker, Billerica, MA, USA) XRD diffractometer by using Cu K$_\alpha$ radiation source and a nickel filter, operated at 40 kV and 40 mA. The step size and scanning range of 2θ for analysis was set to 0.01° and 5–100°, respectively. The present phases were documented using standard powder XRD cards (JCPDS). Catalyst morphology was studied using JEOL JSM-7100F (JEOL, Tokyo, Japan) (field emission scanning electron microscope, equipped with energy-dispersive X-ray spectroscopy (EDXS) for surface elemental analysis.

3.4. Catalytic Perfromance

Methane reforming reaction was accomplished in a fixed-bed tubular stainless-steel micro-reactor (ID = 9 mm) at atmospheric pressure. The reactor system was provided by Process Integral Development (Process Integral Development Eng & Tech SL, Madrid, Spain). Before performing the DRM reaction, a 0.10 g catalyst was activated by H_2 flow of 40 mL/min at 700 °C for 60 min. N_2 gas was then admitted to the reactor for 20 min to remove adsorbed H_2 while the catalyst was kept at reaction temperature (700 °C). Afterwards, feed gases of CH_4, CO_2, and N_2 were injected at flow rates of 30, 30 and 5 mL/min, respectively. The temperature, pressure and reaction variables were inspected through the reactor panel. A GC (Shimadzu Corp., Kyoto, Japan) unit having a thermal conductivity detector and two columns, Porapak Q and Molecular Sieve 5A, was connected in series/bypass connections in order to have a complete analysis of the reaction products. The following equations were used to calculate the CH_4 and CO_2 conversions respectively.

$$\%CH_4 \text{ conversion} = \frac{CH_4 \text{ in} - CH_4 \text{ out}}{CH_4 \text{ in}} \times 100$$

$$\%CO_2 \text{ conversion} = \frac{CO_2 \text{ in} - CO_2 \text{ out}}{CO_2 \text{ in}} \times 100$$

4. Conclusions

This paper investigated the dry reforming of methane, CH_4, over Ti-CAT-V catalyst, and the effects of promoters such as CeO_2 and MgO, on the catalytic activity and stability of the catalyst. The promoter loading was 10.0 wt. % and 1.0 wt. % for CeO_2 and MgO, respectively. Promoted Ti-CAT-V catalyst showed better conversion of both CH_4 and CO_2 than the un-promoted counterpart. Ti-CAT-II had the highest CH_4 and CO_2 conversion of about 55% and 64% respectively, while no reaction was observed for Ti-CAT-VI and Ti-CAT-IV. It can be inferred from the improved performance of the promoted catalysts that the promoters had a positive influence on the textural properties, metal support interaction and reduction behavior of the catalyst. These impacts of promoters were well shown by the characterization techniques used. From the thermogravimetric analysis, un-promoted catalyst gave the lowest carbon deposition. The promoted catalyst, especially by Ce, with higher amounts than 10% was found to have the highest carbon formation. This result implied that the promoters enhanced the activity performance of the catalyst, resulting in the formation of graphitic carbon, and hence, were not effective in boosting the stability via reduction of carbon deposition relative to the un-promoted catalyst. The TPO investigation indicated the types of carbon formed, which include atomic, amorphous, and graphitic carbon.

Ti-CAT-V was selected for further investigation at different GHSVs and subsequently at various CO_2/CH_4 ratios. An inverse relationship between GHSV and catalytic activity was observed. A GHSV of 19,500 $\frac{\text{feed flow rate}}{\text{mass of cat.}}$ $\left(\frac{mL}{g \cdot h} \right)$ and CO_2/CH_4 ratio of 0.5 gave the best results.

Supplementary Materials: The following are available online at http://www.mdpi.com/2073-4344/9/2/188/s1, Figure S1: XRD patterns of fresh catalysts.

Author Contributions: A.S.F., A.H.F, S.O.K., R.A. (Rasheed Alrasheed), R.A. (Rawan Ashamari) and A.B. carried out all experiments and characterization tests as well as shared in the analysis of the data and shared in the writing of the manuscript. A.S.F, S.O.K, A.E.A and A.B. wrote the paper and shared data analysis. A.H.F and A.A.I. contributed in writing the paper and edited it.

Funding: This research was funded by the Deanship of Scientific Research at King Saud University, Project No. RGP-119.

Acknowledgments: The authors would like to express their sincere appreciation to the Deanship of Scientific Research at King Saud University for its funding for this research group project No. (RGP-119).

Conflicts of Interest: The authors declare no conflict of interest.

References

1. Albarazi, A.; Beaunier, P.; Da Costa, P. Hydrogen and syngas production by methane dry reforming on SBA-15 supported nickel catalysts: On the effect of promotion by Ce0.75Zr0.25O$_2$ mixed oxide. *Int. J. Hydrogen Energy* **2013**, *38*, 127–139. [CrossRef]

2. Bao, Z.; Lu, Y.; Han, J.; Li, Y.; Yu, F. Highly active and stable Ni based bimodal pore catalysts for dry reforming of methane. *Appl. Catal. A Gen.* **2015**, *491*, 116–126. [CrossRef]

3. Buelens, L.C.; Galvita, V.V.; Poelman, H.; Detavernier, C.; Marin, G.B. Super-dry reforming of methane intensifies CO$_2$ utilization via Le Chatelier's principle. *Science* **2016**, *354*, 449–452. [CrossRef] [PubMed]

4. Julián-Durán, L.M.; Ortiz-Espinoza, A.P.; El-Halwagi, M.M.; Jiménez-Gutiérrez, A. Techno-economic assessment and environmental impact of shale gas alternatives to methanol. *ACS Sustain. Chem. Eng.* **2014**, *2*, 2338–2344.

5. Schulz, H. Short history and present trends of Fischer-Tropsch synthesis. *Appl. Catal. A Gen.* **1999**, *186*, 3–12. [CrossRef]

6. Sheu, E.J.; Mokheimer, E.M.A.; Ghoniem, A.F. A review of solar methane reforming systems. *Int. J. Hydrogen Energy* **2015**, *40*, 12929–12955. [CrossRef]

7. Barelli, L.; Bidini, G.; Gallorini, F.; Servili, S. Hydrogen production through sorption-enhanced steam methane reforming and membrane technology. *Energy* **2008**, *33*, 554–570. [CrossRef]

8. Hyunjin, J.; Junghun, L.; Eunyeong, C.; Ilsung, S. Hydrogen production from steam reforming using an indirect heating method. *Int. J. Hydrogen Energy* **2018**, *43*, 3655–3663.

9. Cyril, P.; Wenhao, F.; Mickaël, C.; Sébastien, P.; Louise, J.D. Steam reforming, partial oxidation and oxidative steam reforming for hydrogen production from ethanol over cerium nickel based oxyhydride catalyst. *Appl. Catal. A Gen.* **2016**, *518*, 78–86.

10. Yousri, M.A.W.; Mohamed, M.E.G.; Nader, R.A. Steam and partial oxidation reforming options for hydrogen production from fossil fuels for PEM fuel cells. *AEJ* **2012**, *51*, 69–75.

11. Devin, M.W.; Sandra, L.P.; John, T.W.; John, N.K. Synthesis gas production to desired hydrogen to carbon monoxide ratios by tri-reforming of methane using Ni–MgO–(Ce,Zr)O$_2$ catalysts. *Appl. Catal. A Gen.* **2012**, *445–446*, 61–68.

12. Linus, A.S.; Lea, C.S.K.; Karla, H.D.; Stephan, A.S.; Johannes, A.L. On the coke deposition in dry reforming of methane at elevated pressures. *Appl. Catal. A Gen.* **2015**, *504*, 599–607.

13. Drif, A.; Bion, N.; Brahmi, R.; Ojala, S.; Pirault-Roy, L.; Turpeinen, E.; Seelam, P.K.; Keiski, R.L.; Epron, F. Study of the dry reforming of methane and ethanol using Rh catalysts supported on doped alumina. *Appl. Catal. A Gen.* **2015**, *504*, 576–584. [CrossRef]

14. Jabbour, K.; El Hassan, N.; Casale, S.; Estephane, J.; El Zakhem, H. Promotional effect of Ru on the activity and stability of Co/SBA-15 catalysts in dry reforming of methane. *Int. J. Hydrogen Energy* **2014**, *39*, 7780–7787. [CrossRef]

15. Du, X.; France, L.J.; Kuznetsov, V.L.; Xiao, T.; Edwards, P.P.; AlMegren, H.; Bagabas, A. Dry Reforming of Methane Over ZrO$_2$-Supported Co–Mo Carbide Catalyst. *Appl. Petrochem. Res.* **2014**, *4*, 137–144. [CrossRef]

16. France, L.J.; Du, X.; Almuqati, N.; Kuznetzov, V.L.; Zhao, Y.; Zheng, J.; Xiao, T.; Bagabas, A.; Almegren, H.; Edwards, P.P. The Effect of Lanthanum Addition on The Catalytic Activity of γ-Alumina Supported Bimetallic Co–Mo Carbides for Dry Methane Reforming. *Appl. Petrochem. Res.* **2014**, *4*, 145–156. [CrossRef]

17. Tao, X.; Wang, G.; Huang, L.; Li, X.; Ye, Q. Effect of Cu-Mo Activities on the NiCuMo/Al$_2$O$_3$ Catalyst for CO$_2$ Reforming of methane. *Catal. Lett.* **2016**, *146*, 2129–2138. [CrossRef]

18. Budiman, A.W.; Song, S.-H.; Chang, T.-S.; Shin, C.-H.; Choi, M.-J. Dry reforming of methane over cobalt catalysts: A literature review of catalyst development. *Catal. Surv. Asia* **2012**, *16*, 183–197. [CrossRef]

19. Park, J.H.; Lee, D.; Lee, H.C.; Park, E.D. Steam reforming of liquid petroleum gas over Mn-promoted Ni/γ-Al$_2$O$_3$ catalysts. *Korean J. Chem. Eng.* **2010**, *27*, 1132–1138. [CrossRef]

20. Gao, X.; Bare, S.R.; Fierro, J.L.G.; Banares, M.A.; Wachs, I.E. Preparation and in-Situ Spectroscopic Characterization of Molecularly Dispersed Titanium Oxide on Silica. *J. Phys. Chem. B* **1998**, *102*, 5653–5666. [CrossRef]

21. Klein, S.; Thorimbert, S.; Maier, W.F. Amorphous microporous titania–silica mixed oxides: Preparation, characterization, and catalytic redox properties. *J. Catal.* **1996**, *163*, 476–488. [CrossRef]

22. Miller, J.B.; Ko, E.I. Control of mixed oxide textural and acidic properties by the sol-gel method. *Catal. Today* **1997**, *35*, 269–292. [CrossRef]

23. Niu, F.; Li, S.; Zong, Y.; Yao, Q. Catalytic Behavior of Flame-Made Pd/TiO$_2$ Nanoparticles in Methane Oxidation at Low Temperatures. *J. Phys. Chem. C* **2014**, *118*, 19165–19171. [CrossRef]

24. Tauster, S.J.; Fung, S.C.; Baker, R.T.K.; Horsely, J.A. Partial Oxidation of Methane to Synthesis Gas over Ru/TiO$_2$ Catalysts: Effects of Modification of the Support on Oxidation State and Catalytic Performance. *Science* **1981**, *211*, 1121–1125. [CrossRef]

25. Shamskar, F.R.; Meshkani, F.; Rezaei, M. Preparation and characterization of ultrasound-assisted co-precipitated nanocrystalline La-, Ce-, Zr–promoted Ni-Al$_2$O$_3$ catalysts for dry reforming reaction. *J. CO$_2$ Util.* **2017**, *22*, 124–134. [CrossRef]

26. Jang, W.J.; Jung, Y.S.; Shim, J.O.; Roh, H.S.; Yoon, W.L. Preparation of a Ni-MgO-Al$_2$O$_3$ catalyst with high activity and resistance to potassium poisoning during direct internal reforming of methane in molten carbonate fuel cells. *J. Power Sources* **2018**, *378*, 597–602. [CrossRef]

27. Ahmed, S.A.A.; Anis, H.F.; Ahmed, E.A. Effects of Selected Promoters on Ni/Y-Al$_2$O$_3$ Catalyst Performance in Methane Dry Reforming. *Chin. J. Catal.* **2011**, *32*, 1604–1609.

28. Zhu, F.; Zhang, H.; Yan, X.; Yan, J.; Ni, M.; Li, X.; Tu, X. Plasma-catalytic reforming of CO$_2$-rich biogas over Ni/γ-Al$_2$O$_3$ catalysts in a rotating gliding arc reactor. *Fuel* **2017**, *199*, 430–437. [CrossRef]

29. Mei, D.; Ashford, B.; He, Y.-L.; Tu, X. Plasma-catalytic reforming of biogas over supported Ni catalysts in a dielectric barrier discharge reactor: Effect of catalyst supports. *Plasma Process. Polym.* **2017**, *14*, 1600076. [CrossRef]

30. Akbari, E.; Alavi, S.M.; Rezaei, M. Synthesis gas production over highly active and stable nanostructured NiMgOAl$_2$O$_3$ catalysts in dry reforming of methane: Effects of Ni contents. *Fuel* **2017**, *194*, 171–179. [CrossRef]

31. Sudarsanam, P.; Hillary, B.; Deepa, D.K.; Amin, M.H.; Mallesham, B.; Reddy, B.M.; Bhargava, S.K. Highly efficient cerium dioxide nanocube-based catalysts for low temperature diesel soot oxidation: The cooperative effect of cerium- and cobalt-oxides. *Catal. Sci. Technol.* **2015**, *5*, 3496–3500. [CrossRef]

32. Li, R.; Qian, R.; Jiang, Z.-T. Synthesis of high specific surface area titania monolith by addition of poly (ethylene oxide). *Mater. Res. Innov.* **2015**, *19*, S8–S136. [CrossRef]

33. Liu, Y.; Lan, K.; Bagabas, A.A.; Zhang, P.; Gao, W.; Wang, J.; Sun, Z.; Fan, J.; Elzatahry, A.A.; Zhao, D. Ordered Macro/Mesoporous TiO$_2$ Hollow Microspheres with Highly Crystalline Thin Shells for High-Efficiency Photoconversion. *Small* **2016**, *12*, 860–867. [CrossRef] [PubMed]

34. Pudukudy, M.; Yaakob, Z.; Akmal, Z.S. Direct decomposition of methane over SBA-15 supported Ni, Co and Fe based bimetallic catalysts. *Appl. Surf. Sci.* **2015**, *330*, 418–430. [CrossRef]

35. Siew, K.W.; Lee, H.C.; Gimbun, J.; Chin, S.Y.; Khan, M.R.; Taufiq-Yap, Y.H.; Cheng, C.K. Syngas production from glycerol-dry (CO$_2$) reforming over La-promoted Ni/Al$_2$O$_3$ catalyst. *Renew. Energy* **2015**, *74*, 441–447. [CrossRef]

36. Wang, W.; Wang, S.P.; Ma, X.B.; Gong, J.L. Recent advances in catalytic hydrogenation of carbon dioxide. *Chem. Soc. Rev.* **2011**, *40*, 3703–3727. [CrossRef] [PubMed]

37. Wu, J.; Fang, Y.; Wang, Y.; Zhang, D.K. Combined coal gasification and methane reforming for production of syngas in a fluidized-bed reactor. *Energy Fuel* **2005**, *19*, 512–516. [CrossRef]

38. Hao, Z.; Zhu, Q.; Jiang, Z.; Hou, B.; Li, H. Characterization of aerogel Ni/Al$_2$O$_3$ catalysts and investigation on their stability for CH$_4$-CO$_2$ reforming in a fluidized bed. *Fuel Process. Technol.* **2009**, *90*, 113–121. [CrossRef]

39. Lalit, S.G.; Guido, S.J.S.; Valero-Romero, M.J.; Mallada, R.; Santamaria, J.; Stankiewicz, A.I.; Stefanidis, G.D. Synthesis, characterization, and application of ruthenium-doped SrTiO$_3$ perovskite catalysts for microwave-assisted methane dry reforming. *Chem. Eng. Process. Process Intensif.* **2018**, *127*, 178–190.

 *catalysts*

Article

Ni-Mo Sulfide Semiconductor Catalyst with High Catalytic Activity for One-Step Conversion of CO_2 and H_2S to Syngas in Non-Thermal Plasma

Xiaozhan Liu [1,2], Lu Zhao [2,*], Ying Li [2], Kegong Fang [2,*] and Minghong Wu [1]

[1]   School of Environmental and Chemical Engineering, Shanghai University, Shanghai 200444, China; liuxiaozhan0810@163.com (X.L.); mhwu@shu.edu.cn (M.W.)

[2]   State Key Laboratory of Coal Conversion, Institute of Coal Chemistry, Chinese Academy of Sciences, Taiyuan 030001, Shanxi, China; liying163@mails.ucas.ac.cn

*   Correspondence: zhaolu@sxicc.ac.cn (L.Z.); kgfang@sxicc.ac.cn (K.F.); Tel.: +86-(0)351-4041153 (L.Z.)

Received: 24 May 2019; Accepted: 11 June 2019; Published: 12 June 2019

Abstract: Carbon dioxide (CO_2) and hydrogen sulfide (H_2S) ordinarily coexist in many industries, being considered as harmful waste gases. Simultaneously converting CO_2 and H_2S into syngas (a mixture of CO and H_2) will be a promising economic strategy for enhancing their recycling value. Herein, a novel one-step conversion of CO_2 and H_2S to syngas induced by non-thermal plasma with the aid of Ni-Mo sulfide/Al_2O_3 catalyst under ambient conditions was designed. The as-synthesized catalysts were characterized by using XRD, nitrogen sorption, UV-vis, TEM, SEM, ICP, and XPS techniques. Ni-Mo sulfide/Al_2O_3 catalysts with various Ni/Mo molar ratios possessed significantly improved catalytic performances, compared to the single-component catalysts. Based on the modifications of the physical and chemical properties of the Ni-Mo sulfide/Al_2O_3 catalysts, the variations in catalytic activity are carefully discussed. In particular, among all the catalysts, the 5Ni-3Mo/Al_2O_3 catalyst exhibited the best catalytic behavior with high CO_2 and H_2S conversion at reasonably low-energy input in non-thermal plasma. This method provides an alternative route for syngas production with added environmental and economic benefits.

Keywords: syngas production; hydrogen sulfide; carbon dioxide; Ni-Mo sulfide semiconductor; non-thermal plasma

1. Introduction

Carbon dioxide (CO_2) and hydrogen sulfide (H_2S) are often considered as harmful waste gases, coexisting in many industries. These massive acid gases must be harmlessly treated for environmental improvement. Particularly, converting CO_2 and H_2S acid gases into value-added products will bring about more environmental and economic benefits. However, CO_2 is an extremely stable molecule that commonly needs to be activated at high temperature. Hence, converting CO_2 into valuable products, such as chemicals and fuels, is a global challenge [1]. Several methods for CO_2 conversion have been reported. Traynor et al. [2] revealed that using solar energy could directly reduce CO_2. They found that the high-temperature solar irradiance system provided strong heating of CO_2 with the resultant dissociation. Huh [3] reported the catalytic cycloaddition reaction of CO_2 into organic epoxides to produce cyclic carbonates using MOFs material as efficient catalysts for this reaction. Furthermore, in recent years, photocatalytic reduction of CO_2 has been also an attractive approach [4–6]. A series of new transition-metal-centered electrocatalysts has been developed for the electrocatalytic reduction of CO_2 to produce value-added C_1 or C_2 chemicals [7]. Among the aforementioned methods, the catalytic CO_2 conversion seems to be a promising process for its utilization due to the ambient operating conditions.

Hydrogen sulfide is a highly toxic pollutant, and a major source of acid rain when oxidized in the atmosphere. In industry, H_2S is usually removed by the Claus process, in which it is partially oxidized to produce water and elemental sulfur [8]. Additionally, Li et al. [9] studied the oxidation process of H_2S on activated carbon (AC) to simultaneously capture H_2S and SO_2. The results indicate that H_2S was adsorbed on the AC surface and combined with oxygen-containing functional groups to form sulfate (SO_4^{2-}) in the absence of O_2. Palma et al. [10] investigated the H_2S thermal oxidative decomposition at different operating conditions. The results show that the reaction temperature of 1100 °C and a O_2/H_2S ratio equal to 0.2 allowed to achieve the highest H_2S conversion and the lowest selectivity to SO_2. They also prepared cordierite-honeycomb-structured catalysts for H_2S oxidative decomposition at high temperature. It revealed that the optimal washcoat percentage of 30 wt% for the catalysts could obtain high H_2S conversion and H_2 yield [11]. Previously, we demonstrated that the semiconductor catalysts synergistically working with non-thermal plasma could exhibit excellent performance in H_2S decomposition [12–14]. The photons and electric fields generated by the plasma could excite the semiconductor catalyst to generate electron–hole pairs, which dramatically enhanced H_2S decomposition.

The above-mentioned studies have been reported for the separate conversion of H_2S or CO_2. The one-step conversion of CO_2 and H_2S acid gas to syngas (a mixture of CO and H_2) is expected to provide an alternative route to reduce CO_2 emissions and detoxify H_2S with added environmental and economic benefits. In the present work, we demonstrated a low-temperature and novel non-thermal plasma method aided by Ni-Mo sulfide/Al_2O_3 catalysts for syngas production from the simultaneous conversion of CO_2 and H_2S. A series of Ni-Mo sulfide/Al_2O_3 catalysts with different Ni/Mo molar ratios was prepared. The effects of the chemical and physical properties on the catalytic behaviors of the as-prepared catalysts were carefully investigated by various characterization methods such as XRD, nitrogen sorption, UV-vis, TEM, SEM, ICP, and XPS. Some intensive understandings for the optimizations and designs of catalysts were also provided through studying the structure–performance correlations.

2. Results and Discussion

2.1. XRD Analysis

Figure 1a shows the XRD patterns of the Ni-Mo sulfide/Al_2O_3 catalysts with different Ni/Mo molar ratios. The characteristic diffraction peaks of each catalyst observed at 2θ about 14.4°, 32.7°, and 58.4° were attributed to the (0 0 2), (1 0 0), and (1 1 0) planes of MoS_2 (JCPDS#65-1951), respectively. Similarly, the peaks located at 2θ of 27.2°, 31.5°, 35.3°, 38.8°, 45.1°, 53.5°, 56.1°, 58.6°, and 61.0° were assigned to the (1 1 1), (2 0 0), (2 1 0), (2 1 1), (2 2 0), (3 1 1), (2 2 2), (0 2 3), and (3 2 1) crystal surfaces of the NiS_2 phase (JCPDS#65-3325), respectively. Meanwhile, the diffraction peaks of Al_2O_3 were also detectable. Especially, the XRD patterns also exhibited a variation trend correlated to the chemical compositions. With an increase in the Ni/Mo molar ratio, the diffraction peaks shifted to slightly higher 2θ values, as seen in Figure 1b. The Ni^{2+} ions replaced the position of Mo ions or entered the gap position of MoS_2 to form the Ni-Mo-S_x phase [15]. Since the radius of Ni^{2+} ion is bigger than the radius of Mo^{4+} ion, the lattice parameters of Ni-Mo sulfides increased with increasing Ni content. Therefore, it can be deduced that the weak diffraction peaks of MoS_2 with an increase in the content of Ni may be related to the formation of the Ni-Mo-S_x phase. In addition, the other Ni and Mo species were in the states of NiS_2 and MoS_2, respectively. These sulfides were uniformly mixed, owing to the metal ions being uniformly dispersed on the support. In other words, Ni-Mo-S_x, NiS_2, and MoS_2 were well mixed and highly dispersed on the support.

In addition, the weak and broad peaks illustrate that the sulfide particles were highly dispersed with particles in small nanoscale size. Based on the Scherrer equation, the average particle sizes of the Ni-Mo sulfide/Al_2O_3 catalysts were calculated. As shown in Table 1, the average particle sizes were estimated to be 7.8, 9.1, 10.2, 11.5, and 13.9 nm for the Ni-Mo sulfide/Al_2O_3 catalysts of which

Ni/Mo molar ratios were 2/6, 3/5, 4/4, 5/3, and 6/2, respectively, which is agreement with the sizes of the nanoparticles in the TEM analysis. Moreover, as the Ni/Mo molar ratio increased, the average particle sizes gradually increased. The change in the average particle size can be explained by an increase in the Ni content. In addition, the surface areas of the various Ni-Mo sulfide/Al_2O_3 catalysts are compared in Table 1. It seems that the Ni/Mo molar ratio did not significantly affect the surface areas.

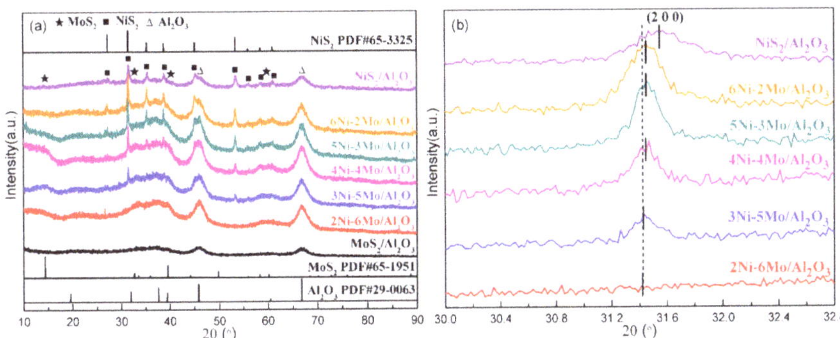

Figure 1. The XRD patterns of the Ni-Mo sulfide/Al_2O_3 catalysts with different Ni/Mo molar ratios (**a**) scanning angle of 10–90°; (**b**) scanning angle of 30–32.8°.

Table 1. BET surface areas, particle sizes, and band gaps of the various Ni-Mo sulfide/Al_2O_3 catalysts.

Catalyst	BET Surface Area (m²/g)	Particle Size (nm)	Band Gap (eV)
NiS_2/Al_2O_3	243	14.3	1.01
6Ni-2Mo/Al_2O_3	248	13.9	1.30
5Ni-3Mo/Al_2O_3	250	11.5	1.56
4Ni-4Mo/Al_2O_3	253	10.2	1.81
3Ni-5Mo/Al_2O_3	259	9.1	2.00
2Ni-6Mo/Al_2O_3	261	7.8	2.17
MoS_2/Al_2O_3	265	6.2	2.32

2.2. UV-Visible Analysis

Figure 2a shows the UV-visible spectra of the Ni-Mo sulfide/Al_2O_3 catalysts with different Ni/Mo molar ratios, compared with NiS_2 and MoS_2. From the UV-visible spectra, the band edge of MoS_2 was located around 600–800 nm, which belonged to the absorption of visible-light. Compared with MoS_2, with increasing the Ni/Mo molar ratio, the absorption boundaries of the Ni-Mo sulfide/Al_2O_3 catalysts were gradually red shifted. Continuous shift of the absorption boundaries suggests that the band gaps of the Ni-Mo sulfide/Al_2O_3 catalysts can be controllably adjusted through changing the Ni/Mo molar ratio. The relationship between the incident photon energy and the absorption coefficient of a semiconductor can be determined by the Kubelka–Munk equation [16,17]:

$$\alpha(h\nu) = C(h\nu - Eg)^{n/2}, \tag{1}$$

where α is the absorption coefficient and its value can be achieved by the equation: $\alpha = (1 - R)^2/2R$; R is the diffuse reflectance and its relationship with absorbance can be defined by $R = 10^{-A}$; A is absorbance. ν is frequency, h is Planck's constant, and C is a constant. For a direct transition semiconductor, n = 1; for an indirect transition semiconductor, n = 4. The nature of transition is possible to be determined through plotting the graph of $(\alpha h\nu)^2$ versus $h\nu$; therefore, the band gap energies can be deduced by extrapolating the straight-linear portions of the plot to intersect the photon energy axis. As shown in Figure 2b and Table 1, the band gaps obtained in such a way were 2.17, 2.00, 1.81, 1.56, and 1.30 eV for the Ni-Mo sulfide/Al_2O_3 catalysts, of which Ni/Mo molar ratios were 2/6, 3/5, 4/4, 5/3, and 6/2, respectively. For all the catalysts, the influence of the chemical compositions of the Ni-Mo sulfide/Al_2O_3 catalysts on the band gap can be observed. When the Ni/Mo molar ratio increased, the band gap decreased gradually. This indicates that changing the Ni/Mo molar ratio can significantly adjust the band gaps of the Ni-Mo sulfide/Al_2O_3 catalysts. Meanwhile, the changes in band gaps also illustrate that the relative redox abilities of the Ni-Mo sulfide/Al_2O_3 catalysts were effectively changed.

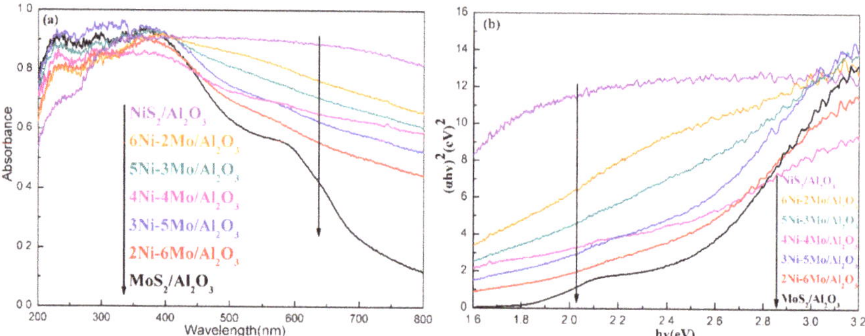

Figure 2. UV-vis diffuse reflection spectra for the Ni-Mo sulfide/Al_2O_3 catalysts with different Ni/Mo molar ratios (**a**) relationship of absorbance and wavelength; (**b**) relationship of absorption coefficient and incident photon energy.

2.3. TEM Analysis

The morphologies and microstructures of the Ni-Mo sulfide/Al_2O_3 catalyst with Ni/Mo of 5/3 are presented in Figure 3. The TEM images suggest that the highly distributed nanoparticles were dispersed on the supports, which can prevent their grain growth. Furthermore, it also reveals that the average particle size was about 10 nm, which was in agreement with the particle size estimated from Scherrer equation (shown in Table 1). As shown in Figure 3d, the observed interlayer spacing of 6.1 Å, which was identical to the lattice fringe of Ni-Mo-S_x, was bigger than that of the (0 0 2) plane of MoS_2 (6.0 Å) [18]. It can be attributed to the bigger radius of Ni^{2+} ions. Meanwhile, the interplanar spacings were about 2.8 and 3.2 Å, which correspond to the (2 0 0) and (1 1 1) planes of NiS_2, respectively [19,20]. These results of TEM analysis are in good agreement with the XRD analysis.

In addition, the porosity of the catalyst also plays an important role in the generation of an electric field in non-thermal plasma. As shown in Table 1, the obtained Ni-Mo sulfide/Al_2O_3 catalysts had high surface areas (>200 m^2/g). As pointed by Fridman [21], the porous material in the gap refracts the electric field, enhancing the local field by a factor of over 10 depending on the porosity of the materials. The electric field can excite the Ni-Mo sulfide semiconductor to generate electron–hole pairs, which plays an extremely important role in converting CO_2 and H_2S. Simultaneously, the strong electric field is beneficial for delaying the recombination of electron–hole pairs, thereby extending their lifetime.

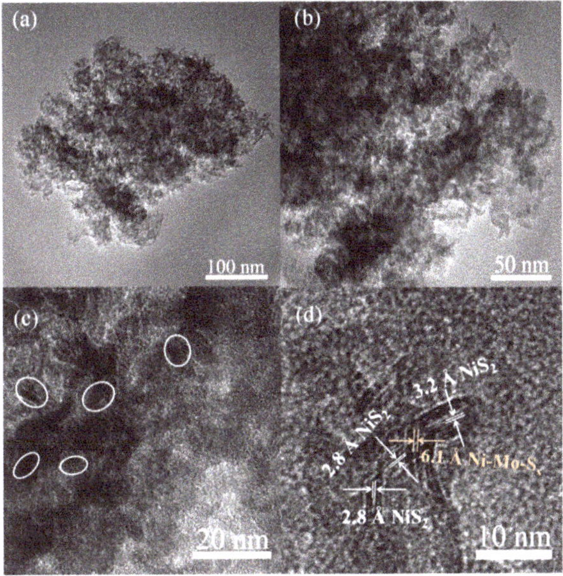

Figure 3. TEM and HR-TEM images of the 5Ni-3Mo/Al$_2$O$_3$ catalyst (**a**) scaleplate of 100 nm; (**b**) scaleplate of 50 nm; (**c**) scaleplate of 20 nm; (**d**) scaleplate of 10 nm.

2.4. XPS Analysis

X-ray photoelectron spectroscopy (XPS) analysis was carried out to study the chemical state and surface ratio of MoS$_2$, NiS$_2$, and Ni-Mo sulfide/Al$_2$O$_3$ catalysts with different Ni/Mo molar ratios. Figure 4 shows the Mo 3d and Ni 2p spectra of the various catalysts. For Mo element, as shown in Figure 4a, the observed binding energy (BE) of Mo 3d$_{5/2}$ was about 229.0 eV, indicating that the Mo species were Mo^{4+} [22]. In Figure 4b, for the case of Ni, the main peaks at the BE of about 855.0 eV can be attributed to the Ni 2p$_{3/2}$ peaks of Ni^{2+} [22]. However, as the Ni/Mo molar ratio increased, the peak position of Mo 3d$_{5/2}$ gradually shifted toward the peak position of the lower BE, accompanied by the BE shift of Ni 2p$_{3/2}$. This phenomenon indicates the increased electron density in Mo 3d$_{5/2}$, resulting from the electron donating property of Ni 2p$_{3/2}$. Therefore, a strong electron interaction between Ni and Mo occurs on the catalyst surface, wherein electrons likely transfer from the Ni species to the Mo species in the Ni-Mo sulfide/Al$_2$O$_3$ catalysts.

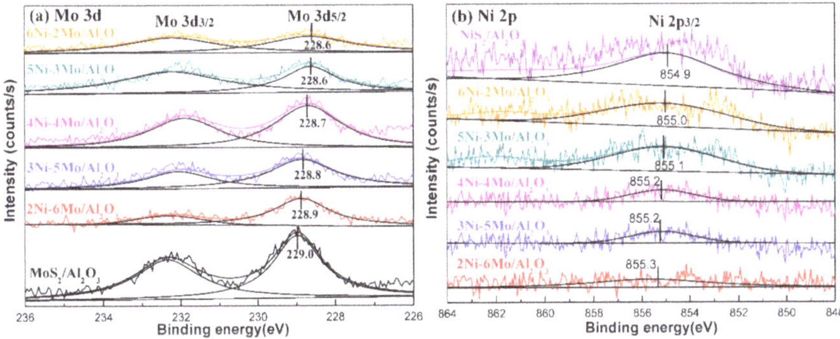

Figure 4. XPS spectra of the Ni-Mo sulfide/Al$_2$O$_3$ catalysts with different Ni/Mo molar ratios: (**a**) Mo 3d, (**b**) Ni 2p.

Table 2 shows the surface and total Ni/Mo ratio of Ni-Mo sulfide/Al_2O_3 catalysts with different Ni/Mo molar ratios. As presented, the total Ni/Mo molar ratio was consistent with the theoretical ratio. However, these total Ni/Mo ratios (0.31İ2.97) were higher than the surface Ni/Mo ratio (0.18-2.41). This is because Ni^{2+} ions were intercalated into the gap position of the MoS_2 lattice, and a large number of Mo vacancies could be generated. Therefore, the surface of the Ni-Mo sulfide/Al_2O_3 catalysts became slightly Ni-depleted.

Table 2. The surface and total Ni/Mo atomic ratios of the Ni-Mo sulfide/Al_2O_3 catalysts with different Ni/Mo molar ratios.

Sample	Surface and Total Ni/Mo Atomic Ratios	
	Surface Ni/Mo Atomic Ratio [1]	Total Ni/Mo Atomic Ratio [2]
6Ni-2Mo/Al_2O_3	2.41	2.97
5Ni-3Mo/Al_2O_3	1.37	1.74
4Ni-4Mo/Al_2O_3	0.78	0.95
3Ni-5Mo/Al_2O_3	0.49	0.63
2Ni-6Mo/Al_2O_3	0.18	0.31

[1] By XPS. [2] By ICP.

2.5. Catalytic Evaluation for the One-Step Conversion of CO_2 and H_2S to Syngas

The catalytic performances of the various Ni-Mo sulfide/Al_2O_3 catalysts were evaluated through converting CO_2 and H_2S into syngas in non-thermal plasma. For comparison, the performances of NiS_2/Al_2O_3 and MoS_2/Al_2O_3 were also investigated. As seen in Figure 5a,b, all the Ni-Mo sulfide/Al_2O_3 catalysts possessed better activities in CO_2 and H_2S conversion than NiS_2/Al_2O_3 and MoS_2/Al_2O_3 catalysts, and the CO_2 and H_2S conversions could reach high levels. The experimental results show that the Ni/Mo molar ratio had a great influence on the conversion of CO_2 and H_2S. As the Ni/Mo molar ratio increased, the catalytic activity presented a primary enhancement followed by a decline. The CO_2 and H_2S conversions were strongly dependent on the SEI (Specific energy input). At SEI of 60.0 kJ/L, CO_2 conversions were 25.1%, 45.0%, 46.2%, 46.9%, 47.7%, 56.3%, and 49.0%, and H_2S conversions were 87.8%, 93.7%, 94.8%, 95.7%, 96.4%, 98.9%, and 97.3% when NiS_2/Al_2O_3, MoS_2/Al_2O_3, 2Ni-6Mo/Al_2O_3, 3Ni-5Mo/Al_2O_3, 4Ni-4Mo/Al_2O_3, 5Ni-3Mo/Al_2O_3, and 6Ni-2Mo/Al_2O_3 were filled in the gap, respectively. Especially, the 5Ni-3Mo/Al_2O_3 catalyst exhibited the best catalytic performance and achieved relatively high CO_2 and H_2S conversions with the lowest SEI.

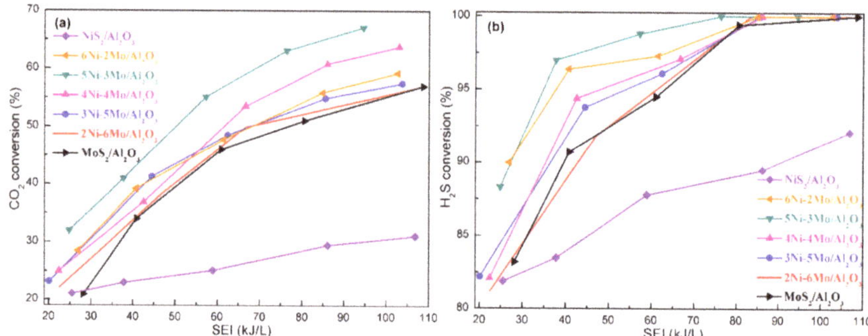

Figure 5. The conversions of CO_2 (a) and H_2S (b) as a function of SEI on the Ni-Mo sulfide/Al_2O_3 catalysts with various Ni/Mo molar ratios. Reaction conditions: feed: H_2S/CO_2 molar ratio = 20:15; flow rate: 35 mL/min; catalyst bed volume: 15.0 mL.

In addition, as seen in Figure 6a,b, the major products were CO and H_2 in the CO_2 and H_2S conversion. CO and H_2 concentrations were in line with SEI, which indicates that the behavior for CO_2 and H_2S conversion had relatively stronger dependence on the energy input. An increase of SEI could generate more active H species and obviously promote CO_2 activation and CO production, together with the decrease in H_2 yields. Meanwhile, very small amounts of light hydrocarbons (others: CH_4, C_2H_4, and C_2H_6) were also generated. The selectivity to light hydrocarbons was very low (<2%) during the reaction. Furthermore, there were not any C_{3+} hydrocarbons. Therefore, this novel method may produce clean syngas. Additionally, it was also found that SEI strongly affected the H_2/CO ratio. In Figure 7, when SEI was changed from 20 to 110 kJ/L, the H_2/CO ratio considerably decreased from about 4.5 to 1.0, which illustrates that the H_2/CO ratio strongly depends on the energy input. An increase of SEI could induce the decrease in the H_2/CO ratio. Hence, the H_2/CO ratio can be controllably adjusted on a large scale through varying SEI by this method.

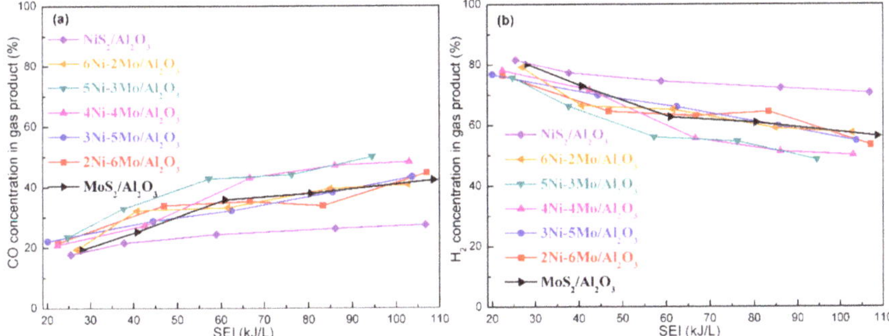

Figure 6. CO concentration (**a**) and H_2 concentration (**b**) in gas product as a function of SEI in the plasma-induced CO_2 and H_2S conversion on the Ni-Mo sulfide/Al_2O_3 catalysts with various Ni/Mo molar ratios. Reaction conditions: feed: H_2S/CO_2 molar ratio = 20:15; flow rate: 35 mL/min; catalyst bed volume: 15.0 mL.

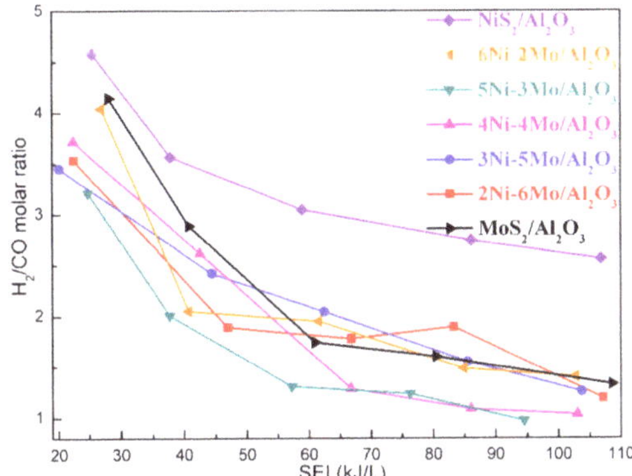

Figure 7. H_2/CO molar ratio as a function of SEI in the plasma-induced CO_2 and H_2S conversion on the Ni-Mo sulfide/Al_2O_3 catalysts with various Ni/Mo molar ratios. Reaction conditions: feed: H_2S/CO_2 molar ratio = 20:15; flow rate: 35 mL/min; catalyst bed volume: 15.0 mL.

A series of characterizations of the Ni-Mo sulfide/Al_2O_3 catalysts displays that the Ni/Mo molar ratio had a significant effect on the physical and chemical properties of the catalyst. We have reported that the synergistic effects of semiconductor catalyst and non-thermal plasma in the H_2S decomposition [12]. In the present work, the Ni-Mo sulfide/Al_2O_3 catalyst in non-thermal plasma can be excited by both the strong electric field and UV-visible light irradiation, and thus generate highly active hole–electron pairs. The hole–electron pairs will react with the adsorbed surface species, thereby accelerating the conversion of CO_2 and H_2S. Hence, since the generated hole–electron pairs are sufficiently reactive to convert CO_2 and H_2S to H_2 and CO, the rate of CO_2 and H_2S conversion depends on the number of electron–hole pairs generated on the surface of the Ni-Mo sulfide/Al_2O_3 catalyst. A higher number of hole–electron pairs may be linked to the relatively higher behavior on CO_2 and H_2S conversion. From the results of UV-vis spectra (shown in Figure 2 and Table 1), the change in the Ni/Mo molar ratio affects the optical properties of the Ni-Mo sulfide/Al_2O_3 catalyst. With increasing the Ni/Mo molar ratio, a monotonous variation in the absorption in visible light region and band gap of Ni-Mo sulfide could be clearly found. For a semiconductor catalyst with a narrower band gap, less energy for electrons is required to jump from valence band (VB) to conduction band (CB). Therefore, a decrease in band gap can lead to the increase in the amount of hole–electron pairs. Moreover, the other optical properties of semiconductor catalyst, such as conduction band position and valence band position, are also related to its chemical compositions. According to the XRD and TEM results, the Ni-Mo sulfide possessed the layer structure, the Ni^{2+} ions can replace the position of Mo ions or enter the gap position of MoS_2 to form Ni-Mo-S_x phase. Hence, the suitable impurity energy level could be provided through a proper doping amount of Ni^{2+} ions into MoS_2. The presence of impurity levels leads to the easy injection of the excited electrons from VB to CB of MoS_2.

In addition, all the Ni-Mo sulfide/Al_2O_3 catalysts exhibited relatively high BET surface areas (shown in Table 1). The high surface area facilitates photon absorption, provides more active sites, and reduces the distance of generated carriers from the catalyst surface [23]. Moreover, the average particle size was around 10 nm. The small nanoparticles with low crystallinity are favorable for the fast electron transportation from bulk to surface, which prevent the recombination of the generated electrons and holes of the catalyst [24]. Therefore, the reduction in the particle size of the Ni-Mo sulfide/Al_2O_3 catalyst also contributes to the improvement of the catalytic activity.

Additionally, the Ni^{2+} ions can be evenly incorporated into the MoS_2 lattice to form Ni-Mo-S_x phase, which would bring about the Mo vacancies formation. Ideally, the incorporation of two Ni^{2+} ions may generate one Mo vacancy. Therefore, the incorporation of Ni^{2+} ions can produce a large amount of Mo vacancies. Mo vacancies favor the separation of the energy-induced electrons and holes, which induce the high catalytic performance in CO_2 and H_2S conversion, compared to the MoS_2/Al_2O_3 catalyst. Nevertheless, the 6Ni-2Mo/Al_2O_3 catalyst with higher Ni content possessed relatively stronger visible light absorption capacity than other Ni-Mo sulfide/Al_2O_3 catalysts, but the CO_2 and H_2S conversions were lower. The reason for the low catalytic activity over the 6Ni-2Mo/Al_2O_3 catalyst may be that the excessive amount of Ni would result in the unevenly distributed Ni^{2+} ions. Owing to the higher concentration of Ni, the probability of electron–hole recombination was regarded to become comparably high. Consequently, a superfluous increase in the Ni/Mo molar ratio not only encumbered the light absorption, but also offered more recombination sites for hole–electron pairs, so the catalytic activity was repressed. In particular, when the Ni/Mo molar ratio was 5/3, the Ni-Mo sulfide/Al_2O_3 catalyst exhibited the best catalytic activity for CO_2 and H_2S conversion with the most proper optical and structural properties.

Figure 8 presents the CO_2 and H_2S conversion, and H_2/CO molar ratio variations over 5Ni-3Mo/Al_2O_3 during the long-term test. The results demonstrate that the catalytic activity did not exhibit loss in the runs. The XPS spectra and SEM images were taken before and after evaluation, as shown in Figures S1 and S2 (Supplementary Materials), respectively. There was no obvious difference detected in the spent 5Ni-3Mo/Al_2O_3 catalyst after the reaction test. Moreover, Figure S3 (Supplementary Materials) shows a comparison between the fresh 5Ni-3Mo/Al_2O_3 catalyst and the spent one after reaction tests in the XRD patterns. According to the Scherrer equation, the average particle size of the 5Ni-3Mo/Al_2O_3 catalyst increased from 11.5 to 13.2 nm after the 50 h long-term test, proving high stability of the active phases on the 5Ni-3Mo/Al_2O_3 catalyst in the CO_2 and H_2S conversion process. Furthermore, the surface area decreased only from about 250 to 231 m^2/g. The active phases were highly dispersed in Al_2O_3 support, which could also prevent the agglomeration formation and inhibit the growth of the particles. Hence, it is clear that Ni-Mo sulfide underwent no variation in non-thermal plasma and can maintain stable structures in the plasma-induced CO_2 and H_2S conversion.

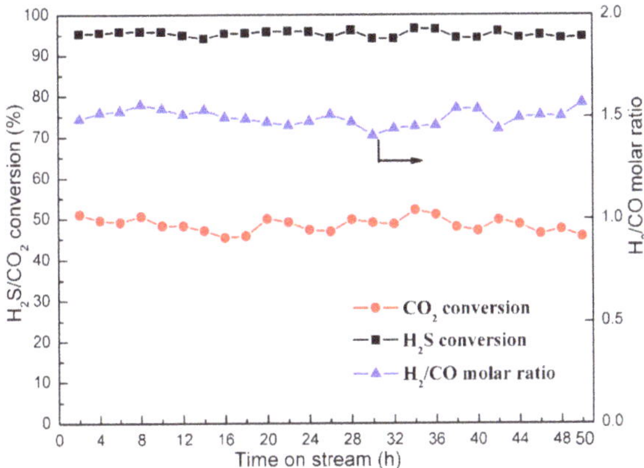

Figure 8. Variations of CO_2 and H_2S conversion and H_2/CO molar ratio with time in the plasma-induced conversion over 5Ni-3Mo/Al_2O_3. Reaction conditions: H_2S/CO_2 molar ratio = 20:15; flow rate: 35 mL/min; catalyst bed volume: 15.0 mL; SEI: 50.0 kJ/L.

3. Materials and Methods

3.1. Catalyst Preparation

A series of Al_2O_3 supported Ni-Mo sulfide catalysts with different Ni/Mo molar ratios was synthesized. Al_2O_3 with surface area about 310 m^2/g was grinded and then sieved to 40–60 mesh. In order to prepare the precursors of Ni-Mo sulfide supported on Al_2O_3, Al_2O_3 was impregnated by a mixed solution of nickel nitrate and ammonium molybdate at room temperature. The total loading of the oxide precursor was 5 wt%. After the impregnation, the samples were dried at 120 °C for 8 h, and then calcined at 500 °C for 3 h. After calcination, the samples were sulfided in 60 vol% H_2S/H_2 flow (80 mL/min) for 3 h. The catalysts were denoted as xNi(8 − x)Mo/Al_2O_3, where x/(8 − x) was the Ni/Mo molar ratio.

3.2. Catalyst Evaluation for Plasma-Induced CO_2-H_2S Conversion

A dielectric barrier discharge (DBD) reactor was used to generate non-thermal plasma at atmospheric pressure in this work. The configuration of reactor has been illustrated in detail in our previous study [12,13]. A high voltage generator was applied to supply a voltage from 0 to 20 kV with a sinusoidal waveform at a frequency of about 10 kHz. This reactor consisted of one quartz tube and two electrodes. The discharge volume of DBD reactor is 15 mL. The stainless-steel rod was used as a high voltage electrode on the axis of the tubular reactor and connected to the plasma generator. The aluminum foil was used as a grounding electrode, and wrapped around the quartz tube and grounded by wires. An amount of 15 mL of the Ni-Mo sulfide/Al_2O_3 catalyst with 40–60 mesh was placed in the gap between the quartz tube and the high voltage electrode. The discharge power was determined by the Q-V Lissajous pattern, which was measured by the digital oscilloscope. In the conversion of CO_2 and H_2S to syngas, elemental sulfur was produced. In order to prevent sulfur deposition on the catalyst surface, the reactor was immersed in the oil bath at 120 °C to turn the generated sulfur into liquid phase and out of the catalyst bed. The feed gas was flowed through the catalyst bed while the non-thermal plasma was generated by high-voltage discharge.

A cold trap was placed at the exit of the DBD reactor for the condensation of any liquid products. The gas products were analyzed by a two-channel gas chromatography equipped with two thermal conductivity detections (TCD). The first channel contained a Porapak Q column for the measurement of CO_2, H_2S, and C_2-C_4 hydrocarbons, while the second channel was equipped with a Molecular Sieve 5A column for the separation of H_2, CO, and CH_4. The gas chromatography was calibrated for a wide range of concentrations for each gaseous component using reference gas mixtures from the calibrated gas mixes.

For the H_2S and CO_2 conversion, the conversions of H_2S (x_{H2S}) and CO_2 (x_{CO2}) were defined as:

$$x_{H_2S} = \frac{H_2S \text{ converted}}{H_2S \text{ input}} \times 100\%, \tag{2}$$

$$x_{CO_2} = \frac{CO_2 \text{ converted}}{CO_2 \text{ input}} \times 100\%, \tag{3}$$

The H_2/CO molar ratios were defined as:

$$H_2/CO = \frac{H_2 \text{ produced}}{CO \text{ produced}}, \tag{4}$$

The gas product distributions (C, %) were calculated by the selectivity (S, %) of the products:

$$S_{CO} = \frac{CO \text{ produced}}{CO_2 \text{ converted}} \times 100\%, \tag{5}$$

$$S_{H_2} = \frac{H_2 \text{ produced}}{H_2S \text{ converted}} \times 100\%, \tag{6}$$

$$S_{CxHy} = \frac{x \times C_xH_y \text{ produced}}{CO_2 \text{ converted}} \times 100\%, \tag{7}$$

$$C_{H2} = \frac{H_2 \text{ produced}}{H_2 \text{ produced } + CO \text{ produced} \times (1 + \sum \frac{S_{CxHy}}{x \times S_{CO}})} \times 100\%, \tag{8}$$

$$C_{CO} = \frac{CO \text{ produced}}{H_2 \text{ produced } + CO \text{ produced} \times (1 + \sum \frac{S_{CxHy}}{x \times S_{CO}})} \times 100\%, \tag{9}$$

$$C_{others} = \frac{CO \text{ produced} \times \sum \frac{S_{CxHy}}{x \times S_{CO}}}{H_2 \text{ produced } + CO \text{ produced} \times (1 + \sum \frac{S_{CxHy}}{x \times S_{CO}})} \times 100\%, \tag{10}$$

The sulfur and carbon balances were defined as:

$$B_{sulfur} = \frac{[H_2S]_{out} + [Sulphur]_{out}}{[H_2S]_{into}} \times 100\%, \tag{11}$$

$$B_{carbon} = \frac{[CO_2]_{out} + [CO]_{out} + [C_xH_y]_{out}}{[CO_2]_{into}} \times 100\%, \tag{12}$$

The sulfur and carbon balances were based on sulfur-atom and carbon-atom, respectively. The error of the balances were within 5% and typically better than this.

The area of the Lissajous diagram measures the energy dissipated in the discharge during one period of the voltage. The charge was determined by measuring the voltage across the capacitor of 0.47 μF connected in series to the ground line of the plasma reactor. The discharge power was calculated from the area of charge–voltage parallelogram and the frequency of discharge. Specific energy input (SEI, J/L), which measures the energy input in the plasma process, was calculated by:

$$SEI = \frac{P}{V},$$ (13)

where P is the discharge power (W), and V is the gas flow rate (L/s).

3.3. Catalyst Characterization

The X-ray diffraction (XRD) profiles of the catalysts were recorded using a Rigaku D/Max-RA diffractometer (Tokyo, Japan) with Cu Kα radiation operated at 40 kV. The specific surface areas of the catalysts were carried out on a Micromeritics ASAP-2000 instrument (Norcross, GE, USA). UV-visible spectroscopy was obtained on a Jasco V-550 spectrophotometer (Tokyo, Japan). Transmission electron microscope (TEM) images were conducted by a Tecnai G2 F20 S-Twin microscope operating (Hillsboro, OR, USA) at 200 kV. The morphology of the catalyst was investigated using a JEOL JSM-7001F scanning electron microscope (SEM). The total chemical compositions of the catalysts were determined using inductively coupled plasma-atomic emission spectroscopy (ICP-AES) with an Optima 2000 DV spectrometer (PerkinElmer, Waltham, MA, USA). X-ray photoelectron spectroscopy (XPS) was performed using a Thermo VGESCALAB250 X-ray photoelectron spectrometer (Waltham, MA, USA) with an Al Kα source. All binding energies were referenced to the C 1s peak at 284.6 eV.

4. Conclusions

In summary, a series of Ni-Mo sulfide/Al_2O_3 catalysts with different molar ratios of Ni/Mo were prepared and applied for the one-step conversion of CO_2 and H_2S into syngas in non-thermal plasma. According to the characterization results, the optical and structural properties of the as-synthesized catalysts were significantly influenced by the Ni/Mo molar ratio. The Ni/Mo molar ratio can effectively adjust the particle size and band gap. Additionally, the generated Mo vacancies are also favorable for the transfer and separation of holes and electrons. The experimental results indicated that the Ni-Mo sulfide/Al_2O_3 catalysts possessed excellent catalytic activities for CO_2 and H_2S conversion, compared to the single-component NiS_2/Al_2O_3 and MoS_2/Al_2O_3 catalysts. In particular, the Ni-Mo sulfide/Al_2O_3 catalyst with Ni/Mo molar ratio of 5/3 showed relatively higher CO_2 and H_2S conversions, and exhibited good stability in the long-term test for CO_2 and H_2S conversion to syngas.

Supplementary Materials: The following are available online at http://www.mdpi.com/2073-4344/9/6/525/s1, Figure S1: XPS spectra of Mo 3d (**a**) and Ni 2p (**b**) for the 5Ni-3Mo/Al_2O_3 catalyst before and after reaction; Figure S2: SEM images of the 5Ni-3Mo/Al_2O_3 catalyst before (**a**) and after (**b**) reaction; Figure S3: XRD patterns of the 5Ni-3Mo/Al_2O_3 catalyst before and after reaction.

Author Contributions: The project was conceived by L.Z.; X.L. performed the experiments and drafted the paper under the supervision of L.Z. and K.F.; Y.L. and M.W. helped to collect and analyze some characterization data. The manuscript was revised and checked through the comments of all authors. All authors have given approval for the final version of the manuscript.

Funding: This research was funded by the National Key R&D Program of China (2018YFB0604700), the National Natural Science Foundation of China (21603255, 21473230), and the Natural Science Foundation of Shanxi (201601D021052).

Conflicts of Interest: The authors declare no conflict of interest.

References

1. Song, C. Global challenges and strategies for control, conversion and utilization of CO_2 for sustainable development involving energy, catalysis, adsorption and chemical processing. *Catal. Today* **2006**, *115*, 2–32. [CrossRef]

2. Traynor, A.J.; Jensen, R.J. Direct solar reduction of CO_2 to fuel: first prototype results. *Ind. Eng. Chem. Res.* **2002**, *41*, 1935–1939. [CrossRef]

3. Huh, S. Direct catalytic conversion of CO_2 to cyclic organic carbonates under mild reaction conditions by metal-organic frameworks. *Catalysts* **2019**, *9*, 34. [CrossRef]
4. Olowoyo, J.O.; Kumar, M.; Jain, S.L.; Shen, S.-H.; Zhou, Z.-H.; Mao, S.S.; Vorontsov, A.V.; Kumar, U. Reinforced photocatalytic reduction of CO_2 to fuel by efficient S-TiO_2: Significance of sulfur doping. *Int. J. Hydrogen Energy* **2018**, *43*, 17682–17695. [CrossRef]
5. Reithmeier, R.; Bruckmeier, C.; Rieger, B. Conversion of CO_2 via visible light promoted homogeneous redox catalysis. *Catalysts* **2012**, *2*, 544–571. [CrossRef]
6. Tasbihi, M.; Schwarze, M.; Edelmannová, M.; Spöri, C.; Strasser, P.; Schomäcker, R. Photocatalytic reduction of CO_2 to hydrocarbons by using photodeposited Pt nanoparticles on carbon-doped titania. *Catal. Today* **2019**, *328*, 8–14. [CrossRef]
7. Feng, D.-M.; Zhu, Y.-P.; Chen, P.; Ma, T.-Y. Recent advances in transition-metal-mediated electrocatalytic CO_2 Reduction: from homogeneous to heterogeneous systems. *Catalysts* **2017**, *7*, 373. [CrossRef]
8. Lagas, J.A.; Borsboom, J.; Berben, P.H. Selective-oxidation catalyst improves Claus process. *Oil Gas J. (United States)* **1988**, *10*, 68.
9. Li, Y.-R.; Lin, Y.-T.; Xu, Z.-C.; Wang, B.; Zhu, T.-Y. Oxidation mechanisms of H_2S by oxygen and oxygen-containing functional groups on activated carbon. *Fuel Process. Technol.* **2019**, *189*, 110–119. [CrossRef]
10. Palma, V.; Vaiano, V.; Barba, D.; Colozzi, M.; Palo, E.; Barbato, L.; Cortese, S. H_2 production by thermal decomposition of H_2S in the presence of oxygen. *Int. J. Hydrogen Energy* **2015**, *40*, 106–113. [CrossRef]
11. Palma, V.; Barba, D.; Vaiano, V.; Colozzi, M.; Palo, E.; Barbato, L.; Cortese, S.; Miccio, M. Honeycomb structured catalysts for H_2 production via H_2S oxidative decomposition. *Catalysts* **2018**, *8*, 488. [CrossRef]
12. Zhao, L.; Wang, Y.; Jin, L.; Qin, M.-L.; Li, X.; Wang, A.-J.; Song, C.-S.; Hu, Y.-K. Decomposition of hydrogen sulfide in non-thermal plasma aided by supported CdS and ZnS semiconductors. *Green Chem.* **2013**, *15*, 1509–1513. [CrossRef]
13. Zhao, L.; Wang, Y.; Li, X.; Wang, A.-J.; Song, C.-S.; Hu, Y.-K. Hydrogen production via decomposition of hydrogen sulfide by synergy of non-thermal plasma and semiconductor catalysis. *Int. J. Hydrogen Energy* **2013**, *38*, 14415–14423. [CrossRef]
14. Zhao, L.; Wang, Y.; Sun, Z.-L.; Wang, A.-J.; Li, X.; Song, C.-S.; Hu, Y.-K. Synthesis of highly dispersed metal sulfide catalysts via low temperature sulfidation in dielectric barrier discharge plasma. *Green Chem.* **2014**, *16*, 2619–2626. [CrossRef]
15. Hillerová, E.; Sedláček, J.; Zdražil, M. Bimetallic sulphide catalysts Ni-MS_x/SiO_2 prepared by unconventional method involving thiourea complexes. *Collect. Czech. Chem. Comm.* **1987**, *52*, 1748–1757. [CrossRef]
16. Butler, M.A. Photoelectrolysis and physical properties of the semiconducting electrode WO_2. *J. Appl. Phys.* **1977**, *48*, 1914–1920. [CrossRef]
17. Domen, K.; Kudo, A.; Onishi, T. Mechanism of photocatalytic decomposition of water into H_2 and O_2 over NiO-$SrTiO_3$. *J. Catal.* **1986**, *102*, 92–98. [CrossRef]
18. Nath, M.; Govindaraj, A.; Rao, C.N.R. Simple synthesis of MoS_2 and WS_2 nanotubes. *Adv. Mater.* **2001**, *13*, 283–286. [CrossRef]
19. Wu, X.-L.; Yang, B.; Li, Z.-J.; Lei, L.-C.; Zhang, X.-W. Synthesis of supported vertical NiS_2 nanosheets for hydrogen evolution reaction in acidic and alkaline solution. *RSC Adv.* **2015**, *5*, 32976–32982. [CrossRef]
20. Yin, L.-S.; Yuan, Y.-P.; Cao, S.-W.; Zhang, Z.-Y.; Xue, C. Enhanced visible-light-driven photocatalytic hydrogen generation over g-C_3N_4 through loading the noble metal-free NiS_2 cocatalyst. *RSC Adv.* **2014**, *4*, 6127–6132. [CrossRef]
21. Fridman, A. *Plasma Chemistry*; Cambridge University Press: New York, NJ, USA, 2008.
22. Wagner, C.D.; Riggs, W.M.; Davis, L.E.; Moulder, J.F.; Muilenberg, G.E. *Handbook of X-ray Photoelectron Spectroscopy*; Perkin-Elmer Corporation Physical Electronics Division: Eden Prairie, MN, USA, 1978.
23. Hu, J.-S.; Ren, L.-L.; Guo, Y.-G.; Liang, H.-P.; Cao, A.-M.; Wan, L.-J.; Bai, C.-L. Mass production and high photocatalytic activity of ZnS nanoporous nanoparticles. *Angew. Chem. Int. Ed.* **2005**, *44*, 1269–1273. [CrossRef] [PubMed]

24. Anpo, M.; Shima, T.; Kodama, S.; Kubokawa, Y. Photocatalytic hydrogenation of CH$_3$CCH with H$_2$O on small-particle TiO$_2$: size quantization effects and reaction intermediates. *J. Phys. Chem.* **1987**, *91*, 4305–4310. [CrossRef]

Article

Bench-Scale Steam Reforming of Methane for Hydrogen Production

Hae-Gu Park [1,2], Sang-Young Han [1,3], Ki-Won Jun [1,3], Yesol Woo [4], Myung-June Park [4,5,]* and Seok Ki Kim [1,3,]*

[1] Carbon Resources Institute, Korea Research Institute of Chemical Technology, Daejeon 34114, Korea
[2] Department of Chemical and Biomolecular Engineering, Korea Advanced Institute of Science and Technology (KAIST), Daejeon 34141, Korea
[3] Advanced Materials and Chemical Engineering, School of Science, Korea University of Science and Technology (UST), Yuseong, Daejeon 305-333, Korea
[4] Department of Energy Systems Research, Ajou University, Suwon 16499, Korea
[5] Department of Chemical Engineering, Ajou University, Suwon 16499, Korea
* Correspondence: mjpark@ajou.ac.kr (M.-J.P.); skkim726@krict.re.kr (S.K.K.);
  Tel.: +82-31-219-2895 (M.-J.P.); +82-42-860-7530 (S.K.K.)

Received: 5 July 2019; Accepted: 17 July 2019; Published: 20 July 2019

Abstract: The effects of reaction parameters, including reaction temperature and space velocity, on hydrogen production via steam reforming of methane (SRM) were investigated using lab- and bench-scale reactors to identify critical factors for the design of large-scale processes. Based on thermodynamic and kinetic data obtained using the lab-scale reactor, a series of SRM reactions were performed using a pelletized catalyst in the bench-scale reactor with a hydrogen production capacity of 10 L/min. Various temperature profiles were tested for the bench-scale reactor, which was surrounded by three successive cylindrical furnaces to simulate the actual SRM conditions. The temperature at the reactor bottom was crucial for determining the methane conversion and hydrogen production rates when a sufficiently high reaction temperature was maintained (>800 °C) to reach thermodynamic equilibrium at the gas-hourly space velocity of 2.0 L $CH_4/(h \cdot g_{cat})$. However, if the temperature of one or more of the furnaces decreased below 700 °C, the reaction was not equilibrated at the given space velocity. The effectiveness factor (0.143) of the pelletized catalyst was calculated based on the deviation of methane conversion between the lab- and bench-scale reactions at various space velocities. Finally, an idling procedure was proposed so that catalytic activity was not affected by discontinuous operation.

Keywords: methane steam reforming; hydrogen production; bench scale; effectiveness factor

---

1. Introduction

The demand for hydrogen has traditionally been high because hydrogen has been widely used as a chemical raw material in various refineries, as it is essential for the Fischer–Tropsch process and methanol synthesis [1,2]. Hydrogen is also expected to play an important role as a carbon-free energy carrier in the future [3,4]. Various methods for producing hydrogen with renewable energies have been proposed over the past few decades [5–13]. However, large-scale commercialization of hydrogen production using renewable energy to meet the massive demand for hydrogen remains challenging [14–16]. Until hydrogen production technology using renewable energy is sufficiently mature to facilitate the implementation of a sustainable hydrogen economy, a large amount of hydrogen is required to construct and operate the infrastructure for its storage, transportation, and utilization. Currently, steam reforming of fossil fuels or biomass is the most realistic option for producing large amounts of hydrogen [17]. Among various resources, natural gas is abundant and inexpensive compared to other sources, and its reforming technologies are widely used on commercial scales [18].

Methane constitutes the majority of natural gas, but it is very stable and requires a significant energy input for utilization. The steam reforming of methane (SRM 1 and 2) is a strongly endothermic reaction, as shown in the reaction Equations (1) and (2), and it is usually operated at ≥ 800 °C. Here, the ratio of steam/methane is stoichiometrically 1, but steam is practically supplied at a ratio of ≥ 2.5 to prevent carbon deposition and improve the long-term stability of the catalyst. In addition, if excess water is supplied, a water–gas shift (WGS, Equation (4)) occurs despite its moderately exothermic nature, resulting in additional hydrogen production. As can be seen from Equations (1) and (2), the SRM is a volumetric expansion reaction, so the process is often operated at low pressure as it is thermodynamically preferred. However, to reduce the size of the reactor and facilitate the overall operation, the reactor is operated at a pressure of >0.5 MPa. Therefore, it is necessary to derive the optimal operating conditions according to the composition, amount of the desired product, and the process scale. Due to the small amount of CO_2 produced during the reaction, dry reforming of methane (DRM, Equation (3)) can also occur.

$$\text{SRM1 (Steam reforming of methane): } CH_4 + H_2O \rightarrow CO + 3H_2 \ (\Delta H_{298K} = 205.9 \text{ kJ/mol}), \quad (1)$$

$$\text{SRM2 (Steam reforming of methane): } CH_4 + 2H_2O \rightarrow 2CO_2 + 4H_2 \ (\Delta H_{298K} = 164.7 \text{ kJ/mol}), \quad (2)$$

$$\text{DRM (Dry reforming of methane): } CH_4 + CO_2 \rightarrow 2CO + 2H_2 \ (\Delta H_{298K} = 247.0 \text{ kJ/mol}), \quad (3)$$

$$\text{WGS (Water-gas shift) } CO + H_2O \rightarrow CO_2 + H_2 \ (\Delta H_{298K} = -41.1 \text{ kJ/mol}). \quad (4)$$

To date, most studies on SRM catalysts have focused on their activity and stability, which include studying the effect of the type and amount of active metal on catalyst performance and identifying the causes of deactivation, which include sintering of metallic species and coke deposition [19,20]. These studies have been performed in lab-scale reactors using powdered catalysts from a microscopic point of view. However, to increase the scale of the process, the catalyst must be pelletized to a certain size and shape considering the heat and mass transfer as well as the pressure drop in the reactor. Accordingly, the reactor and operating conditions must be properly engineered [21]. For catalysts used in commercial-scale reactors, their physicochemical properties must be first evaluated in a bench-scale process (or larger), and appropriate operating conditions must be derived. However, few studies have been performed on bench-scale reactions [22]. Herein, a commercial Ni-based catalyst was tested in lab- and bench-scale reactors, wherein powder- and pellet-type catalysts were used, respectively. We focused on determining the crucial factors of reactor design, especially for commercializing methane reforming reactions, by conducting a series of experiments under various conditions, including idling for intermittent operations.

2. Results and Discussion

2.1. Methane Steam Reforming Reaction in a Lab-Scale Reactor

Preliminary lab-scale reactions were performed using a powder-type catalyst obtained by grinding a commercial pellet-type catalyst and sieving it through a 16–20 size mesh. In the lab-scale reactor, the effects of reaction temperature, steam/methane ratio, and reaction pressure on SRM performance were studied. It should be noted that the temperature at the catalyst bed reported here was somewhat underestimated compared to the overall reactor system, so the experimental values could exceed the equilibrium values calculated based on the temperature of the catalyst bed. Figure 1a shows the effect of reaction temperature on methane conversion. For this reaction, the gas-hourly space velocity (GHSV) was fixed at 4.8 L $CH_4/(h \cdot g_{cat})$, the pressure was fixed at 1 MPa, and the steam/methane ratio was fixed at 3. As expected from the highly endothermic nature of the SRM (Equation (1)), methane conversion increased with reaction temperature. The experimental values of methane conversion were close to equilibrium, indicating that the SRM reaction rate was not limited by the kinetics of the catalyst, but by the overall system thermodynamics.

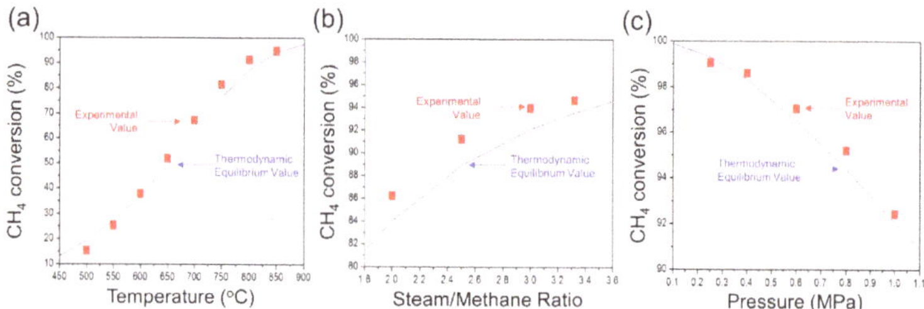

Figure 1. Effect of (**a**) reaction temperature, (**b**) steam/methane ratio, and (**c**) reaction pressure on methane conversion in the lab-scale reactor.

The effect of the steam/methane ratio on the methane conversion is shown in Figure 1b. For this reaction, the reaction temperature, pressure, and GHSV were fixed at 830 °C, 1 MPa, and 4.8 L CH_4/(h·g_{cat}), respectively. If the SRM is the only reaction taking place in the reactor, excessive steam does not necessarily affect methane conversion according to Equation (1). However, the increase in methane conversion as a function of steam/methane ratio suggests that an additional WGS (Equation (2)) also occurs, resulting in a shift in the SRM equilibrium so that methane consumption is accelerated at a higher steam/methane ratio [23].

Figure 1c shows the effect of the reaction pressure on methane conversion. For this reaction, the reaction temperature was fixed at 830 °C, the steam/methane ratio was fixed at 3, and GHSV was fixed at 4.8 L CH_4/(h·g_{cat}). The decreased methane conversion with increasing reaction pressure was in good agreement with the thermodynamic equilibrium conversion, indicating that the reaction rate was thermodynamically limited under the reaction conditions tested herein.

2.2. Methane Steam Reforming Reaction in a Bench-Scale Reactor

2.2.1. Effect of Reaction Temperature

The bench-scale SRM reaction was performed using a fixed-bed reactor, as shown in Figure 2. The reactor temperature was controlled by three heaters placed continuously. The inner diameter and length of the reactor were 32.52 mm and 110 cm, respectively. The temperature gradient along the vertical distance of the reactor was monitored using five thermocouples (TCs). The position of the TCs are shown in Figure 3.

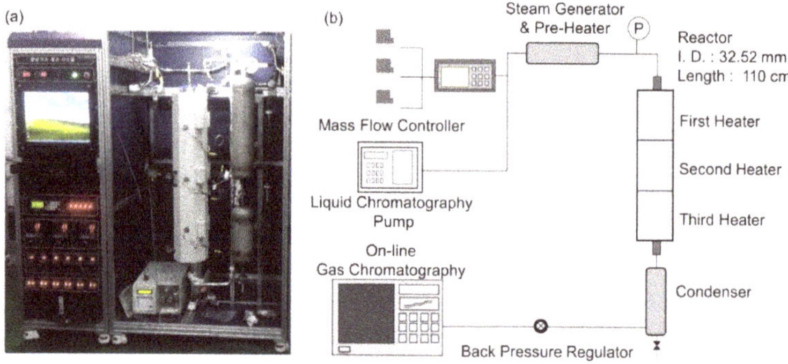

Figure 2. Bench-scale reaction system for the steam reforming of methane (hydrogen production rate of 10 L/min): (**a**) photograph of the unit and (**b**) schematic diagram of the unit.

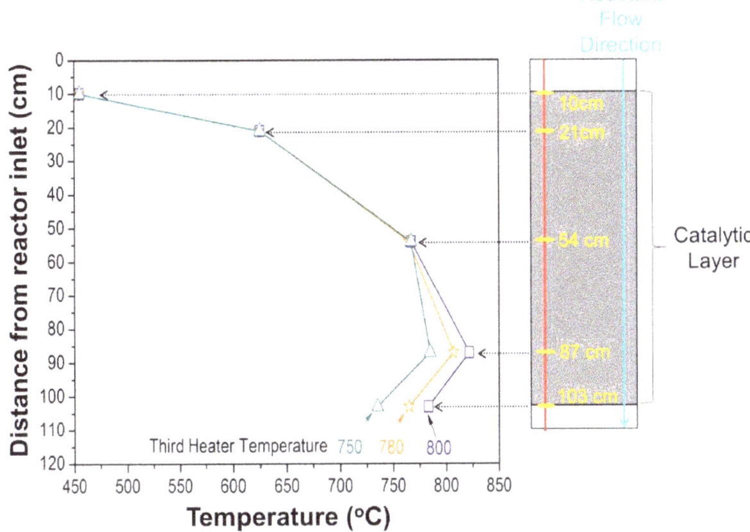

Figure 3. Temperature gradient along the catalyst bed depending on the third heater temperature (catalytic layer: 93 cm length, 96 g catalyst weight, 1386 g dilution agent weight, physical mixing). Schematic diagram of the reactor with the thermocouple (TC) positions (yellow bars) shown on the right-hand side.

The bench-scale reaction conditions were designed based on the lab-scale reaction results. The reaction was performed at a heater temperature of 800 °C, steam/methane ratio of 3, reaction pressure of 0.6 MPa, and GHSV of 2.0 L CH_4/(h·g_{cat}). To confirm whether the reaction set was close to the equilibrium state of the reforming reaction, the temperature of the bottom heater was changed to 800, 780, and 750 °C. The temperature profiles along the reactor distance and corresponding methane conversions are shown in Figure 3 and Table 1, respectively.

Table 1. Catalyst bed temperature gradient and CH_4 conversion at various third heater temperatures.

(a) Experimental Value					
Temperature (°C)				CH_4 Conversion (%)	Hydrogen Production Rate (L/min)
3rd Heater	4th TC	5th TC	Mean Value (between the 4th and 5th TC)		
800	822	783	802.5	94.07	10.76
780	807	764	785.5	92.43	10.68
750	785	738	761.5	89.63	10.53
(b) Thermodynamic Equilibrium Value					
Temperature (°C)				CH_4 Conversion (%)	
750				87.10	
760				89.00	
770				90.67	
780				92.16	
790				93.46	
800				94.57	

Reaction conditions: steam/methane ratio = 3, feed composition of $CH_4/H_2O/N_2$ = 1/3/1, reaction pressure = 0.6 MPa, and gas-hourly space velocity (GHSV) = 2.0 L CH_4/(h·g_{cat}).

The results show that methane conversion decreased with decreasing bottom heater temperature. When the temperature of the bottom heater was maintained at 800, 780, and 750 °C, the methane conversions were 94.07%, 92.43%, and 89.63%, respectively. Accordingly, the hydrogen production rates were 10.76, 10.68, and 10.53 L/min, respectively. The methane conversion obtained for each condition is similar to the equilibrium conversion calculated based on the mean value of the temperature measured between the 4th and 5th TCs. For instance, when the mean temperature of 4th and 5th TCs was 802.5 °C, the bench reaction exhibited a methane conversion of 94.07% (Table 1a), which is close to the equilibrium conversion calculated at 800 °C (94.57%, in Table 1b). These results indicated that the SRM reaction rate was limited by the thermodynamic state that can be determined under the bench-reaction conditions. In addition, these results highlight the importance of the temperature at the bottom part of the reactor when the reaction is close to equilibrium.

The above reaction results were obtained under conditions that were sufficient to reach system equilibrium. However, in a commercial process, a more rapid temperature gradient would be expected depending on reactor design and climate conditions. Herein, two cases for the rapid temperature gradients that could be caused by heater malfunction were tested. In the first case, heat was assumed to be intensively supplied at the middle of the reactor. This is a typical scenario that can occur when the commercial side-fired reactor is operated in cold regions. The heater temperatures of those located at of the top, middle, and bottom were set to 600, 800, and 600 °C, respectively. In the second case, only the bottom heater was heated intensively, but the temperatures of the top and middle heaters were lowered to simulate another abnormal situation, where the extensive endothermic reaction takes place beyond the capacity of heaters. The temperatures of the top, middle, and bottom heaters were set to 500, 650, and 800 °C, respectively, for this scenario. These temperature profiles along with the catalyst bed distances are shown in Figure 4.

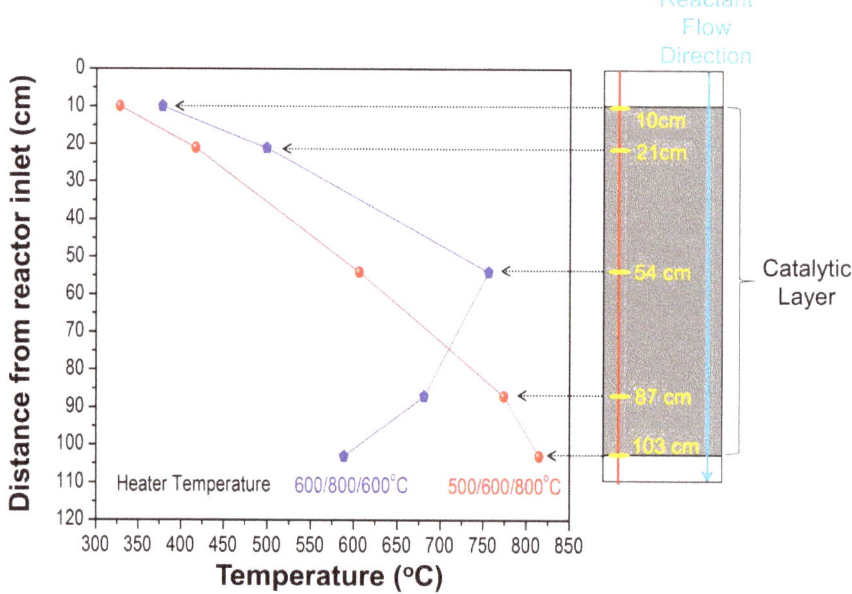

Figure 4. Temperature gradient along the catalyst bed under two abnormal conditions (catalytic layer: 93 cm length, 96 g catalyst weight, 1386 g dilution agent weight, and physical mixing).

The reaction results obtained under the two abnormal conditions are listed in Table 2. Methane conversion in the second scenario was 57.27%, which was ~7% lower than that obtained in the first scenario (64.13%). Given that the temperature of the bottom heater in the second scenario was higher than that of the first, both SRM reactions were not under equilibrium conditions. This is in contrast to the cases listed in Table 2, where the SRMs were under equilibrium conditions as the top and middle heater temperatures were maintained at 800 °C.

Table 2. Catalyst bed temperature gradient and CH_4 conversion under abnormal conditions.

(a) Experimental Value			
Temperature (°C)			CH_4 Conversion (%)
1st Heater	2nd Heater	3rd Heater	
500	650	800	57.27
600	800	600	64.13

(b) Thermodynamic Equilibrium Value	
Temperature (°C)	CH_4 Conversion (%)
620	53.40
630	56.30
640	59.20
650	62.20
660	65.20
670	68.20

Reaction conditions: steam/methane ratio = 3, feed composition of $CH_4/H_2O/N_2$ = 1/3/1, reaction pressure = 0.6 MPa, and GHSV = 2.0 L CH_4/(h·g_{cat}).

The higher methane conversion in the first scenario was due to the wider region of the effective catalyst bed, which sufficiently maintained the reaction rate (>600 °C). However, in the second scenario where only the bottom heater temperature increased, the allowance for maintaining rapid catalysis was reduced. As a result, the space velocity was increased in the effective catalyst layer, consequently preventing the system from reaching equilibrium. To summarize, operation of the SRM reaction under equilibrium conditions can be achieved when supplying sufficient heat to the catalyst bed in as wide a manner as possible.

2.2.2. Effect of Space Velocity

As shown above, the reaction could not reach equilibrium if the reactor exhibited a sufficiently large temperature gradient because the space velocity was too fast for the catalyst to participate in the reaction. The effects of space velocity for different types of catalysts for the SRM reaction were investigated using lab- and bench-scale reactors, as shown in Figure 5. First, 0.15 g of a powdered catalyst 850 to 1250 μm in size was used in the lab-scale reactor, while 12 catalyst pellets (7.34 g) were used in the bench-scale reactor. For the latter reactor, the catalyst pellets were evenly distributed with 1440 g of alumina balls, and the length of the catalyst bed was 93.5 cm. The reaction was performed under various GHSV conditions at 800 °C for the three heaters, steam/methane ratio of 3, and reaction pressure of 0.6 MPa.

For both lab- and bench-scale reactions, methane conversion decreased with increasing space velocity, but the latter showed a larger decrease. That is, at a GHSV of 2.0 L CH_4/(h·g_{cat}), both experiments showed similar methane conversions (94.73% for the lab-scale and 91.52% for the bench-scale), but at a GHSV of 7.5 L CH_4/(h·g_{cat}), while the lab-scale reaction still showed a comparable methane conversion of 89.33%, that of the bench-scale reaction significantly decreased to 53.58%. This indicates that when the GHSV is ≥2.0 L CH_4/(h·g_{cat}), penetration of the reaction gas through the wall of the catalyst pellet does not occur to a sufficient extent, and the active component of the catalyst is not fully utilized, compared to the powdered catalyst. Accordingly, we suggest that the appropriate space velocity for hydrogen production in the bench-scale reaction should be ≤2.0 L CH_4/(h·g_{cat}). Based on the above results, the effectiveness factors of the pelletized catalysts were derived and the results reported in Section 2.2.3.

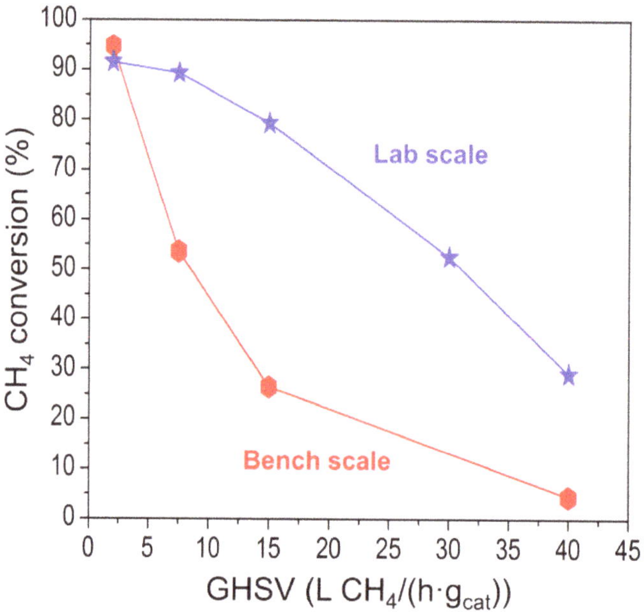

Figure 5. Methane conversion as a function of space velocity in the lab-scale and bench-scale reactors (reaction conditions: steam/methane ratio = 3, feed composition of CH$_4$/H$_2$O/N$_2$ = 1/3/1, reaction pressure = 0.6 MPa, and GHSV = 2.0–40.0 L CH$_4$/(h·g$_{cat}$).

2.2.3. Determination of the Effectiveness Factor

Determining an optimal catalyst loading in the reactor to maximize participation in the reaction is important when designing a large-scale reactor using pelletized catalysts. The effectiveness factor is defined as the ratio of the apparent reaction rate of the catalyst pellet to the intrinsic reaction rate, which provides information on the fraction of the catalyst pellet that participates in the reaction [24]. Based on experimental results for the lab- and bench-scale reactions, the effectiveness factor for the pelletized catalyst was determined. Instead of deriving all effectiveness factors for each individual reaction, the overall effectiveness factor ($\eta_{overall}$), as well as CO (SRM1) and CO$_2$ (SRM2) production from methane by steam reforming, dry reforming of methane (DRM), and the water–gas-shift reaction (WGS) were calculated as follows:

$$\left(r_{apparent}\right)_i = \eta_{overall}\left(r_{intrinsic}\right)_i \qquad i = \text{SRM1, SRM2, DRM, WGS.} \tag{5}$$

To calculate the reaction rates for the commercial catalysts, $\left(r_{apparent}\right)_i$, the reaction rates and kinetic parameters from our previous work were used without modification [25]:

$$r_{\text{SRM 1}} = \frac{k_{\text{SRM1}}\left(f_{\text{CH}_4}f_{\text{H}_2\text{O}} - f_{\text{H}_2}^3 f_{\text{CO}}/K_{\text{pSRM1}}\right)/f_{\text{H}_2}^{2.5}}{\left[1 + K_{\text{CO}}f_{\text{CO}} + K_{\text{H}_2}f_{\text{H}_2} + K_{\text{CH}_4}f_{\text{CH}_4} + K_{\text{H}_2\text{O}}\left(f_{\text{H}_2\text{O}}/f_{\text{H}_2}\right)\right]^2}; \tag{6}$$

$$r_{\text{SRM 1}} = \frac{k_{\text{SRM1}}\left(f_{\text{CH}_4}f_{\text{H}_2\text{O}} - f_{\text{H}_2}^3 f_{\text{CO}}/K_{\text{pSRM1}}\right)/f_{\text{H}_2}^{2.5}}{\left[1 + K_{\text{CO}}f_{\text{CO}} + K_{\text{H}_2}f_{\text{H}_2} + K_{\text{CH}_4}f_{\text{CH}_4} + K_{\text{H}_2\text{O}}\left(f_{\text{H}_2\text{O}}/f_{\text{H}_2}\right)\right]^2}; \tag{7}$$

$$r_{\text{DRM}} = \frac{k_{\text{DRM}}\left(f_{\text{CH}_4}f_{\text{CO}_2} - f_{\text{H}_2}^2 f_{\text{CO}}^2/K_{\text{pDRM}}\right)}{\left(1 + K_{\text{CH}_4}f_{\text{CH}_4} + K_{\text{CO}}f_{\text{CO}}\right)\left(1 + K_{\text{CO}_2}f_{\text{CO}_2}\right)}; \tag{8}$$

$$r_{WGS} = \frac{k_{WGS}\left(f_{CO}f_{H_2O} - f_{H_2}f_{C0_2}/K_{pWGS}\right)/f_{H_2}}{\left[1 + K_{CO}f_{CO} + K_{H_2}f_{H_2} + K_{CH_4}f_{CH_4} + K_{H_2O}\left(f_{H_2O}/f_{H_2}\right)\right]^2};$$ (9)

where k_i and K_i denote the reaction rate constant and adsorption equilibrium constants, respectively, for species i. Fugacity (f) was calculated using the generalized correlations for the fugacity coefficient, as described in the literature [26]. The symbol Kp represents the reaction equilibrium constant, which was obtained from the process simulator UniSim Design Suite R400 (Honeywell Inc., Charlotte, NC, USA)

Because the inert fraction of the catalyst bed was extremely high (7.34 g of catalyst pellet and 1440 g of inert materials) in the bench-scale reactor, a catalyst pellet was considered to be a single reactor module in the process simulator, as shown in Figure 6a, 12 of which were connected consecutively over the entire packing of the reactor. Figure 6b shows a comparison of the methane conversion between the experimental data and simulated results, where the simulated values coincide with the observed data satisfactorily (mean of absolute relative residuals (MARRs) and relative standard deviation were 26.7% and 2.23%, respectively, for GHSV values of 7.5 and 15.0 L CH_4/(h·g_{cat}), when the value at 40.0 mL CH_4/(h·g_{cat}) was excluded as an outlier).

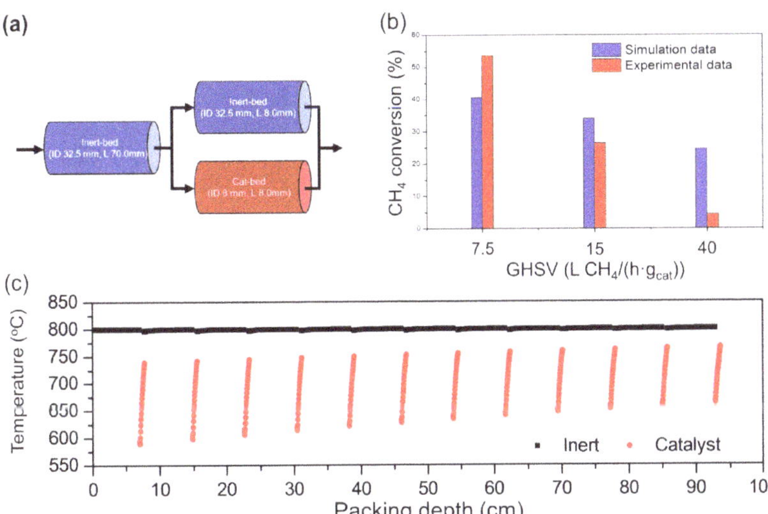

Figure 6. (a) Schematic of the bench-scale reactor (hydraulic diameter of the catalyst pellet was used in the Cat-bed module), (b) comparison of the CH_4 conversion for various space velocities [mL CH_4/(h·g_{cat})], and (c) temperature profile in the reactor at 7.5 L CH_4/(h·g_{cat}). Wall temperature = 800 °C, pressure = 0.6 MPa, overall heat transfer coefficient =100 W/(m²·K), and feed composition of CH_4/H_2O/N_2 = 1/3/1.

For the lab-scale reactor, a single plug flow reactor (PFR) was used in the simulator, and a reaction rate of $(r_{intrinsic})_i = (r_{apparent})_i / \eta_{overall}$ was used. Figure 7a shows the MARR values for CH_4 conversion as a function of $\eta_{overall}$, where the optimal value of $\eta_{overall}$ was 0.143 for the minimum MARR (18.8%). Figure 7b shows that the simulated values of CH_4 conversion agreed well with the experimental data for various space velocities. The temperature profile was also estimated, as shown in Figure 7c. The reaction temperature decreased to ~600 °C at the initial part of the catalyst bed and increased gradually due to heat transfer from the wall, resulting in the exit temperature being close to that of the wall.

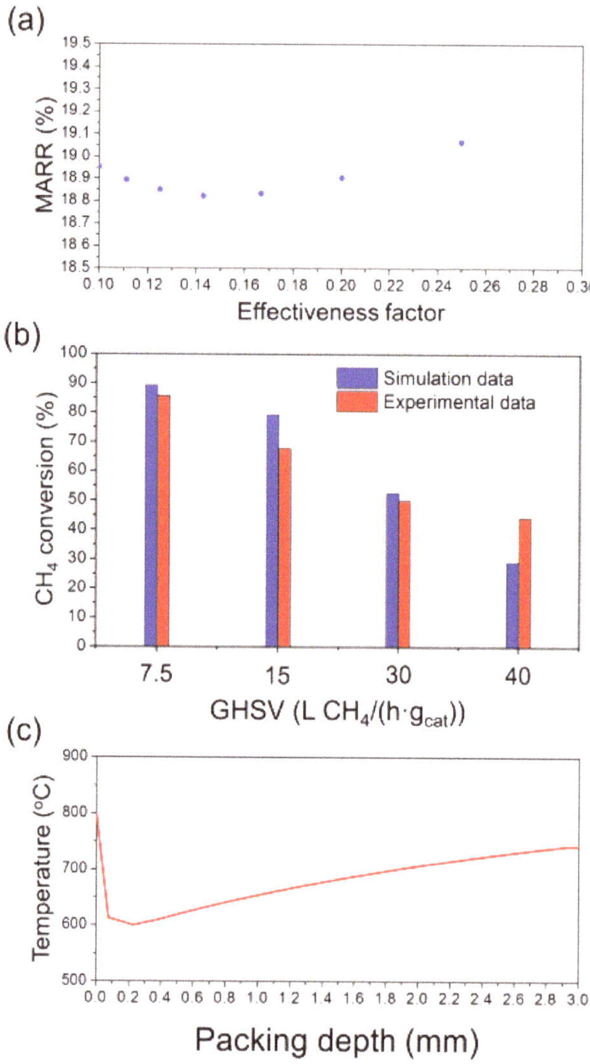

Figure 7. (**a**) Mean of absolute relative residuals (MARRs) values with respect to the overall effectiveness factor, (**b**) comparison of CH_4 conversion for various space velocities [mL CH_4/(h·g_{cat})], (**c**) temperature profile in the reactor at 7500 mL CH_4/(h·g_{cat}). Wall temperature = 800 °C, pressure = 0.6 MPa, overall heat transfer coefficient = 100 W/m^2·K, and feed composition of CH_4/H_2O/N_2 = 1/3/1.

2.3. Idling Conditions

In addition to the extreme temperature gradient of the heaters, a stable idling condition was also simulated under the assumption of discontinuous power supply. The activity of the catalyst can be maintained by maintaining stable idling conditions. By applying an effective idling operation to the process, the reaction may not be completely terminated, which would shorten the preparation time for restarting the operation.

Figure 8 shows temperatures recorded along the SRM reaction followed by idling operation and the restart process. In a typical starting procedure, the reactor containing a reduced catalyst was heated to the reaction temperature (800 °C, region (1) in Figure 8b) prior to feeding the reactants.

After stabilizing the reactor temperature, the SRM reaction was initiated by feeding the reactants (region (2)). After completion of the reaction, the three heaters were maintained at 500 °C, and nitrogen flowed inside the reactor at a rate of 1 L/min (region (3)). When the reforming reaction proceeded again, the reactor temperature was heated (region ((4)) followed by feeding of the reactants (region (5)). As shown in Table 3, no significant changes in catalytic performance were observed before or after the idling operation.

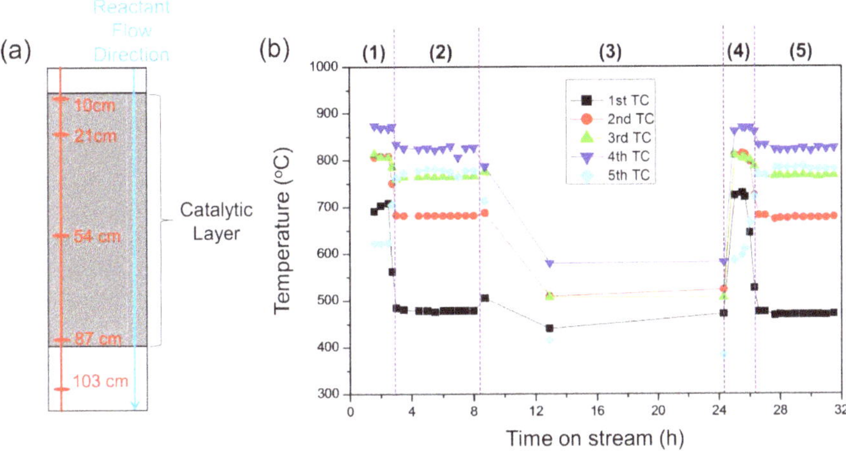

Figure 8. (**a**) Schematic of a catalytic layer for idling operation, (**b**) temperature recorded before (1, 2), during (3), and after idling (4, 5).

Table 3. Comparison of temperature, methane conversion, and hydrogen production before and after idling.

	Temperature (°C)					CH$_4$ Conversion (%)	Rate of Hydrogen Production (L/min)
	1st TC	2nd TC	3rd TC	4th TC	5th TC		
Before idling	478	682	766	826	781	92.95	6.67
After idling	469	677	767	825	782	92.81	6.65

Reaction conditions: steam/methane ratio = 3, feed composition of CH$_4$/H$_2$O/N$_2$ = 1/3/1, reaction pressure = 0.6 MPa, GHSV = 2.0 L CH$_4$/(h·g$_{cat}$), catalytic layer = 82 cm length, catalyst weight = 80 g, dilution agent weight = 1220 g, and physical mixing.

3. Materials and Methods

3.1. Catalyst Characterization

As a preliminary study for applying a pelletized catalyst to a commercial process, reactions were performed using a commercial Ni-based catalyst suitable for mass production of hydrogen. The textural properties and composition of the catalyst are listed in Table 4.

Table 4. Textural properties of the catalyst used herein.

Parameters	Data
Shape	1–hole cylinder
Size	O.D. 8.17 mm, I.D. 2.85 mm, Height 7.21 mm
Composition	Ni 20 wt.%, CaO-Al$_2$O$_3$ 80 wt.%
Density	1.80 g/cm^3
Packing Density (in bench reactor)	0.718 g/cm^3
Surface Area	21.26 m^2/g
Micropore Area	13.07 m^2/g
Pore Volume	0.033 cm^3/g
Pore Size	111 Å

For structural characterization, the commercial catalyst was ground and meshed to a size of ≤250 μm. The Brunauer–Emmett–Teller (BET) specific surface area, micropore area, pore volume, and pore size distribution of the powdered catalyst were estimated from the N$_2$ adsorption and desorption isotherm obtained at −195.7 °C using a constant-volume adsorption apparatus (Micromeritics, ASAP-2020, Norcross, GA, USA). The pore volumes were determined at a relative pressure (P/Po) of 0.99. The catalyst was degassed at 300 °C for 4 h before the measurements. The pore size distributions of the samples were calculated using the Barrett–Joyner–Halenda (BJH) model.

3.2. Steam Reforming Reaction

3.2.1. Methane Steam Reforming Reaction in the Lab-Scale Reactor

The catalytic activity of the powdered catalyst for the methane steam reforming reaction was tested in a fixed-bed tubular Inconel reactor (ID = 10 mm). Prior to feeding the reactants, the catalyst was activated by flowing H$_2$ at a rate of 50 mL/min at 800 °C for 120 min. A TC was placed at the center of the catalyst bed to monitor the reaction temperature, and the feed flow was controlled using a mass flow controller (Brooks, 5850E, Hatfield, PA, USA). The gas products were analyzed using an online gas chromatograph (GC) (Young Lin Acme 6000, Gyeonggi-do, Korea) with a 40/60 carboxen-1000 column (2.0 ft × 1/8 in. × 2.1 mm) and a thermal conductivity detector (TCD). Nitrogen was used as an internal standard gas to verify the composition of the analytical gas (methane) as a volume or half volume.

The activity data shown in Figure 1 were collected by varying the reaction temperature (500 to 850 °C), steam/methane ratio (2 to 3.3), and reaction pressure (0.2 to 1 MPa). The methane conversions shown in Figure 5 (lab-scale) were evaluated under the following reaction conditions: heater temperature = 800 °C; pressure = 0.6 MPa; steam/methane ratio = 3; feed composition of CH$_4$/H$_2$O/N$_2$ = 1/3/1; and gas hourly space velocity (GHSV) = 2.0–40.0 L CH$_4$/(h·g$_{cat}$).

The equilibrium conversion was calculated using "HSC chemistry" software (Outotec, Espoo, Finland).

3.2.2. Methane Steam Reforming Reaction in the Bench-Scale Reactor

The bench-scale reactor consisted of three heaters, a stainless-steel reactor with an inner diameter of 32.52 mm and length of 110 cm, and five TCs. Figure 9 shows the bench reactor in detail. The TCs of the three heaters were located 22, 55, and 88 cm from the reactor inlet. Five TCs were located inside the reactor to monitor the temperature of the catalyst bed, at positions of 10, 21, 54, 87, and 103 cm. To prevent localization of heat and mass, the reactor was filled with a mixture of a pellet-type catalyst and spherical diluent (alpha-alumina). Prior to the SRM reaction, the catalyst mixture was activated by flowing H$_2$ at a rate of 1 L/min at 800 °C for 120 min. The experiment was conducted under the conditions mentioned above, and the analysis method was the same as that of the lab-scale reaction.

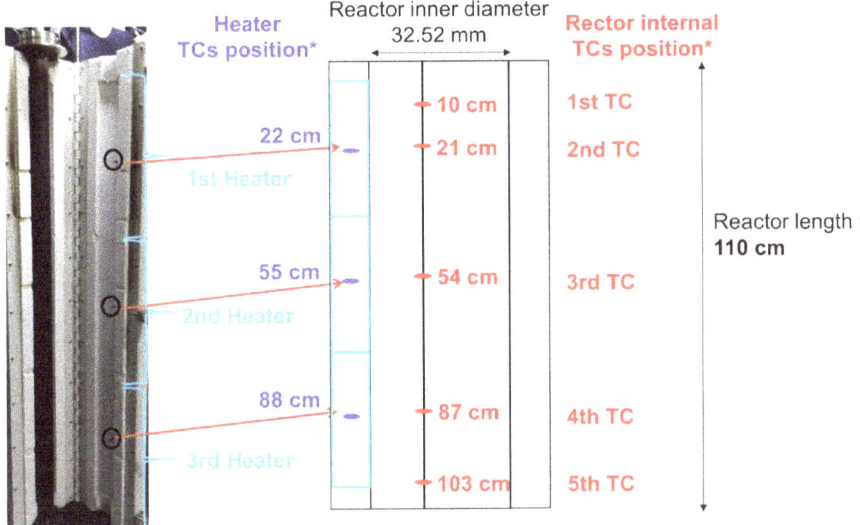

Figure 9. Detailed position of the heaters and TCs in the bench-scale reactor.

4. Conclusions

A series of lab- and bench-scale SRM experiments were performed to identify and solve the problems that could occur during scale-up. In the lab-scale reaction, the effects of temperature, pressure, and steam/methane ratio on methane conversion were investigated in comparison to the corresponding equilibrium conversion. Based on the lab-scale experiments, a bench-scale reaction was designed. The methane conversion in the bench-scale reactor was >90%, and the hydrogen production was >10 L/min if the three consecutive heaters sufficiently supplied the heat required to reach the equilibrium (>800 °C). Under these conditions, the heater temperature positioned at the bottom of the reactor (outlet side) largely governed the methane conversion. Under abnormal reactor temperature conditions, where the catalyst bed was not heated sufficiently (<650 °C), the reaction was not equilibrated simply by maintaining the temperature of the bottom heater at 800 °C. This is similar to the case where the space velocity is relatively high (>10.0 L CH_4/(h·g_{cat})). Using kinetic data obtained from the lab- and bench-scale reactions, the effective factor (0.143) for the pelletized catalysts was calculated. Finally, we proposed effective idling operating conditions that prevented catalyst deactivation during process downtime and reduced the time and costs involved in restarting the process.

Author Contributions: H.-G.P., K.-W.J., and S.K.K. conceived and designed the experiments; H.-G.P. and S.-Y.H. performed the experiments; Y.W. and M.-J.P. analyzed and simulated the data; S.K.K. administrated the project; H.-G.P. wrote original draft; M.-J.P. and S.K.K. reviewed and modified the manuscript.

Funding: This work was supported by the Korea Gas Corporation R&D program (Project No. IPT17-12) and Korea Research Institute of Chemical Technology (Project No. SI1911-60).

Acknowledgments: We thank Jin-Mo Park and Hyung-Sik Kim of Korea Gas Corporation for helpful discussion on this work.

Conflicts of Interest: The authors declare no conflict of interest.

References

1. Iglesia, E. Design, synthesis, and use of cobalt-based Fischer-Tropsch synthesis catalysts. *Appl. Catal. A Gen.* **1997**, *161*, 59–78. [CrossRef]
2. Zhang, C.; Jun, K.-W.; Kwak, G.; Lee, Y.-J.; Park, H.-G. Efficient utilization of carbon dioxide in a gas-to-methanol process composed of CO_2/steam mixed reforming and methanol synthesis. *J. CO_2 Util.* **2016**, *16*, 1–7. [CrossRef]
3. Prater, K.B. Polymer electrolyte fuel cells: A review of recent developments. *J. Power Sources* **1994**, *51*, 129–144. [CrossRef]
4. Minh, N.Q. Ceramic Fuel Cells. *J. Am. Ceram. Soc.* **1993**, *76*, 563–588. [CrossRef]
5. Nagaoka, K.; Eboshi, T.; Takeishi, Y.; Tasaki, R.; Honda, K.; Imamura, K.; Sato, K. Carbon-free H_2 production from ammonia triggered at room temperature with an acidic RuO_2/gamma-Al_2O_3 catalyst. *Sci. Adv.* **2017**, *3*, e1602747. [CrossRef] [PubMed]
6. Hill, A.K.; Torrente-Murciano, L. Low temperature H_2 production from ammonia using ruthenium-based catalysts: Synergetic effect of promoter and support. *Appl. Catal. B Environ.* **2015**, *172*, 129–135. [CrossRef]
7. Cheng, C.; Shi, J.; Hu, Y.; Guo, L. WO_3/g-C_3N_4 composites: One-Pot preparation and enhanced photocatalytic H_2 production under visible-light irradiation. *Nanotechnology* **2017**, *28*, 164002. [CrossRef]
8. Ouyang, S.; Tong, H.; Umezawa, N.; Cao, J.; Li, P.; Bi, Y.; Zhang, Y.; Ye, J. Surface-alkalinization-induced enhancement of photocatalytic H_2 evolution over $SrTiO_3$-based photocatalysts. *J. Am. Chem. Soc.* **2012**, *134*, 1974–1977. [CrossRef]
9. Qin, J.; Huo, J.; Zhang, P.; Zeng, J.; Wang, T.; Zeng, H. Improving the photocatalytic hydrogen production of Ag/g-C_3N_4 nanocomposites by dye-sensitization under visible light irradiation. *Nanoscale* **2016**, *8*, 2249–2259. [CrossRef]
10. Lindgren, M.; Panas, I. Confinement dependence of electro-catalysts for hydrogen evolution from water splitting. *Beilstein. J. Nanotechnol.* **2014**, *5*, 195–201. [CrossRef]
11. Yildiz, B.; Kazimi, M.S. Efficiency of hydrogen production systems using alternative nuclear energy technologies. *Int. J. Hydrogen Energy* **2006**, *31*, 77–92. [CrossRef]
12. Nagasawa, K.; Davidson, F.T.; Lloyd, A.C.; Webber, M.E. Impacts of renewable hydrogen production from wind energy in electricity markets on potential hydrogen demand for light-duty vehicles. *Appl. Energy* **2019**, *235*, 1001–1016. [CrossRef]
13. Sarma, S.J.; Brar, S.K.; Sydney, E.B.; Le Bihan, Y.; Buelna, G.; Soccol, C.R. Microbial hydrogen production by bioconversion of crude glycerol: A review. *Int. J. Hydrogen Energy* **2012**, *37*, 6473–6490. [CrossRef]
14. Turner, J.A. Sustainable hydrogen production. *Science* **2004**, *305*, 972–974. [CrossRef] [PubMed]
15. Szima, S.; Nazir, S.M.; Cloete, S.; Amini, S.; Fogarasi, S.; Cormos, A.-M.; Cormos, C.-C. Gas switching reforming for flexible power and hydrogen production to balance variable renewables. *Renew. Sustain. Energy Rev.* **2019**, *110*, 207–219. [CrossRef]
16. Rashid, M.M.; Al Mesfer, M.K.; Naseem, H.; Danish, M. Hydrogen production by water electrolysis: A review of alkaline water electrolysis, PEM water electrolysis and high temperature water electrolysis. *Int. J. Eng. Adv. Technol.* **2015**, *4*, 80–93.
17. Barelli, L.; Bidini, G.; Gallorini, F.; Servili, S. Hydrogen production through sorption-enhanced steam methane reforming and membrane technology: A review. *Energy* **2008**, *33*, 554–570. [CrossRef]
18. Zamaniyan, A.; Ebrahimi, H.; Mohammadzadeh, J.S.S. A unified model for top fired methane steam reformers using three-dimensional zonal analysis. *Chem. Eng. Process. Process. Intensif.* **2008**, *47*, 946–956. [CrossRef]
19. Wu, H.; Parola, V.L.; Pantaleo, G.; Puleo, F.; Venezia, A.M.; Liotta, L.F. Ni-Based Catalysts for low temperature methane steam reforming: Recent results on Ni-Au and comparison with other bi-metallic systems. *Catalysts* **2013**, *3*, 563–583. [CrossRef]
20. Liu, C.-J.; Ye, J.; Jiang, J.; Pan, Y. Progresses in the preparation of coke resistant Ni-based catalyst for steam and CO_2 reforming of methane. *Chem. Cat Chem.* **2011**, *3*, 529–541. [CrossRef]
21. Mohammadzadeh, J.S.S.; Zamaniyan, A. Catalyst shape as a design parameter—Optimum shape for methane-steam reforming catalyst. *Chem. Eng. Res. Des.* **2002**, *80*, 383–391. [CrossRef]
22. Saric, M.; Delft, Y.C.; Sumbharaju, R.; Meyer, D.F.; Groot, A. Steam reforming of methane in a bench-scale membrane reactor at realistic working conditions. *Catal. Today* **2012**, *193*, 74–80. [CrossRef]

23. Wang, Y.; Chin, Y.H.; Rozmiarek, R.T.; Johnson, B.R.; Gao, Y.; Watson, J.; Tonkovich, A.Y.L.; Vander Wiel, D.P. Highly active and stable Rh/MgO–Al$_2$O$_3$ catalysts for methane steam reforming. *Catal. Today* **2004**, *98*, 575–581. [CrossRef]
24. Baek, S.M.; Kang, J.H.; Lee, K.-J.; Nam, J.H. A numerical study of the effectiveness factors of nickel catalyst pellets used in steam methane reforming for residential fuel cell applications. *Int. J. Hydrogen Energy* **2014**, *39*, 9180–9192. [CrossRef]
25. Park, N.; Park, M.-J.; Baek, S.-C.; Ha, K.-S.; Lee, Y.-J.; Kwak, G.; Park, H.-G.; Jun, K.-W. Modeling and optimization of the mixed reforming of methane: Maximizing CO$_2$ utilization for non-equilibrated reaction. *Fuel* **2014**, *115*, 357–365. [CrossRef]
26. Smith, J.M.; Van Ness, H.C.; Abbott, M.M. *Introduction to Chemical Engineering Thermodynamics*, 7th ed.; McGraw-Hill: New York, NY, USA, 2005.

Article

Recent Advances in Industrial Sulfur Tolerant Water Gas Shift Catalysts for Syngas Hydrogen Enrichment: Application of Lean (Low) Steam/Gas Ratio

Bonan Liu [1,2,*], Liang Zhao [1], Zhijie Wu [1], Jin Zhang [2], Qiuyun Zong [2], Hamid Almegren [3], Feng Wei [4], Xiaohan Zhang [1], Zhen Zhao [1], Jinsen Gao [1] and Tiancun Xiao [5,*]

[1] State Key Laboratory of Heavy Oil Processing, China University of Petroleum (Beijing), 18 Fuxue Road, Changping District, Beijing 102249, China; liangzhao@cup.edu.cn (L.Z.); zhijiewu@cup.edu.cn (Z.W.); zxh19950925@126.com (X.Z.); zhenzhao@cup.edu.cn (Z.Z.); jsgao@cup.edu.cn (J.G.)

[2] Industrial Engineering Laboratory of Sulfur Tolerant Water Gas Shift Catalyst Subjected to China Petroleum and Chemical Industry Federation (CPCIF), Qingdao Lianxin Catalyst Company, Yunxi Road, Jiaozhou 266300, Shandong Province, China; zhangjin19870227@163.com (J.Z.); zqy1959@163.com (Q.Z.)

[3] Materials Research Institute, KACST, PO Box 6086, Riyadh 11442, Saudi Arabia; almegren@kacst.edu.sa

[4] State Key Laboratory of Advanced Materials for Smart Sensing, General Research Institute for Nonferrous Metals, 11 Xingkedong Street, Huairou District, Beijing 101402, China; weifeng@grinm.com

[5] KACST-Oxford Petrochemical Research Centre, Inorganic Chemistry Laboratory, University of Oxford, South Parks Road, Oxford OX1 3QR, UK

* Correspondence: liubonan@cup.edu.cn (B.L.); xiao.tiancun@chem.ox.ac.uk (T.X.)

Received: 16 July 2019; Accepted: 10 September 2019; Published: 14 September 2019

Abstract: A novel sulfur tolerant water gas shift (SWGS) catalyst has been developed for the applications under lean (low) steam/gas ratio conditions, which has been extensively used for H_2/CO adjustment of syngas and H_2 enrichment in the world since 2000s with safer operation and lower steam consumption. Technology design and catalyst performances under different lean steam/gas conditions were comprehensively reported. Industrial data were collected from several large scale running plants with a variety of served catalysts characterized and precisely re-examined in the laboratory. It is shown that the developed Mo–Co/alkali/Al_2O_3 SWGS catalyst can operate very steadily even with the steam/gas ratio as low as 0.2–0.3, and the main deactivation factors are accidental caking, sintering, as well as poisoning impurities, such as As or Cl. The adoption of lean steam/gas SWGS catalyst can significantly improve the plant efficiency & safety and remarkably reduce the actual steam consumption for H_2 production, which can decrease CO_2 emission correspondingly. The work helps to evaluate how specially designed SWGS catalysts performed under applied lean steam/gas conditions, providing important references for researchers and industry.

Keywords: Sulfur tolerant water gas shift catalyst; steam/gas ratio; Mo–Co/alkali/Al_2O_3 catalyst; catalyst deactivation; syngas; H_2 production

---

1. Introduction

Water gas shift (WGS) reaction is a key step of hydrogen (H_2) enrichment in many important industrial sectors where syngas (CO + H_2) is routinely generated from biomass, residue oil, coal and natural gas [1,2]. Such a process plays a unique role in the adjustment of syngas CO/H_2 ratio, syngas quality upgrade by H_2 enrichment, and more importantly bringing in a diversity for downstream productions [1–4]. The reaction stands on a *mars-van Krevelen* mechanism according to which one oxygen atom is transferred from H_2O to CO leading to the liberation of H_2 and the formation of CO_2 [5].

Explorations on WGS never stopped with many new catalysts continuously developed in the laboratory, such as the recent advances in supported gold catalyst for low-temperature WGS [6–10].

However, the huge cost of noble metal inevitably limits the scaling up of such novel catalysts [6,8]. In fact, as the purification of syngas, e.g., sulfur removal, is always expensive, $Mo–Co/Al_2O_3$-based catalysts with sulfur-tolerance are still being used as the major industrial WGS catalyst, especially for sulfur-containing syngas, referred as sulfur tolerant (sour) water gas shift (SWGS) catalyst [11,12].

Reaction conditions are undoubtedly crucial for the industrial SWGS process leading to many important improvements in technology to decrease the actual production costs, e.g., water usage, and corresponding changes in the designed catalysts. In this work, the application of newly developed lean (low) steam/gas (ratio) technology for SWGS as well as the performance of designed catalysts are comprehensively reported, as one of the most important recent advances in the syngas industry [13,14].

2. Lean Steam/gas (Ratio) Sulfur Tolerant Water Gas Shift

The WGS system mainly contains five general compounds, i.e., CO, CO_2, H_2, H_2O and CH_4, which are supposed to maintain a balance under given reaction conditions. Two reactions proceed in a parallel way to decide the final product composition.

(1)   Water gas shift as the main reaction

$$CO\ (g) + H_2O\ (g) \leftrightarrow CO_2\ (g) + H_2\ (g)\ (\Delta H = -41.1\ kJ/mol)$$

(2)   Methanation of CO as the core side reaction

$$CO\ (g) + 3H_2\ (g) \leftrightarrow CH_4\ (g) + H_2O\ (g)\ (\Delta H = -206.3\ kJ/mol)$$

WGS reaction is mildly exothermic while methanation is commonly known as strongly exothermic. Intensive methanation consumes H_2; in addition, it often leads to an unpredictable, suddenly raised catalyst bed temperature which causes the damage of catalyst structure. In the real industrial process of syngas refining, to obtain a satisfied CO/H_2 ratio, methanation should be strictly avoided in the WGS section [1,2,15,16]. One useful method is to overfeed the content of water (steam) in the reactant stream which helps to selectively promote the WGS main reaction while effectively suppressing methanation with the reaction equilibrium; thus, a high steam/gas (dry syngas without water) ratio should be applied. For SWGS in the coal/natural gas industry, the steam/gas ratio employed by a pre-reactor (R1) is routinely 1.6 or higher [12,13,17].

Water (steam) consumption became another challenge when a high steam/gas ratio was applied. On the other hand, the overfed steam used to avoid methanation may also overpromote WGS reaction with an equilibrium temperature interval up to several hundred degrees. In such case, the control of WGS reaction depth became undesirably difficult and therefore the catalyst bed temperature rapidly went up. The only way to avoid superfluous heat was reducing the actual catalyst loading for each reactor [13,17].

With modified anti-methanation and water-capture features, SWGS catalysts can be used under lean (lower) steam/gas ratio (0.5 or lower) conditions. It was until 2003 that the world's first industrialization of SWGS with a lean steam/gas ratio was successfully applied in China [13,14]. Some plants have been smoothly running for more than 10 years with an average catalyst lifetime of 3–4 years. The adoption of such catalysts requires changes in technological process as explained in Figure 1 [17].

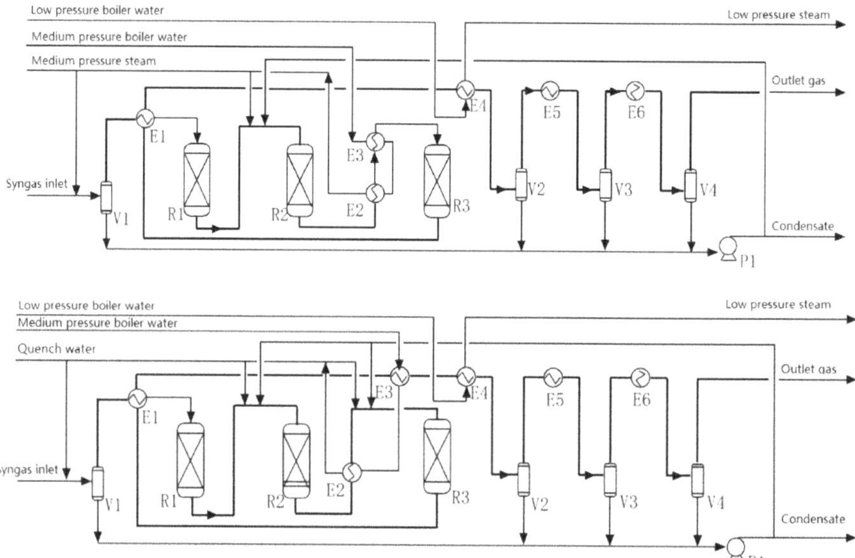

Figure 1. High (1.6 or higher) steam/gas ratio WGS (**top**) vs. Low steam/gas ratio WGS (**bottom**). E1 preheater; E2 medium pressure waste heat boiler; E3 preheater of medium pressure boiler water; E4 low pressure waste heat boiler; E5 desalted water preheater; E6 recycling water cooler; P1 condensation pump; V1–V4 separators; R1 pre WGS reactor; R2 the 1st WGS reactor; R3 the 2nd WGS reactor; figures adjusted from other researchers' work [17].

When applied in the SWGS process (Figure 1), high steam/gas ratio technology (HSGRT) does not require water complementation before a 2nd WGS reactor (R3); however, medium pressure steam used for a reaction will be added into the pre WGS reactor (R1) and 1st WGS reactor (R2), respectively. Instead of steam, lean steam/gas ratio technology (LSGRT) possesses three quench water injections in front of R1, R2 and R3, respectively. In each reactor quench, water vaporized into steam with heat is released from the reaction (there is no need to use out heat), and such improvement has been proven to effectively reduce the actual plant steam consumption and energy cost with the WGS (SWGS) reaction rate being better controlled by the injected quench water amount [17].

Table 1 presents the real industrial data obtained from a 0.5 million ton/year methanol synthesis project (Zhong Yuan Da Hua Coal Chemistry Group, Henan Province, China) employed Shell gasification for syngas production.

As the original steam/gas ratio (~0.3 or higher) of crude syngas had already satisfied the requirement for LSGRT, there was no water complementary, either in terms of steam or quench water. The steam cost (35–60 ton/hour) fell to nearly zero by technological upgrade from HSGRT to LSGRT. On the other hand, the condensed water generation correspondingly reduced from 30–55 ton/hour to 0–5 ton/hour. Noteworthily, with no extra steam added, the steam/gas ratio of each reactor could be maintained at a uniform (0.23–0.26) by LSGRT as a significant advance. At the same time, with the steam (water) amount inside SWGS system steadily controlled, methanation was effectively suppressed and the undesired catalyst bed temperature 'flying away' (the suddenly occurred temperature increasement due to methanation) was also prohibited.

Industrial data from the SWGS plants of a 4 million ton/year Coal-to-Liquid project (Shenhua Ningxia Coal Industry Group, Ningxia Province, China) are more encouraging. The project employed GSP gasification (pressurized pulverized coal gasification) to give an initial steam/gas ratio of ~1.0 for the raw syngas, which already exceeded the steam usage requirement for LSGRT. After upgrading to LSGRT, there was no need to add extra steam into the SWGS process; for a single SWGS reactor

R2, the overall saved cost on steam was approximately 1.04×10^9 RMB, equivalent to 1.52×10^8 USD every year.

Table 1. Industrial data from SWGS plants of a 0.5 million ton/year methanol synthesis project (Zhong Yuan Da Hua Coal Chemistry Group, Henan Province, China);. Catalytic performances were compared between the high (1.6 or higher) steam/gas ratio technology and low (0.3 or lower) steam/gas ratio technology.

SWGS	Rector	Steam/Gas		Temperature, °C		Steam Cost t/h	Condensed Water t/h
		Inlet	Outlet	Inlet	Hot Spot		
HSGRT	R1	1.82	0.79	286	502	35–60	30–55
	R2	0.79	0.49	260	373		
	R3	0.49	0.35	240	336		
LSGRT	R1	0.26	0.03	210	417	0	0–5
	R2	0.23	0.04	210	371		
	R3	0.26	0.03	200	317		

Data were collected by the catalyst inventors from Qingdao Lianxin Catalyst Company who are the co-authors of the current work.

3. Catalysts Used for Lean Steam/Gas (Ratio) Sulfur Tolerant Water Gas Shift

Syngas from the gasification and partial oxidation of heavy oil, tar sands, coal, coke or biomass contain much larger concentrations of CO (up to 50 mol%) and sulfur levels of 5–50 ppm, as a result of which both sulfur tolerance and effective CO activation are crucial for conventional SWGS catalysts. A standard commercial SWGS catalyst composition is based on Co (Ni)-Mo/alumina, here the core catalytic function relies on the effective CO activation of sulfurized Mo sites while other metal species, e.g., Cobalt, are co-catalytic sites or promoters helping in the effective electron transportations [10,11,18–20]. The accelerated CO conversion requires more water molecules to finish the reaction. It is, hence, necessary to either operate at higher steam/gas (or H_2O/CO) ratios or use extra volume of catalysts to support the desired CO conversion [10,11,18–21].

One question is to what extent the water molecules used by high steam/gas conditions are utilized to successfully convert CO into CO_2, or whether the enriched water concentration would affect the WGS reaction in practice. Herein, the interaction between catalytic sites and water molecules becomes a key point. The secret behind LSGRT for SWGS is its *enhanced water (steam) capture* by employed catalyst [13]. It is well known that the WGS catalyst promotes both water gas shift and methanation reactions. In a closed system reaction equilibrium works to draw the final product composition; however, in a constant flow system, e.g., the WGS reactor, the utilization of catalytic sites by different reactants dominates the product output [12,18,21]. By applying a specially designed SWGS catalyst with better water-capture ability, H_2O molecules were selectively enriched on/near the catalytic sites to meet the requirement of accelerated CO conversions, while the H_2 adsorption on catalytic sites was effectively weakened. The first effect greatly promotes the WGS reaction and enables step-wised control of WGS reaction depth with even very limited water amount; the second effect better suppresses the methane formation.

Inclusion of alkali oxides into the Mo–Co/alumina composition brings the above enhanced water capture ability to LSGRT SWGS catalysts. The increase of K_2O content in $MgAl_2O_4$ spinel modified alumina supported CoMoOx catalyst (general form of conventional Mo–Co/Al_2O_3 catalyst for SWGS) led to higher CO conversions under the lean steam/gas conditions while effectively suppressing the formation of CH_4 [13,14]. It was thought that K_2CO_3 helped in the capture of water molecules while the alkalinity of K_2CO_3 also effectively depressed methane formation over the MoS_2 active sites.

A general composition of industrial Mo–Co/alkali/Al_2O_3 catalyst for LSGRT is given as below (based on the evaluation of X-ray fluorescence spectroscopy, atom weight):

Na 1–3%, Mg 0–1%, Al 65–80%, K 0–15%, Ca 0–5%, Mo 5–10%, and Co 0–5%.

In this paper, we have selected representative samples of a laboratory-developed LSGRT SWGS catalyst (marked as CAT) that has been and is being employed in the SWGS plant of many large scale coal/natural-gas chemistry projects in the Chinese industry. Samples of different running time, and various deactivations were characterized and re-evaluated in laboratory.

4. Catalyst Performances and Characterizations

Residual catalytic properties of served/deactivated samples were measured by rationally designed laboratory experiments using as-prepared (sulfurized in factory) CAT for reference. The experimental system (Figure 2) employed two fixed-bed reactors connected in series in order to better simulate the real SWGS process (R1 used for pre-sulfurization was removed to show the authentic residual catalytic property of served/deactivated industrial samples, all samples were vacuumed immediately at the time of unloading; the reactor 1 and reactor 2 in Figure 2 simulated the R2 and R3 SWGS reactors in Figure 1, respectively). Each time, 0.3 ml catalysts ground and sieved into 40–60 mesh were loaded in the isothermal zone of each reactor (quartz wool and quartz sand were used for protection and supporting). N_2 (100 ml/min) was used as a carrier gas to bring steam into the reactors. Water was vaporized in the calibrated two-stage evaporators, and the steam/gas ratio was precisely set at 0.7, a bit higher than the real LSGRT to guarantee the CO conversion over served/deactivated samples. The inlet dry gas (50 ml/min) contained CO of 45 vol%, CO_2 of 5 vol%, H_2 ~50 vol% and H_2S 3000 ppm to make sure a syngas GHSV of 10000 h^{-1}. It should be noted that the composition of industrial syngas as raw materials to SWGS plants does vary a lot due to many issues, such as the coal type, employed gasifiers and the exact gasifying technologies [21–23]. The syngas composition employed in this research, especially the CO content has considered the suggestions from field experts, plant designers, and more importantly the practical CO vol% of several representative commercialized gasifications (typical slurry feed entrained-flow gasifier, e.g., Chevron Texaco Gasifier, gives 30–45 vol% CO; dry feed entrained-flow gasifier, e.g., Shell Gasifier, may give CO percentage of ~60 vol%) [22–25]. In order to extend the catalytic lifetime for deactivated samples and for a safety consideration, industrial 3Mp pressure was not employed and all samples were tested under atmosphere pressure; however, as the reaction is volume constant, the current system causes very limited discrepancy in the catalytic property evaluations.

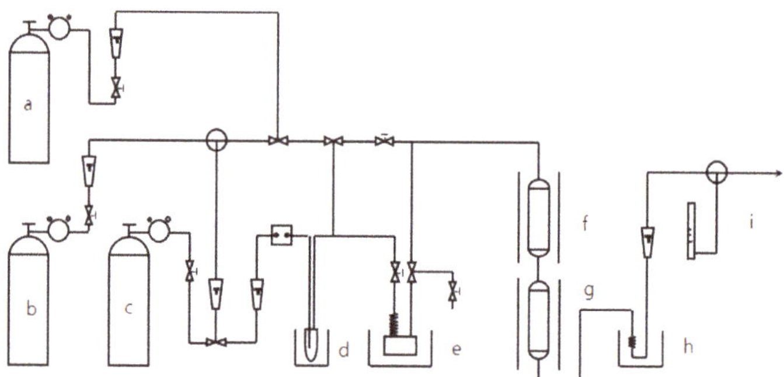

Figure 2. System of laboratory SWGS test: (**a**) N_2 for catalyst stabilization (**b**) Dry gas; (**c**) N_2; (**d**) evaporator 1; (**e**) evaporator 2; (**f**) reactor 1; (**g**) reactor 2; (**h**) gas-liquid separator; (**i**) outlet to GC.

Catalytic performances of different samples at three temperature points (265 °C, 350 °C and 450 °C) are presented in Figures 3 and 4, respectively. At each temperature, the catalyst samples were stabilized in N_2 for 1h ahead of test. The reactor outlet flow was analyzed with on-line gas chromatography

(Agilent 7890, packed column 1.5 m Φ 3 mm, TCD, FID and FPD). The catalytic performance was measured by CO conversion:

$$X_{CO}\% = (Y^0_{CO} - Y_{CO})/(Y^0_{CO}) \times 100$$

$$Y^0_{CO} \text{ - CO inlet vol\%; } Y_{CO} \text{ - CO outlet vol\%}$$

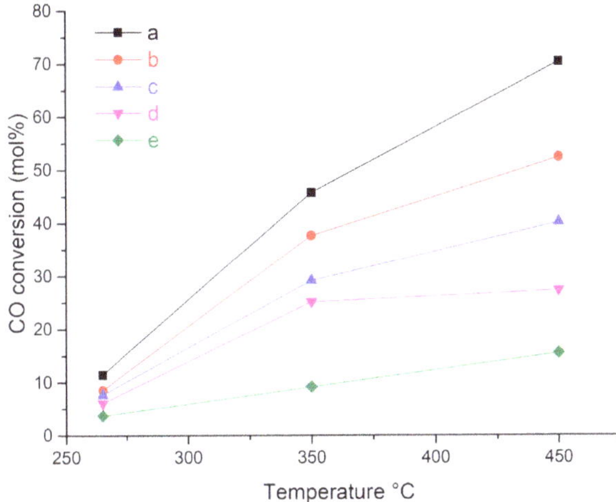

Figure 3. Laboratory test of CO conversion achieved by CATs served in industry for different years: (**a**) CAT as-prepared; (**b**) 1 year served; (**c**) 2 years served; (**d**) 3 year served; (**e**) 4 years served.

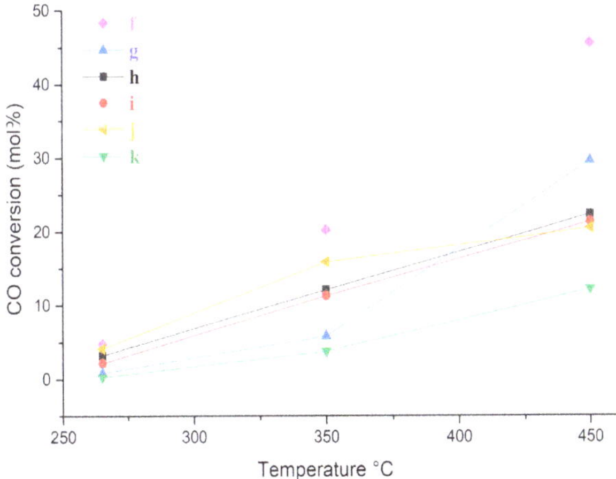

Figure 4. Laboratory test of CO conversion achieved by CATs deactivated for different reasons: (**f**) As poisoned; (**g**) caking 1; (**h**) water soaking; (**i**) caking 2; (**j**) Cl poisoned; (**k**) sintering by hot spot.

It is clearly shown in Figure 3, longer-time served CAT always led to a lower CO conversion in the same laboratory experiment, which is an intuitionistic sign of catalyst 'aging to deactivate'. All samples exhibited very low CO conversion at 265 °C which implies a higher temperature is required for the sufficient catalytic activation. At 450 °C, all samples achieved CO conversion of 10 mol% or much higher (70 mol%). Interestingly, CO conversion at 350 °C showed a huge difference between samples used for 3 and 4 years, while at 450 °C, the difference between these two samples was less. To answer the above question, two notes could be addressed: 1) 310–450 °C is the proper temperature range of industrial High-Temperature WGS (HTS). For HTS, a higher temperature, i.e., 450 °C, possesses more influence on the reaction rate, which could minimize the actual CO conversion difference between differently aged catalysts especially in a stream reactor (*It should be noted that raising catalyst bed temperature is a common way to extend the lifetime of less catalytic, aged catalysts in the chemistry industry although coke formation may also be fastened at the same time) [1,4,12,16,18]. 2) WGS at 350 °C or below relies more on the reserved intrinsic properties of the employed catalyst, thus, the poorer performance of 4 years served CAT was amplified. The above results support recent studies focusing on the development of a lower temperature WGS catalyst; the aim of such exploration is to speed up catalytic performance at 300 °C or lower while taking the advantage of WGS reaction equilibrium (WGS process is moderately exothermic) [1,6–8].

In Figure 4, a series of CAT samples accidentally deactivated were evaluated. We carefully selected samples from several large scale Coal-to-Methanol/Liquid oil/Hydrocarbons industrial plants to make sure they are referential. Except one sample deactivated by caking (~45 mol% CO conversion at 450 °C), most of them were not completely deactivated but quite close to the status of CAT 3–4 years served. Their morphology was presented by scanning electron microscopy (SEM) as shown in Figure 5.

Caking is an early-stage deactivation of alumina-based catalysts serving for years due to which the catalyst particles became loose (not very obviously), further compressed or agglomerated. A general reason for caking could be the high-pressure hydrothermal working conditions, e.g., 3 Mpa for SWGS. As caking is a more 'physical' property change, no obvious morphological difference was found between CATs used for 4 years (Figure 5a), CATs used for 3 years (Figure 5b), CATs deactivated by caking 1 (Figure 5c) and CATs deactivated by caking 2 (Figure 5d). *Water soaking* accidently occurs between the WGS plant operating cycles. There are various reasons for the local enrichment of water/steam in a WGS plant, which leads to the catalyst surface being oversaturated with H_2O which could be another important reason for catalyst caking. From SEM results (Figure 5e), the water-soaked CATs showed very similar morphology as compared to CATs served for years or caked, with only a small difference on the catalyst surface (more obviously seen in Figure 5e3, the surface of CATs presented special 'washing' appearance). CATs deactivated by *sintering* exhibited the worst catalytic performance (Figure 4k) in terms of the lowest CO conversions at the selected three temperatures, indicating the most severe damage of catalytic sites. This is more evidenced by the SEM results (Figure 5f); the apparent crystallization reflected in the picture brought in the dramatic and irreversible reducing of catalyst surface and active sites. It should be mentioned that there were in fact very unusual catalyst deactivations by 'poisoning' in which case poisonous elements (e.g., As and Cl), even in very small amounts, deposited over the catalyst surface (Figure 6). By far, no strong links were found between the cause of WGS deactivation and the above catalyst poisoning. Such accident rarely occurred and could be attributed to the unsuccessful coal preparation before gasification.

More information about the surficial elemental compositions of different samples are listed in Tables 2 and 3. Trends have been found for the 1–4 years served CATs. Fresh (as-prepared) CAT possessed the highest Mo level and Al level, which fell gradually as the serving years increased. The 1 year served CAT became sulfurized, then the sulfur level decreased at a faster rate than Mo and Al. As more Mo elements were continuously lost from the catalyst surface, the surface level (concentration) of Co elements accordingly increased implying they were comparably more stable in the LSGRT SWGS process. No Cl or As was observed on the surface of as-prepared CAT, but their concentrations went up gradually with the time on stream of WGS increasing. For the industrial SWGS process, it was difficult to completely avoid the accumulation of unexpected poisoning elements over the catalyst surface especially during a long-time plant running.

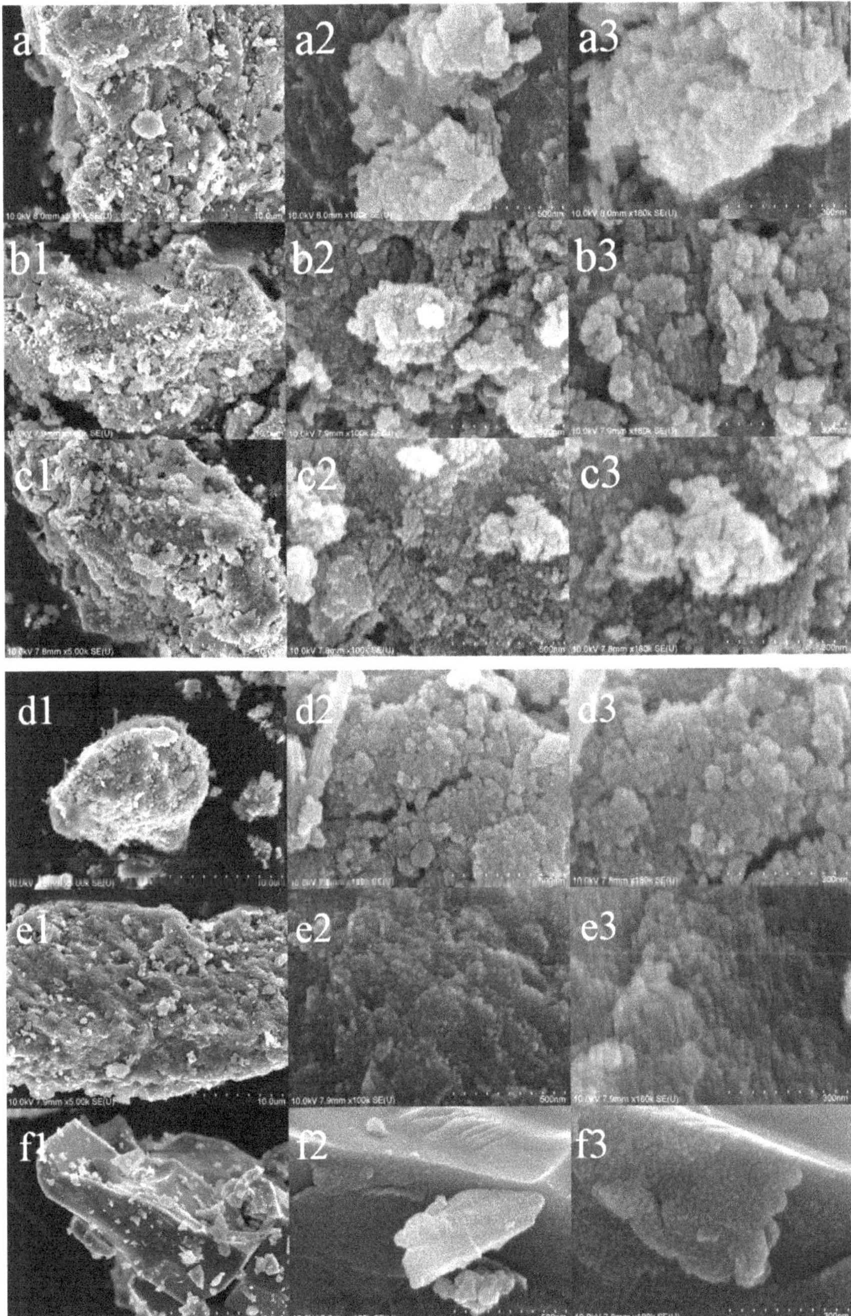

Figure 5. SEM images of CATs unloaded after time on stream for 3–4 years or deactivated for other reasons: (**a**) 4 years; (**b**) 3 years (**c**) caked 1; (**d**) caked 2; (**e**) water soaked (**f**) sintered.

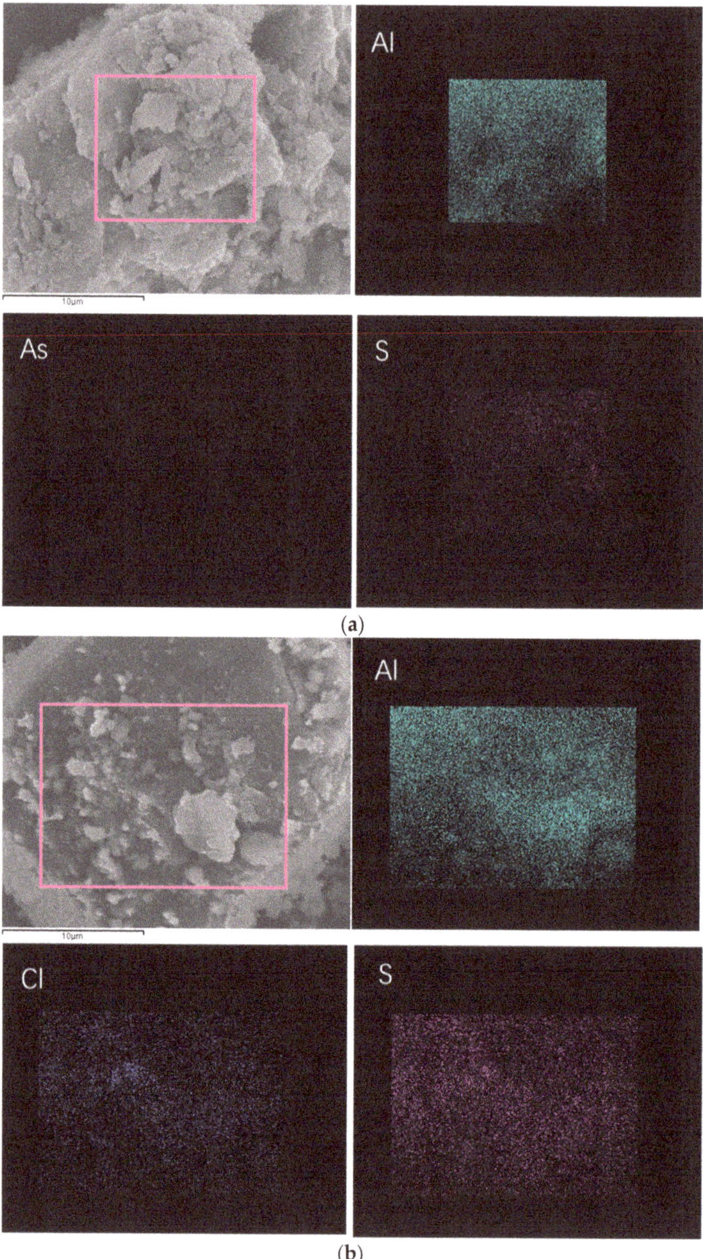

Figure 6. Elemental mapping results of CATs poisoned by As (**a**) and Cl (**b**).

CaO was introduced in the preparation of CAT as a common method to improve the catalyst mechanical strength. As shown in Table 2, even in a very small amount, Ca content over the catalyst surface fell gradually with time on stream extending. However, as shown in Table 3, the sintered CAT and caked CATs exhibited much higher Ca levels than all other samples. The accumulation of Ca over CAT surface might be a direct reason of catalyst particle growth and aggregation, which led to the final deactivation by caking [12,26,27]. In some cases, the slow caking of catalyst particles could be accelerated by high temperatures and finally evolved to the sintering deactivation: 1) The WGS reaction is exothermic, more exothermic is due to higher accompanied methanation side reaction which might occur at aggregated catalyst particles causing locally over-heated sites (hot spot). 2) The local high temperature further accelerates the metal particle growth and finally the aggregation (sintering). 3) The observed high Ca levels of caked and sintered samples supported the above explanations.

Table 2. Elemental distributions of different CATs (fresh–used), the standard deviation of the measured value is set as unity in the last decimal.

Composition	CAT	1 year	2 years	3 years	4 years
Mo (MoO_3)	7.856	6.905	6.885	6.803	6.628
Co (Co_3O_4)	1.985	1.898	2.326	2.659	3.436
Al (Al_2O_3)	75.349	73.334	71.005	70.804	68.801
S(SO_3)	0.222	16.015	13.309	10.315	9.655
Ca (CaO)	0.582	0.565	0.456	0.408	0.249
Cl (Cl)	0	0.22	0.203	0.237	0.274
As (As_2O_3)	0	0.01	0.01	0.01	0.028
C (Organic)	0	0.022	0.035	0.05	0.063

Table 3. Elemental distributions of different CATs deactivated by other reasons, the standard deviation of the measured value is set as unity in the last decimal.

Composition.	Caked 1	Caked 2	Sintered	Cl	As
Mo (MoO_3)	8.323	8.571	9.216	6.905	6.322
Co (Co_3O_4)	2.865	2.015	1.893	1.934	1.751
Al (Al_2O_3)	70.085	72.86	69.008	60.104	71.521
S (SO_3)	10.223	9.05	8.61	14.732	9.772
Ca (CaO)	0.603	0.721	0.859	0.228	0.303
Cl (Cl)	0.15	0.215	0.429	0.725	0.642
As (As_2O_3)	0.01	0.01	0	0.05	0.101
C (Organic)	0.037	0.082	0.097	0.052	0.041

All unloaded CATs showed different levels of carbon deposition over the catalyst surfaces as a straightforward evidence of coking deactivation. However, even the sintered CAT and caked CATs have more cokes; coke formation was still not regarded as the main reason for WGS deactivation; a possible reason could be the H_2 generated in the WGS reaction suppressed the development of catalyst coking.

Crystal structures of different samples were analyzed with X-ray diffraction (XRD) as shown in Figures 7–9. The sharper diffraction peaks of γ-Al_2O_3 in Figure 7 indicate all CATs served for different years have developed further crystalized alumina structures as compared with fresh CAT, which is a common result for alumina supported catalysts long-term served with steam. Diffraction (a peak at ~61°) from the 008 plane of crystal MoS_2 was clearly observed on the 1 year served CAT, as an early sign of the irreversible loss of mono-layer dispersed MoS_2 which is the active sites for catalytic CO conversion [28]. Diffractions from the 105 plane (a peak at ~50°) and 002 plane (a peak at ~14°) of crystal MoS_2 were continuously found on CATs served for 3 years and 4 years [28]. The above results reflected an evolution of the MoS_2 phase supported on CAT when served in the industrial WGS plant under a lean steam/gas ratio, which is supposed to start from a mono-layer distribution and gradually develop into a complex crystal structure. Crystal phases of $CaMoO_4$ and $MgAl_2O_4$ were also found to be present on all served CATs indicating an effective alkali modification.

By comparison (Figure 8), the crystallinity of γ-Al_2O_3 was noted to be less developed on Cl and As poisoned, caked and steam-soaked CATs, which is more similar to the as-prepared CAT. These observations match the fact that CATs in Figure 8 were not used for a long time due to the deactivation. On the other hand, these XRD observations also indicate caking, as the early stage of alumina supported catalyst blocking or sintering, did not show apparent impact on crystallization. Besides, crystal phases of boehmite (precursor of alumina support) were noted to be formed in the caked and Cl poisoned CATs all pointing to an undesired catalyst structure damage.

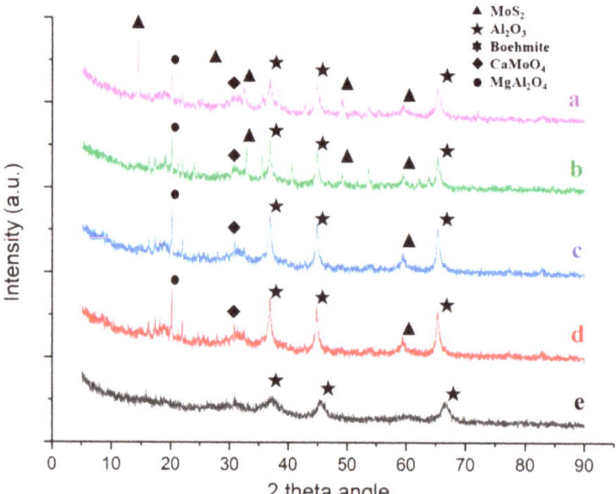

Figure 7. XRD patterns of CATs served for different time (**a**) 4 years; (**b**) 3 years; (**c**) 2 years; (**d**) 1 year; (**e**) as-prepared.

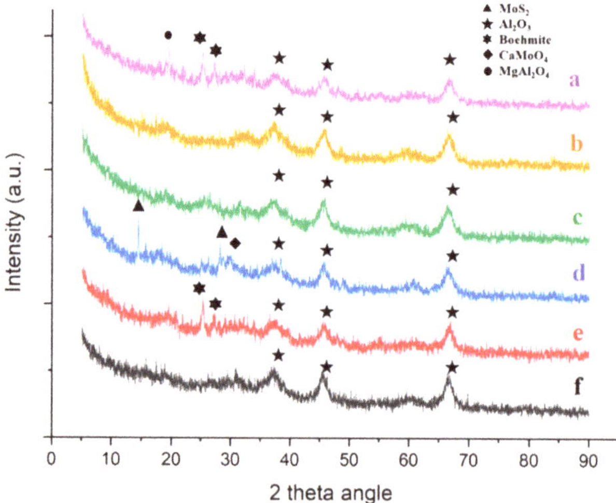

Figure 8. XRD patterns of CATs deactivated for different reasons (**a**) Cl; (**b**) As; (**c**) caking 1; (**d**) caking 2; (**e**) steam-soaked; (**f**) as-prepared.

In operating catalyst beds both sintering and caking could lead to block formation or aggregation of alumina supported catalyst. However, as seen in our results (Figure 9), for CATs served in LSGRT SWGS, only sintering leads to very distinguished and complicated crystalline structures. This is further supported by our SEM images (Figure 5f) since the crystallization of sintered CATs is so apparent and uncontested. In Figure 8 we also noted the formation of crystallized MoS_2 on the caked CAT sample

(Figure 8d, caking 2), which implies that caking might be another reason leading to the loss of mono layered MoS_2 and, therefore, the dramatic reduction of catalytic sites for SWGS.

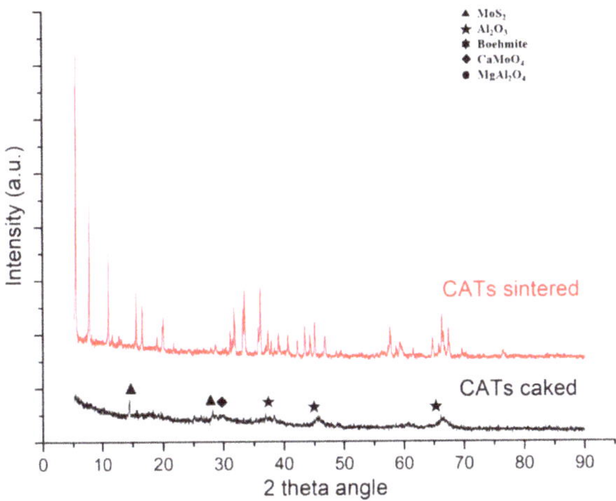

Figure 9. XRD patterns of CATs deactivated by sintering vs. by caking 2.

For further analyses, Raman spectroscopy (Figure 10) is employed to draw the surficial properties of different CATs. MoO_3 species (Raman shifts at ~320, ~910 and 1050 cm^{-1}) are only observed on the as-prepared (un-sulfurized) CAT. All served CATs have formed MoS_2 species on their surface as clearly evidenced by the featured Raman shifts at ~380 and ~410 cm^{-1}, respectively. The MoS_2 signals gradually drop as the serving time of CAT increases. Both the loss of sulfur contents and the MoS_2 crystallization could lead to a reduced MoS_2 catalyst surface dispersion in a long term SWGS process, which can be reflected by the gradually weakened MoS_2 Raman signals.

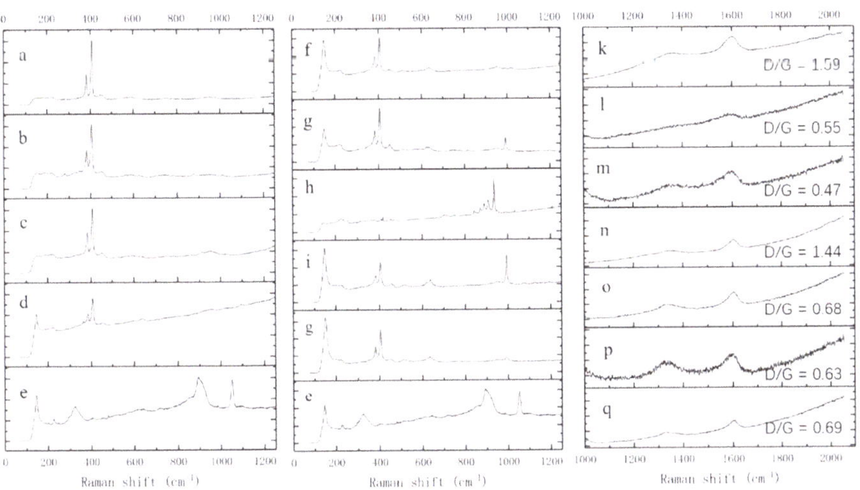

Figure 10. Laser Raman spectra of CATs at different status and coke formations (**a**) served 1 year; (**b**) served 2 years; (**c**) served for 3 years; (**d**) served for 4 years; (**e**) as-prepared; (**f**) caked; (**g**) water soaked; (**h**) sintered; (**i**) Cl; (**g**) As; (**k**) coke-caked; (**l**) coke-water soaked; (**m**) coke-sintered; (**n**) coke-served for 4 years; (**o**) coke-served for 3 years; (**p**) coke-served for 2 years; (**q**) coke-served for 1 year. For D/G ratio the standard deviation of the measured value is set as unity in the last decimal

Caked, Cl-poisoned, As-poisoned CATs also exhibit clear MoS_2 signals in a strong contrast to the sintered CAT which shows only very faint MoS_2 Raman bands. For the sintered CAT, a group of unidentified Raman shifts between 800 and 950 cm^{-1} are observed possibly attributed to the complex interactions between catalyst surface Mo species. Raman shifts between 140 and 150 cm^{-1} only present in the spectra of as-prepared CAT, 4 years served CAT, and those accidentally deactivated CATs (except sintered CAT). A possible reason could be the variational interactions between the catalyst surface Mo species and support alumina; on the other hand, these signals might be part of the background when there is very weak MoS_2 signal. Notably, Cl-poisoned, As-poisoned and water-soaked CATs also exhibited Raman shift of MoO_3 at ~1050cm^{-1}. It is assumed that for these CATs, the sulfurization of MoO_3 was unpredictably disturbed.

Carbon bands (the D peak around 1350 cm^{-1} and the G peak around 1580–1600 cm^{-1}) are also clearly detected by Laser Raman for the caked CAT, water-soaked CAT, sintered CAT and served CATs (Figure 10 k–q). The D/G ratio of carbonaceous species was calculated for each sample to describe the extent of carbon formation [29,30]. The D/G ratio lies between 0.6 and 0.7 for CATs served for 1–3 years indicating a comparably uniform carbonaceous distribution on the catalyst surface. The 4 years served CAT possesses more diamondlike (DLC) carbons of SP_3 bonding which also implies that more hydrogen contents could exist in the deposited carbon compounds. Unlike carbon deposition obtained in a zeolite catalyzed acidic reaction, where the catalyst surface carbonaceous species are normally deeply dehydrogenated, the WGS reaction generates rich amounts of H_2 with steam which might possess an inverse role to hydrogenate the cokes. The D/G ratio of accidentally deactivated CATs varies in the range of 0.47–1.59; the exact reason may require a future investigation with the help of catalyst accidental deactivation simulation in laboratory conditions. Infrared spectroscopy was employed as a complement to the above Raman studies where comparisons were made with the previous IR studies on supported MoS_2 catalysts or others [31,32]. As seen in Figure 11, the as-prepared CAT without sulfur presented apparently divergent spectra in a stark contrast to the served, sulfurized CATs; for this sample (Figure 11a) IR bands only appeared at 1600, 1380, 1280 and 1000 cm^{-1}. The above bands (IR peaks) became eroded (1380, 1280 and 1000cm^{-1}) or slightly shifted (1600 cm^{-1}) in the spectra of served CATs, as a good sign of sulfurization. Meanwhile, new bands at 1450, 1100 (there was even a small peak at 1050 cm^{-1} in the spectra of CAT served 1 year which was 'just sulfurized' compared with the other 3 served CATs) and 890 cm^{-1} appeared; the first two bands were so strong in the spectra that could be a most straightforward evidence showing the occurrence of successful sulfurization. As the serving years of CAT increased (Figure 11b–e), the bands at 1450 and 1100 cm^{-1} became gradually weakened with the WGS catalytic ability accordingly lowered; the 1450 cm^{-1} band completely disappeared in the spectra of 4 years served CAT which was the most severely served CAT.

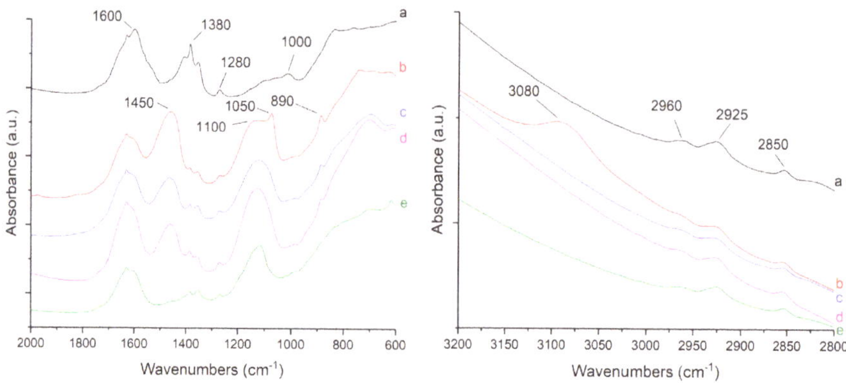

Figure 11. FT-IR spectra of CATs used for different years (**a**) as-prepared; (**b**) served for 1 year; (**c**) served for 2 years; (**d**) served for 3 years; (**e**)served for 4 years.

In the spectra of accidently deactivated CATs, 1450 and 890 cm^{-1} (there was a shift to 840 cm^{-1} in the spectra of sintered CATs) bands were still missing; however, the band at 1100 cm^{-1} was clearly shown despite that signals were much weaker for the As-poisoned CAT (Figure 12f).

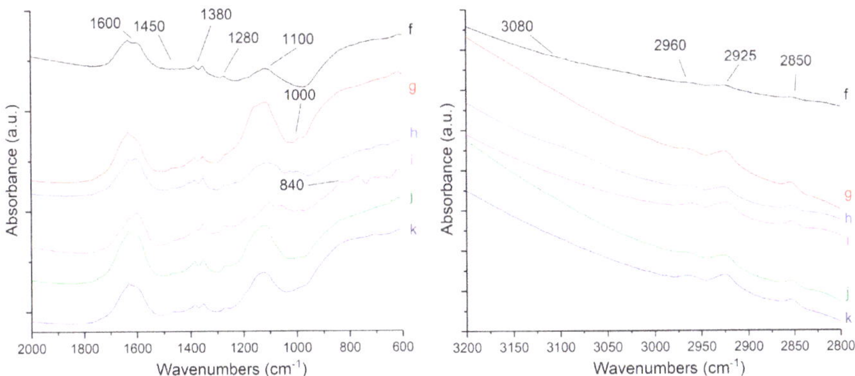

Figure 12. FT-IR spectra of CATs deactivated for different reasons (**f**) As; (**g**) Cl; (**h**) water soaked; (**i**) sintered; (**j**) caked 1; (**k**) caked 2.

Despite the poor resolution in coke analysis, for each sample attempts were also made to capture the possible IR bands of carbonaceous compounds deposited. However, in the commonly used coke domains both aromatic CH stretching modes (2800–3200 cm^{-1}) and paraffinic carbon CH stretching modes (3000–3200 cm^{-1}) were not confidentially detected, whereas the observed bands at 2850, 2925, 2960 and 3080 cm^{-1} are possibly attributed to the catalyst support frameworks [30].

The above IR discussions indicated that CATs in a good sulfurized condition mainly possessed two bands at 1450 and 1100 cm^{-1}, respectively. Both bands might be responsible for the effective sulfur tolerant WGS. However, only the change of 1450 cm^{-1} band was noted more sensitive for those deactivated CATs, and therefore, it is more important to assess the remaining catalytic properties. Notably, this new finding was not reported in the previous studies of high steam/gas ratio sulfur tolerant WGS. Compared with the IR results, although more intensive in signals, Raman spectra only indicated an overall reduction of catalyst surface MoS$_2$ species on the deactivated CATs (Raman shifts at ~380 and ~410 cm^{-1}).

5. Experimental

The CAT samples (Figure 13) were precisely unloaded from the plant and packed in vacuum at the same time. CATs served 1–4 years (Figure 13 c–f) were selected, and all accidentally deactivated CATs (Figure 13 g–l) only served for less than two years.

Experimental sets employed for laboratory sulfur tolerant WGS reaction over selected CATs were discussed in the above.

Scanning electron microscopy (SEM) were taken with a JEOL 840F scanning microscope instrument (JEOL, Peabody, USA). Elemental mapping employed a GeminiSEM 300 scanning microscope instrument (Zeiss, Jena, Germany). Samples were deposited onto dust free platform and treated with gold-spray to enhance the signal before analysis.

X-ray diffraction results were obtained with a PANalytical X'Pert PRO diffractometer (Malvern Panalytical, Royston, UK) using Cu Kα1 radiation in diffraction angle from 5° to 90° (2 theta angular range) and a scan rate of 0.8° min^{-1} in 2θ.

Laser Raman spectra were recorded with a Perkin-Elmer Raman Station 400F Raman Spectrometer (Perkin Elmer, Waltham Mass, USA). Raman shifts were recorded between 50cm and 1250cm^{-1} for all CATs or 1000 and 2050cm^{-1} for selected coked CATs. Powder samples were hold with a piece of clean glass during scan. For D/G ratio calculation LapSpec (Version 5.58, Horiba, Kyoto, Japan) software was used in the peak matching and integration (Gaussian method).

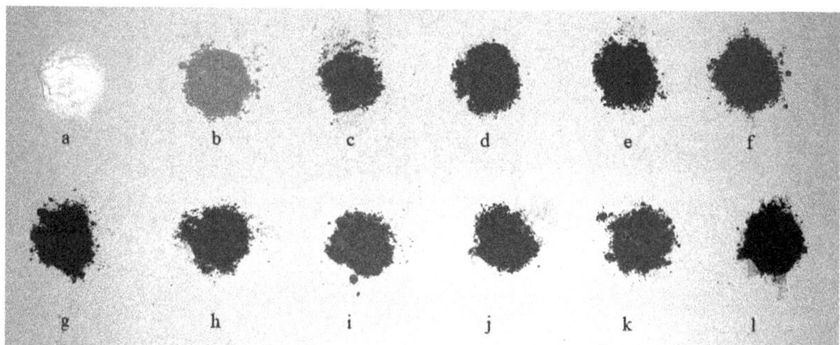

Figure 13. CATs samples investigated (samples in photo were broken into powders for characterizations): (**a**) γ-Al$_2$O$_3$; (**b**) as-prepared; (**c**) 1 year served; (**d**) 2-years served; (**e**) 3-years served; (**f**) 4-years served; (**g**) As; (**h**) Cl; (**i**) water-soaked; (**j**) caked 1; (**k**) caked 2; (**l**) sintered.

Infrared spectra were obtained by a Bruker Tensor II FTIR spectrometer (Bruker, Santa Barbara, USA) in the wavenumber range of 4000–400 cm^{-1}. Samples were mixed with KBr (sample/KBr = 1:50 in weight) and then pressed into pellets for measurements.

6. Summary

The latest development in sulfur-tolerant water gas shift catalysts and its application in low/lean steam/gas ratio technology (LSGRT) in industrial operation were comprehensively reported in this work. LGSRT possessed many improvements as compared with the traditional high steam/gas ratio technology. The LSGRT catalyst does not need extra steam adding in the WGS process; it just requires adding liquid water directly to the raw syngas stream to effectively control the hot spot and reduce the energy input. This can also help to control the WGS reaction depth, as well as the suppression of methanation side reactions.

Catalysts based on Mo–Co/alkali/Al$_2$O$_3$ composition were successfully developed and applied to the LSGRT water gas shift process, and a series of industrial samples (CATs) used under different conditions and lifetime stage were collected, laboratory tested and characterized with the techniques of XRD, SEM and Laser-Raman.

From the above works, many important findings were made: 1) The CO conversion of Mo–Co/alkali/Al$_2$O$_3$ catalyst (CATs in the above discussions) gradually fell after 1–4 years time on stream under the conditions of LSGRT while the difference between medium temperature (350 °C) and high temperature (450 °C) catalytic performances became less as the catalyst was being used. 2) Caking, water-soaking and sintering led to negative effects on the catalytic performance of LSGRT catalysts, whereas sintering gives the most severe damage to catalyst in terms of changes in catalyst crystal structure and surface properties as reflected by the characterizations. 3) Cl and As could be poisonous for LSGRT catalysts and were supposed to disturb the catalyst sulfurization. On the other hand, water-soaking might also lead to a similar result. 4) New spectroscopy evidence was found in the characterizations of served and deactivated LSGRT SWGS catalysts which together with the above experimental results could be a good reference for researchers and industry.

7. Expanded Discussion

Even though LSGRT SWGS catalysts have been successfully employed in many important WGS plants in the past years, there is still an urgent demand for the continuous study and more tests, especially in the real field conditions.

One would argue how the 'accidentally deactivated' samples discussed in this work could be representative for researchers and field experts; besides, these deactivation approaches might be random and therefore, the results only teach less. To make sure our results are representative and referential, one would apply their WGS catalyst under a similar condition, i.e., reaction with a lower steam/gas ratio; the accidental deactivations described in this work have encompassed as many originally unexpected deactivations as the catalyst inventers (part of our authors) could record, since the LSGRT SWGS catalysts were extensively applied:1) LSGRT SWGS catalyst in one plant could be accidentally caked, water-soaked, sintered, Cl poisoned or As poisoned; however, yet no other major reason was reported as newly found accidental deactivation. 2) One type of LSGRT SWGS catalyst accident could occur several times, or in different plants; however, these changes are significant deactivations. Although the exact mechanism of these accidental deactivations is not completely clear, we have given as much reliable explanations (e.g., a general reason for caking could be the high-pressure hydrothermal condition) as the current work could support and we do appeal for more researchers seeking future explorations.

Another discussion could be over a long time period in an environment that is lacking in the normal control which one would use in a lab setting, to which extent we could assure that something besides the operating conditions did not affect the LSGRT SWGS catalytic performance. This question points to not only WGS catalysts but also all other industrially employed catalysts, e.g., zeolites, as the amplification of catalytic performance from a laboratory reaction to industrial plant application does bring in many uncertainties. However, the potential risks could be effectively prohibited and controlled in many ways, such as rational catalytic design, long-term laboratory experiments, pilot test, the real monitoring of plants and more importantly, a continuous study on employed catalysts to further improve its performance [33,34]. While the practical WGS reaction occurring in a plant is a heterogenous, multi-direction process with several side reactions, e.g., methanation, it is first and foremost a chemical phenomenon that can be effectively controlled with the help of plant-integrated monitoring systems [18,21]. The above accidental deactivations may have not been found by the previous WGS laboratory studies, but observed and recorded by the practical plant monitoring; in this work, their significance on real WGS performance was firstly reported with a preliminary evaluation on the causes, which is of great importance to prohibit future deactivations and helps to effectively control plant operating conditions.

Author Contributions: For research articles with several authors, a short paragraph specifying their individual contributions must be provided. The following statements should be used "conceptualization, B.L. and Q.Z.; methodology, B.L., Q.Z. and T.X.; software, X.Z., J.Z. and F.W.; validation, L.Z. and Z.W.; formal analysis, B.L., L.Z., Z.W. and X.Z.; investigation, B.L., L.Z., Z.W., J.Z., Q.Z., and X.Z.; resources, J.Z. and Q.Z.; data curation, B.L., L.Z., Z.W., J.Z. and Q.Z.; writing—original draft preparation, B.L., T.X. and Q.Z.; writing—review and editing, B.L., Q.Z., Z.Z., J.G., and T.X.; visualization, B.L., X.Z. and F.W.; supervision, Q.Z., H.A., Z.Z., J.G. and T.X.; project administration, Q.Z. and J.G.; funding acquisition, B.L., Q.Z., H.A. and T.X.

Funding: The work was supported by the National natural science foundation of China youth program (NSFC code 21808241) and Shandong Provincial Key Research and Development Plan (code 2017CXGC1111).

Acknowledgments: We appreciate all the reviewers for their very useful and constructive comments. More appreciations are given to our group members at the China University of Petroleum (Beijing, China), and the oversea colleagues at the University of Oxford (Oxford, UK) and KACST (Riyadh, Kingdom of Saudi Arabia) for their contributions to the experiments and manuscript. The Industrial Engineering Laboratory of Sulfur Tolerant Water Gas Shift Catalyst subjected to China Petroleum and Chemical Industry Federation (CPCIF) is part of the Qingdao Lianxin Catalyst Company (China). The state Key Laboratory of Advanced Materials for Smart Sensing is a sub-department of General Research Institute for Nonferrous Metals (China). The industrial LSGRT SWGS catalysts were originally invented by Qingdao Lianxin Catalyst Company (China). Industrially served CATs and the as-prepared CATs were provided by the company. More importantly, they also kindly offered the practical plant running data (as the LSGRT SWGS catalyst supplier of those plants, Qingdao Lianxin Catalyst Company supervised and monitored the catalyst loadings as well as the practical plant runnings as part of their product services; data were collected for R&D purposes). (J. Zhang and Q. Zong) from Qingdao Lianxin Catalyst Company also took important roles in the experiments. People (F. Wei) from the General Research Institute for Nonferrous Metals (Beijing, China) helped in the SEM.

Conflicts of Interest: The authors have no commercial interest for the reported results.

References

1. LeValley, T.L.; Richard, A.R.; Fan, M. The progress in water gas shift and steam reforming hydrogen production technologies—A review. *Int. J. Hydrog. Energy* **2014**, *39*, 16983. [CrossRef]
2. Kumar, N.; Spivey, J.J. Direct Conversion of Syngas to Chemicals Using Heterogeneous Catalysts. In *Encyclopedia of Sustainable Technologies*; Abraham, M.A., Ed.; Elsevier: Amsterdam, The Netherlands, 2017; p. 605.
3. Bukur, D.B.; Todic, B.; Elbashir, N. Role of water-gas-shift reaction in Fischer–Tropsch synthesis on iron catalysts: A review. *Catal. Today* **2016**, *275*, 66. [CrossRef]
4. Zhu, M.; Wachs, I.E. ron-Based Catalysts for the High-Temperature Water–Gas Shift (HT-WGS) Reaction: A Review. *ACS Catal.* **2016**, *6*, 722. [CrossRef]
5. Meunier, F.C.; Tibiletti, D.; Goguet, A.; Shekhtman, S.; Hardacre, C.; Burch, R. On the complexity of the water-gas shift reaction mechanism over a Pt/CeO2 catalyst: Effect of the temperature on the reactivity of formate surface species studied by operando DRIFT during isotopic transient at chemical steady-state. *Catal. Today* **2007**, *126*, 143. [CrossRef]
6. Carter, H.J.; Hutchings, J.G. Recent Advances in the Gold-Catalysed Low-Temperature Water–Gas Shift Reaction. *Catalysts* **2018**, *8*, 627. [CrossRef]
7. Yao, S.; Zhang, X.; Zhou, W.; Gao, R.; Xu, W.; Ye, Y.; Lin, L.; Wen, X.; Liu, P.; Chen, B.; et al. Atomic-layered Au clusters on α-MoC as catalysts for the low-temperature water-gas shift reaction. *Science* **2017**, *357*, 389. [CrossRef] [PubMed]
8. Rodriguez, J.A. Gold-based catalysts for the water–gas shift reaction: Active sites and reaction mechanism. *Catal. Today* **2011**, *160*, 3. [CrossRef]
9. Idakiev, V.; Yuan, Z.Y.; Tabakova, T.; Su, B.L. Titanium oxide nanotubes as supports of nano-sized gold catalysts for low temperature water-gas shift reaction. *Appl. Catal. A Gen.* **2005**, *281*, 149. [CrossRef]
10. Fu, Q.; Deng, W.; Saltsburg, H.; Flytzani-Stephanopoulos, M. Activity and stability of low-content gold–cerium oxide catalysts for the water–gas shift reaction. *Appl. Catal. B Environ.* **2005**, *56*, 57. [CrossRef]
11. Nikolova, D.; Edreva-Kardjieva, R.; Gouliev, G.; Grozeva, T.; Tzvetkov, P. The state of (K)(Ni)Mo/γ-Al2O3 catalysts after water–gas shift reaction in the presence of sulfur in the feed: XPS and EPR study. *Appl. Catal. A Gen.* **2006**, *297*, 135. [CrossRef]
12. Newsome, D.S. The Water-Gas Shift Reaction. *Catal. Rev.* **1980**, *21*, 275. [CrossRef]
13. Liu, B.; Zong, Q.; Du, X.; Zhang, Z.; Xiao, T.; AlMegren, H. Novel sour water gas shift catalyst (SWGS) for lean steam to gas ratio applications. *Fuel Process. Technol.* **2015**, *134*, 65. [CrossRef]
14. Liu, B.; Zong, Q.; Edwards, P.P.; Zou, F.; Du, X.; Jiang, Z.; Xiao, T.; AlMegren, H. Effect of Titania Addition on the Performance of CoMo/Al2O3 Sour Water Gas Shift Catalysts under Lean Steam to Gas Ratio Conditions. *Ind. Eng. Chem. Res.* **2012**, *51*, 11674. [CrossRef]
15. Senanayake, S.D.; Evans, J.; Agnoli, S.; Barrio, L.; Chen, T.-L.; Hrbek, J.; Rodriguez, J.A. Water–Gas Shift and CO Methanation Reactions over Ni–CeO2(111) Catalysts. *Top. Catal.* **2011**, *54*, 34. [CrossRef]
16. Xu, J.; Froment, G.F. Methane steam reforming, methanation and water-gas shift: I. Intrinsic kinetics. *AIChE J.* **1989**, *35*, 88. [CrossRef]
17. Ma, L.; Sun, Y.; Wang, J. Application of Low Water-gas Ratio Shift Process in Shell Coal Gasification. *Coal Chem.* **2011**, *39*, 46.
18. Pal, D.; Chand, R.; Upadhyay, S.; Mishra, P. Performance of water gas shift reaction catalysts: A review. *Renew. Sustain. Energy Rev.* **2018**, *93*, 549. [CrossRef]
19. Copperthwaite, R.G.; Gottschalk, F.M.; Sangiorgio, T.; Hutchings, G.J. Cobalt chromium oxide: a novel sulphur tolerant water-gas shift catalyst. *Appl. Catal.* **1990**, *63*, L11. [CrossRef]
20. Hou, P.; Meeker, D.; Wise, H. Kinetic studies with a sulfur-tolerant water gas shift catalyst. *J. Catal.* **1983**, *80*, 280. [CrossRef]
21. Ratnasamy, C.; Wagner, J.P. Water gas shift catalysis. *Catal. Rev.* **2009**, *51*, 325. [CrossRef]
22. Kristiansen, A. *Understanding Coal Gasification*; IEA Coal Research London: London, UK, 1996; Volume 86.
23. Wang, Z.; Yang, J.; Li, Z.; Xiang, Y. Syngas composition study. *Front. Energy Power Eng. China* **2009**, *3*, 369. [CrossRef]
24. Minchener, A.J. Coal gasification for advanced power generation. *Fuel* **2005**, *84*, 2222. [CrossRef]

25. Wilhelm, D.J.; Simbeck, D.R.; Karp, A.D.; Dickenson, R.L. Syngas production for gas-to-liquids applications: technologies, issues and outlook. *Fuel Process. Technol.* **2001**, *71*, 139. [CrossRef]

26. Fuertes, A.; Alvarez, D.; Rubiera, F.; Pis, J.; Marban, G.; Palacos, J. Surface area and pore size changes during sintering of calcium oxide particles. *Chem. Eng. Commun.* **1991**, *109*, 73. [CrossRef]

27. Dias, J.; Assaf, J. Influence of calcium content in Ni/CaO/γ-Al2O3 catalysts for CO2-reforming of methane. *Catal. Today* **2003**, *85*, 59. [CrossRef]

28. Peng, Y.; Meng, Z.; Zhong, C.; Lu, J.; Yu, W.; Yang, Z.; Qian, Y. Hydrothermal Synthesis of MoS2 and Its Pressure-Related Crystallization. *J. Solid State Chem.* **2001**, *159*, 170. [CrossRef]

29. Ferrari, A.C.; Robertson, J. Interpretation of Raman spectra of disordered and amorphous carbon. *Phys. Rev. B* **2000**, *61*, 14095. [CrossRef]

30. Guisnet, M.; Ribeiro, F.R. *Deactivation and Regeneration of Zeolite Catalysts*; World Scientific: Singapore, 2011; Volume 9.

31. Mauge, F.; Lamotte, J.; Nesterenko, N.S.; Manoilova, O.; Tsyganenko, A.A. FT-IR study of surface properties of unsupported MoS2. *Catal. Today* **2001**, *70*, 271. [CrossRef]

32. Van Haandel, L.; Hensen, E.; Weber, T. FT-IR study of NO adsorption on MoS2/Al2O3 hydrodesulfurization catalysts: Effect of catalyst preparation. *Catal. Today* **2017**, *292*, 67. [CrossRef]

33. Hagen, J. *Industrial Catalysis: A Practical Approach*; John Wiley & Sons: Hoboken, NJ, USA, 2015.

34. Leach, B. *Applied Industrial Catalysis*; Elsevier: Amsterdam, The Netherlands, 2012.

Article

Highly Active Catalysts Based on the $Rh_4(CO)_{12}$ Cluster Supported on $Ce_{0.5}Zr_{0.5}$ and Zr Oxides for Low-Temperature Methane Steam Reforming

Andrea Fasolini [1], Silvia Ruggieri [1], Cristina Femoni [1,2] and Francesco Basile [1,2,*]

[1] Department of Industrial Chemistry "Toso Montanari", University of Bologna, V.le Risorgimento, 4, 40136 Bologna, Italy; andrea.fasolini2@unibo.it (A.F.); silvia.ruggieri3@unibo.it (S.R.); cristina.femoni@unibo.it (C.F.)

[2] Consorzio Interuniversitario Reattività e Catalisi (CIRCC), 70126 Bari, Italy

* Correspondence: f.basile@unibo.it; Tel.: +39-0512093663

Received: 8 August 2019; Accepted: 23 September 2019; Published: 25 September 2019

Abstract: Syngas and Hydrogen productions from methane are industrially carried out at high temperatures (900 °C). Nevertheless, low-temperature steam reforming can be an alternative for small-scale plants. In these conditions, the process can also be coupled with systems that increase the overall efficiency such as hydrogen purification with membranes, microreactors or enhanced reforming with CO_2 capture. However, at low temperature, in order to get conversion values close to the equilibrium ones, very active catalysts are needed. For this purpose, the $Rh_4(CO)_{12}$ cluster was synthetized and deposited over $Ce_{0.5}Zr_{0.5}O_2$ and ZrO_2 supports, prepared by microemulsion, and tested in low-temperature steam methane reforming reactions under different conditions. The catalysts were active at 750 °C at low Rh loadings (0.05%) and outperformed an analogous Rh-impregnated catalyst. At higher Rh concentrations (0.6%), the Rh cluster deposited on $Ce_{0.5}Zr_{0.5}$ oxide reached conversions close to the equilibrium values and good stability over long reaction time, demonstrating that active phases derived from Rh carbonyl clusters can be used to catalyze steam reforming reactions. Conversely, the same catalyst suffered from a fast deactivation at 500 °C, likely related to the oxidation of the Rh phase due to the oxygen-mobility properties of Ce. Indeed, at 500 °C the Rh-based ZrO_2-supported catalyst was able to provide stable results with higher conversions. The effects of different pretreatments were also investigated: at 500 °C, the catalysts subjected to thermal treatment, both under N_2 and H_2, proved to be more active than those without the H_2 treatment. In general, this work highlights the possibility of using Rh carbonyl-cluster-derived supported catalysts in methane reforming reactions and, at low temperature, it showed deactivation phenomena related to the presence of reducible supports.

Keywords: Hydrogen; Low Temperature Steam Reforming; $Rh_4(CO)_{12}$ cluster; microemulsion synthesis; CeZr oxide; Zr oxide

---

1. Introduction

Methane steam reforming (SR) is a leading reaction in syngas production with a wide employment on an industrial scale [1–6]. However, in order to allow high conversions of methane, very high temperatures (900 °C) are employed, and this requires feeding the reactor with a large amount of heat, which is provided by burning part of the methane reagent in an external furnace. Therefore, in some conditions, lower operative temperatures (400–500 °C) may be advisable, for instance when very high H_2/CO ratios are required, in what is called Low Temperature Steam Reforming (LTSR) [7–9]. In cases of excess of steam in the process feed, with the right catalyst the two SR and WGS (Water Gas Shift) reactions take place consequently in the same reactor [10,11]. This allows for a drastic decrease in CO content with respect to the classical steam reforming reaction, providing an outlet gas with a high H_2/CO ratio. Moreover, the temperatures employed are compatible with special applications such as use of microreactors, CO_2 capture enhanced reforming, and Pd membrane reactors for pure hydrogen production [12–15]. In the latter case, the steam reforming occurs inside a tubular membrane from which hydrogen is separated. The employment of a membrane requires high operative pressures that reduce the methane conversion. However, the hydrogen removal from the retentate allows to increase the hydrogen yield and methane conversion with regard to a classical reactor, allowing to realize an effective process [11,12]. For these reasons, very active, selective, and stable catalysts are required to efficiently perform LTSR in a membrane reactor. In fact, these must be able to activate methane at low temperatures, thus improving the performances with respect to high active SR and CPO (catalytic partial oxidation) catalysts [16]. Moreover, active-phase stability is important for the application cited above, in order to minimize the operations related to catalyst replacement. For this reason, Rh was selected as active phase rather than the mostly used Ni. In fact, Rh was reported to have both higher activity and stability in steam reforming than other active phases [17–21].

The $Rh_4(CO)_{12}$ cluster [22] had been studied almost 30 years ago for catalyzing different reactions, such as the hydrosilylation of isoprene, cyclohexanone, and cyclohexenone [23], or, more recently, the hydroformylation of cyclopentene co-promoted by $HMn(CO)_5$ [24], all in homogeneous catalysis. $Rh_4(CO)_{12}$-derived catalysts, supported on Al_2O_3, MgO, and CeO_2, had also been tested in 2001 in the CPO process for syngas production [25–27]. However, the CPO conditions and the related generated hot spot do not allow a simple comparison among catalysts. Furthermore, no tests on heterogeneous supported catalysts based on the $Rh_4(CO)_{12}$ cluster have been reported yet.

This work describes the preparation of Rh-based oxide-supported catalysts by exploiting the $Rh_4(CO)_{12}$ carbonyl cluster as source of active phase, as well as their performances on low-temperature methane steam reforming reactions, with the aim of developing a suitable catalyst for future membrane reactor operations. The $Rh_4(CO)_{12}$ cluster was deposited on two different supports, namely $Ce_{0.5}Zr_{0.5}O_2$ and ZrO_2, which were obtained by microemulsion synthesis. Notably, this method gave rise to an improved Rh-supported CeZr-oxide catalyst with respect to classical CeZr ones, in terms of activity and stability in the oxy-reforming reaction [28–31].

The effect of the presence of Ce in the support and its influence on the catalyst deactivation were also investigated. Noteworthy, no deactivation studies of $Rh_4(CO)_{12}$-derived catalysts for syngas production have been reported thus far

2. Results

2.1. Preparation and Characterization of the Supported Catalysts

The CeZr and Zr oxide supports were synthetized by *w/o* microemulsion technique [28] followed by calcination at 750 °C, and were characterized by XRD (X-Ray Diffraction) analyses (Figure 1). The diffractogram of the former showed reflections ascribable to the $Ce_{0.5}Zr_{0.5}O_2$ (CZOm) phase, which is not easy to obtain with other synthetic methods in terms of composition homogeneity and structure. This specific phase possesses higher oxygen mobility and storage properties that enhance the activity in reforming reactions [28]. Debye Scherrer calculations on the peak at lower 2θ showed an average crystallite dimension of 8 nm.

The XRD analysis on the zirconium oxide sample, ZrO_2 (ZOm), showed the formation of the sole tetragonal phase of zirconia, while the pattern related to the monoclinic phase, associated with a lower surface area [32–35], was not present. Debye Scherrer calculations gave an average crystallite dimension of 10 nm.

Table 1 shows the surface area, volume of the pores, and average pore diameter of the CZOm and ZOm samples, determined through Barrett-Joyner-Helenda (BJH) analysis.

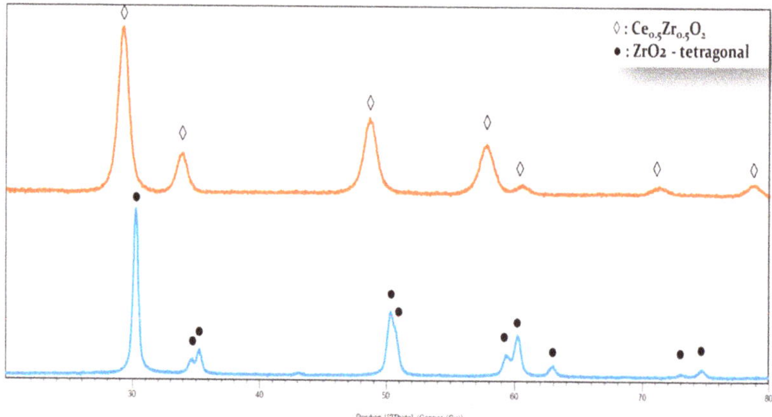

Figure 1. X-Ray Diffraction (XRD) spectra of the $Ce_{0.5}Zr_{0.5}O_2$ (CZOm) and ZrO_2 (ZOm) samples synthetized by microemulsion and followed by calcination at 750 °C.

Table 1. Surface area, pore volume, and average pore diameter of the CZOm and ZOm.

	Surface Area (m²/g)	Pore Vol. (cm³/g)	Avg. Pore Diam (nm)
CZOm	48.8	0.08	5.8
ZOm	40.7	0.07	10.8

The Ce-containing sample showed a higher surface area and a smaller average pore diameter. This is related to the smaller average particle dimension that was obtained for the Ce-containing sample (8 nm) compared to the zirconium oxide one. In any case, high surface and small particles were obtained thanks to the microemulsion technique, in which micelles, used as microreactors, helped to module the crystallite dimensions during the synthesis. Nevertheless, the total pore volume was similar for the two samples, suggesting that a higher number of smaller pores was found in the CZOm sample.

The two supports were also analyzed through IR (Infrared) spectroscopy in the solid state, in nujol mull between NaCl plates, both before and after the deposition with the $Rh_4(CO)_{12}$ cluster, in order to verify that this actually occurred. Initially, both spectra showed peaks at ca. 2100 (w), 2031 (s) and 1892 (w) cm^{-1}, typically associated to the sole supports. The cluster compound was then deposited on CZOm and ZOm (Rh loading of 0.6% wt/wt). The preparation was carried out by mixing a n-hexane solution of $Rh_4(CO)_{12}$ with a slurry of the desired support in the same solvent for 24 hours, under CO atmosphere because of the instability of the neutral cluster in air. In the end, the IR spectra of the n-hexane solutions showed a complete absence of any residual $Rh_4(CO)_{12}$, as indicated by the disappearance of its typical νCO absorptions. Moreover, both supports changed their colors from white to light burgundy after the treatment with the cluster and their IR spectra showed significant changes with respect to the initial ones. More specifically, the spectrum recorded on CZOm exhibited νCO peaks at 2078 (sh), 2051 (s), 2043 (m), 2019 (sh), 1803 (sh) and 1797 (w) cm^{-1}, while the one of ZOm showed νCO peaks at 2068 (w), 2038 (s), 1990 (sh) and 1774 (w) cm^{-1}. These experimental evidences consistently indicate that the cluster deposition on the CZOm and ZOm supports occurred. Figures 2 and 3 show the solid-state IR spectra of the bare and Rh-deposited CZOm and ZOm supports, respectively.

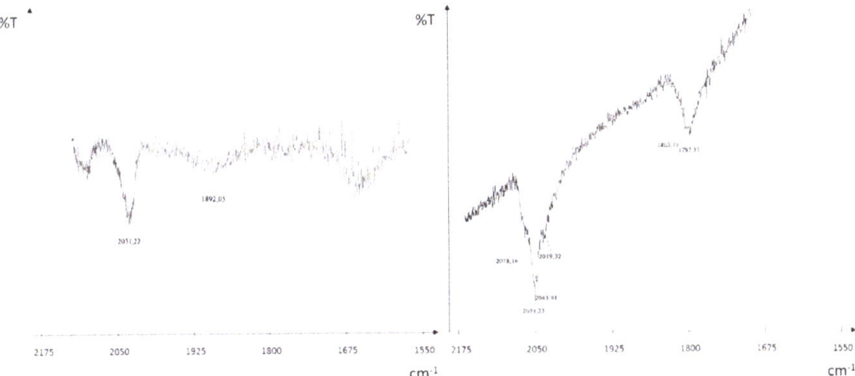

Figure 2. Solid-state infrared (IR) spectra of the bare (**left**) and Rh-deposited (**right**) CZOm support.

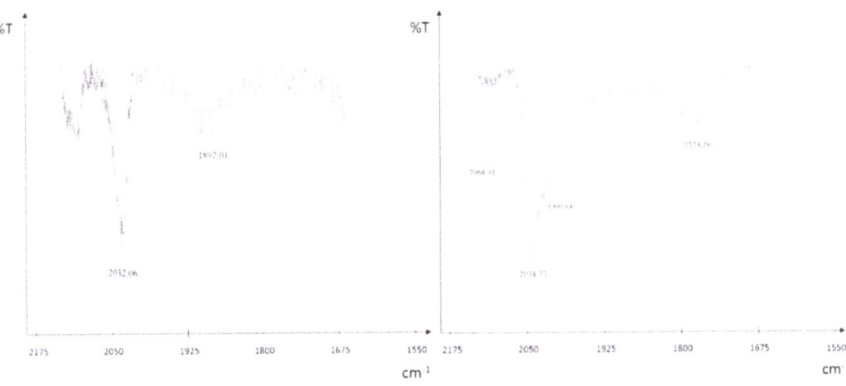

Figure 3. Solid-state IR spectra of the bare (**left**) and Rh-deposited (**right**) ZOm support.

Subsequently, both catalysts were pelleted and treated under N_2 at 500 °C, and new nujol mull spectra were recorded. As illustrated in Figure 4, the ZOm-supported catalyst exhibited residual vCO peaks at 2023 (w) and 1874 (w) cm^{-1}, while the CZOm-supported one displayed the same peaks shown by the support before the deposition [26].

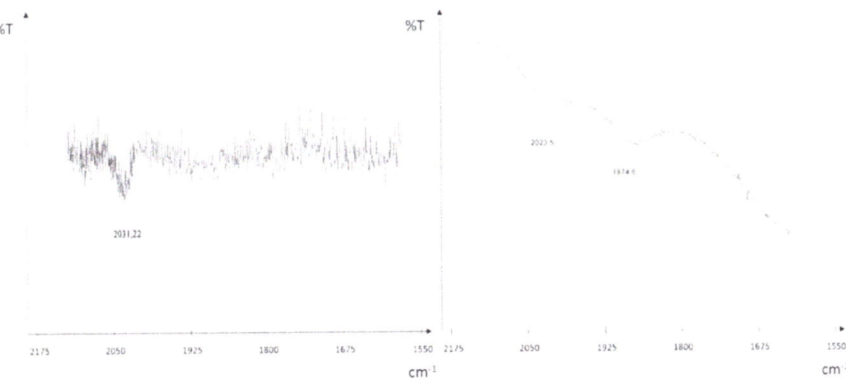

Figure 4. Solid-state IR spectra of the Rh-deposited CZOm (**left**) and ZOm (**right**) supports after treatment under N_2 at 500 °C.

A possible explanation for this different behavior could be that, in the CZOm support, the high mobility of oxygen atoms linked to the Ce(IV)/Ce(III) couple favored the complete oxidation of the Rh carbonyls [36]. After treatment under hydrogen flow at 500 °C, both catalysts only showed IR bands associated to the sole supports.

2.2. Catalytic Tests on Cluster-Derived Catalysts

2.2.1. Cluster-Based Catalyst with High Rh Loading

The cluster-based catalyst with a metal loading of 0.6% deposited on $Ce_{0.5}Zr_{0.5}O_2$ (Rh0.6-CL-R-CZOm) was tested in different conditions, and the effects of the S/C (Steam/Carbon) ratio, temperature, pressure and GHSV (Gas Hourly Space Velocity) were studied. The precursor was charged in the reactor and heated under N_2 at 500 °C, and after cooling down to 200 °C, the sample was heated at 500 °C under H_2 to obtain the final active catalyst. The analyses of the reaction parameters have been carried out with the aim of studying the conditions with which the catalyst could maximize both syngas and hydrogen productions.

At 500 °C, the increase of the S/C ratio had a positive effect on the methane conversion (X CH_4), as shown in Figure 5. In particular, at low S/C ratios, the methane conversion was 19% at S/C 1 and 22% at S/C 1.5, very close to the equilibrium values (19% and 24%, respectively). A further increase in S/C up to 2 and 3 resulted in higher conversions of 23% and 26%, though not enough to meet the increase of the equilibrium values (29% and 36%, respectively), deriving from higher steam concentration, showing a kinetic limitation of the steam reforming reaction. The H_2/CO ratio was also considered to analyze the contribution of the water gas shift reaction (WGSR). In all conditions, the ratio was higher than the stoichiometric value of 3, due to the occurrence of the WGSR. Thus, an advantage of low temperature steam reforming is the possibility of carrying out the two reactions together, as opposed of conducting them in three different reactors (one for steam reforming, a high-temperature shift reactor and a low-temperature shift one), which is the usual procedure. For instance, a H_2/CO ratio of 28 was obtained at S/C 1. When the S/C ratio increased, the H_2/CO raised thanks to an augmented contribution of the WGSR, reaching the values of 35, 43, and 63 at S/C 1.5, 2 and 3, respectively. Indeed, the high amount of steam reacted with CO and produced a major amount of hydrogen. This was also confirmed by the CO selectivity (sCO), which decreased when increasing S/C, falling from 13 (at S/C 1) down to 6 at S/C 3.

Figure 6 highlights the effect of the temperature on the catalytic conversion of methane at 30,000 h^{-1}, S/C 3 and atmospheric pressure. In all conditions, the methane conversions were far from the equilibrium values (indicated by the empty bars). In general, experimental and equilibrium methane conversions increase with temperature as both steam reforming thermodynamics and kinetics are favored at high temperature. In fact, a conversion of 4% was observed at 350 °C, far from the equilibrium value of 11%, which increased to 9% and 17% at 400 and 450 °C, but they were both still far from the equilibrium conversions (18% and 26%, respectively). The equilibrium value was not reached even at 500 °C, where the observed experimental conversion was 26% and the equilibrium one 36%. However, a sharply decreasing trend was observed for the H_2/CO ratio. The WGS and SR reactions are favored in opposite conditions; therefore, at 350 °C water gas shift was predominant, and the as-produced CO was directly converted into hydrogen, resulting in a final gas composition with CO concentration of 75 ppm. Conversely, at 500 °C the water gas shift is less favored, and an increased rate of steam reforming consumed water, further depressing the WGSR. Overall, this resulted in a decrease of the H_2/CO ratio from 1700 at 350 °C to 63 at 500 °C, and in an increase of the CO selectivity with temperature up to 6%.

An increase of the GHSV caused a decrease in the methane conversion from 23% at 30,000 h^{-1}, to 17% at 50,000 h^{-1} and 13% at 100,000 h^{-1}, which also resulted in lower hydrogen production (Figure 7). Moreover, the conversion values were far from the equilibrium ones, which were set at 29%, 28%, and 25%, respectively. This was due to the minor contact time between the gas stream and the catalyst. In these conditions, in fact, the reaction appeared to be strongly limited by kinetics, and the catalyst gave conversion results that were far from the thermodynamic equilibrium. The lower methane conversion is accompanied by a higher water partial pressure, which decreased the CO amount due to WGSR and, overall, increased the H_2/CO ratio from 43 at 30,000 h^{-1}, to 66 at 50,000 h^{-1} and 91 at 100,000 h^{-1}. This catalyst behavior showed the predominance of the WGSR in determining the H_2/CO ratio, suggesting a faster kinetics for the WGS reaction over the steam reforming one. However, it is difficult to offer a complete comparison among the different tests as lower catalytic bed temperatures are developed at higher GHSV, due to decreased conversions. In fact, lower temperatures decrease the equilibrium conversion and steam reforming activities.

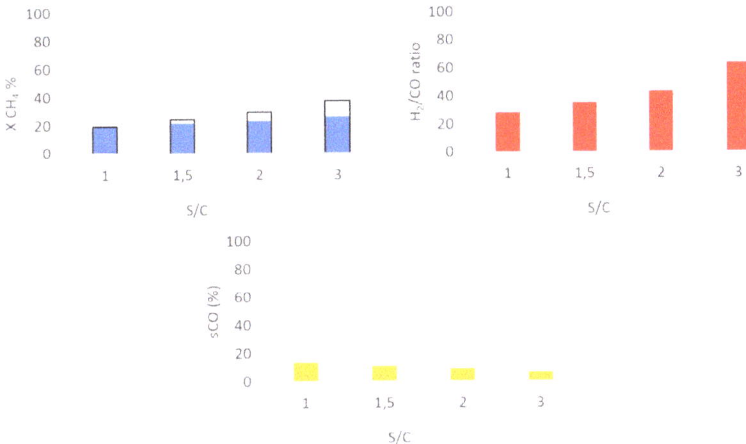

Figure 5. Comparison among tests carried out on Rh0.6-CL-R-CZOm at 30,000 h^{-1}, 1 atm and 500 °C. The graphs report the methane conversion (blue) with respect to the equilibrium values (white bars with black frames), H_2/CO ratio (red), and CO selectivity (yellow) versus different S/C ratios.

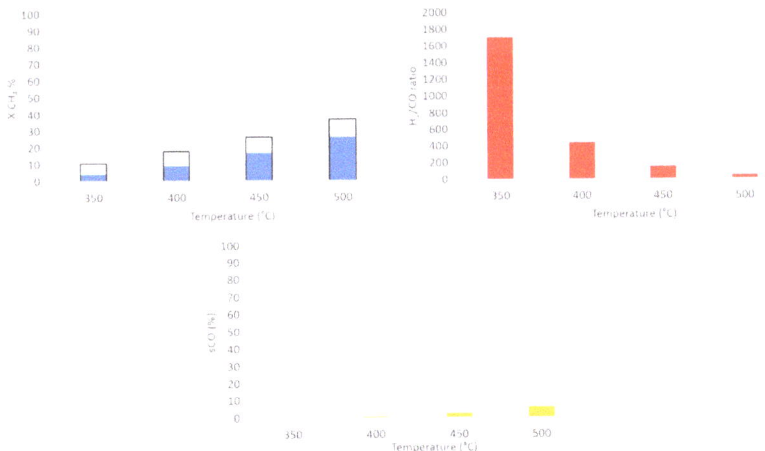

Figure 6. Comparison among tests carried out on Rh0.6-CL-R-CZOm at 30,000 h^{-1}, 1 atm and S/C 3. The graphs report the methane conversion (blue) with respect to the equilibrium values (white bars with black frames), H_2/CO ratio (red), and CO selectivity (yellow) versus different T. The Y scale related to the H_2/CO-ratio graph is very different from the others due to the large values.

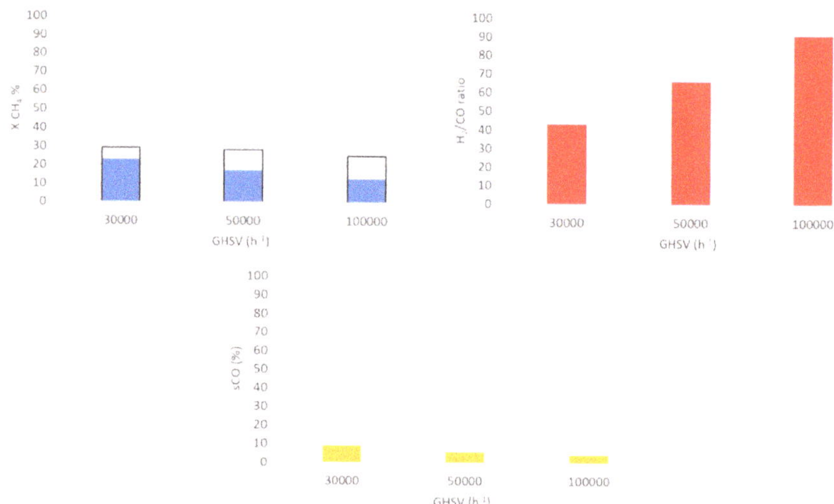

Figure 7. Comparison among tests carried out on Rh0.6-CL-R-CZOm at 1 atm, S/C 2 and 500 °C. The graphs report the methane conversion (blue) with respect to the equilibrium values (white bars with black frames), H_2/CO ratio (red) and CO selectivity (yellow) versus different Gas Hourly Space Velocity (GHSV).

At higher pressure, lower methane conversions were observed (from 29% at 1 atm, to 20%, 17%, and 13% at 3, 5, and 10 atm, respectively) (Figure 8), owing to a decreased thermodynamic equilibrium of the SR reaction. However, the equilibrium conversion was close to the experimental conversion at 5 atm (17%) and reached it at 10 atm (13%). In analogy with what observed above, a lower methane conversion is associated with a higher residual amount of water, which enhances the WGSR and, in turn, the H_2/CO ratio, which raised from 43 at 1 atm to 108 at 10 atm.

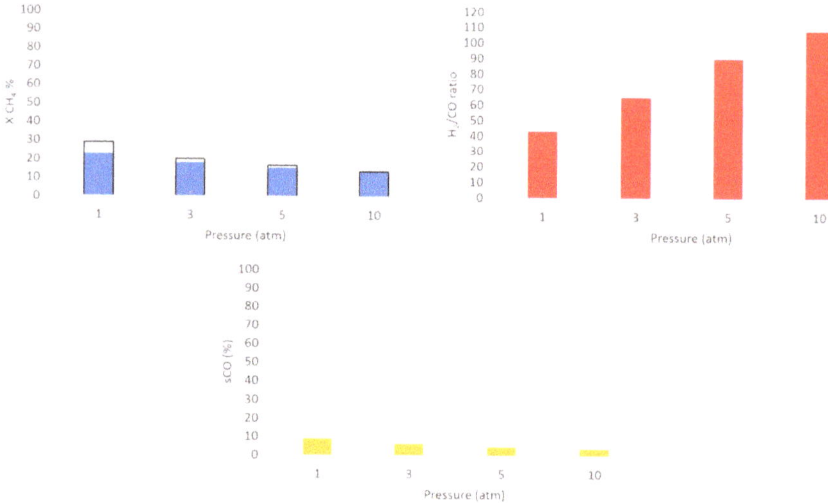

Figure 8. Comparison among tests carried out on Rh0.6-CL-R-CZOm at 30,000 h^{-1}, S/C 2 and 500 °C. The graphs report the methane conversion (blue) with respect to the equilibrium values (white bars with black frames), H_2/CO ratio (red) and CO selectivity (yellow) versus different GHSV. The Y scale related to the H_2/CO-ratio graph is slightly different from the others due to higher values.

The effect of the S/C ratio was investigated at 500 °C and 10 atm (Figure 9). In these conditions, the methane conversion increased with S/C, reaching the equilibrium values up to S/C 2 (9% at S/C 1, 11% at S/C 1.5 and 13% at S/C 2), and the experimental conversion (15%) close to equilibrium value (17%) at S/C 3. The highest conversion was observed at S/C 3 thanks to the increased equilibrium values at higher S/C ratios.

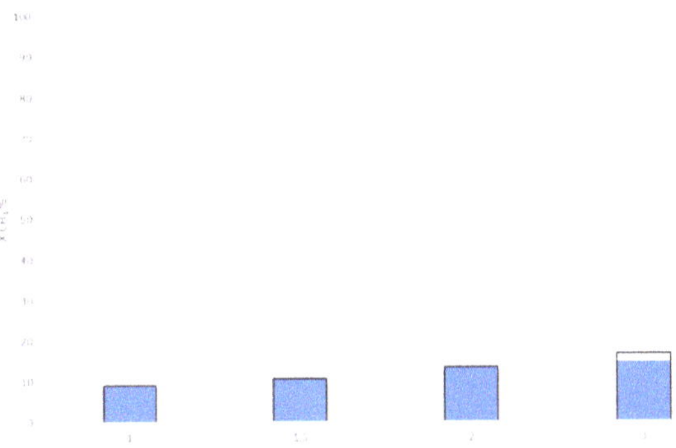

Figure 9. Comparison among tests carried out on Rh0.6-CL-R-CZOm at 30,000 h^{-1}, 10 atm and 500 °C. The graph reports the methane conversion (blue) with respect to the equilibrium values (empty bars with black frame) versus different S/C ratios.

The same tests were also conducted at lower temperature, namely 450 °C (Figure 10). In these operative conditions, the experimental methane conversion was close to the equilibrium values up to a S/C ratio of 2, however, it remained distant from it a S/C 3 due to the lower catalytic activity at this temperature. In particular, conversions of 5%, 6%, 7%, and 8% were, respectively, observed by increasing the S/C, while equilibrium conversions concurrently increased from 5% to 7%, then 9% and 12%.

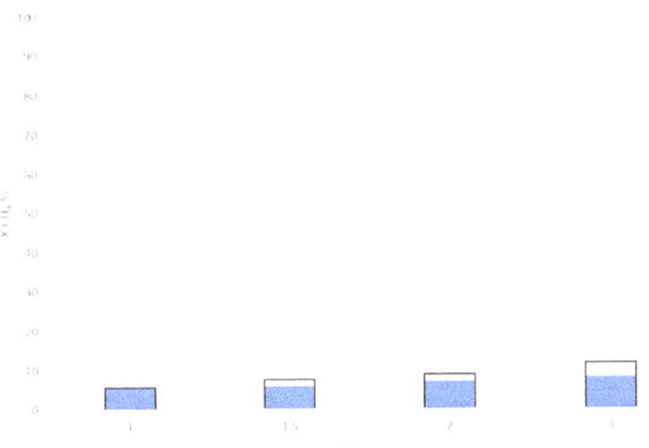

Figure 10. Comparison among tests carried out on Rh0.6-CL-R-CZOm at 30,000 h^{-1}, 10 atm and 450 °C. The graph reports the methane conversion (blue) with respect to the equilibrium values (empty bars with black frames) versus different S/C ratios.

Finally, in order to evaluate the stability of the catalysts in the operative conditions, a test with the same parameters as the first one (500 °C, 30,000 h^{-1}, 1 atm and S/C 1.5, Figure 11) was periodically conducted after every three tests carried out in different conditions, giving rise to the so-called return tests. The results indicate a rather high catalyst stability towards methane conversion. However, a conversion drop of 4% was observed in the last return test, which was carried out after those conducted at low temperature and high pressure.

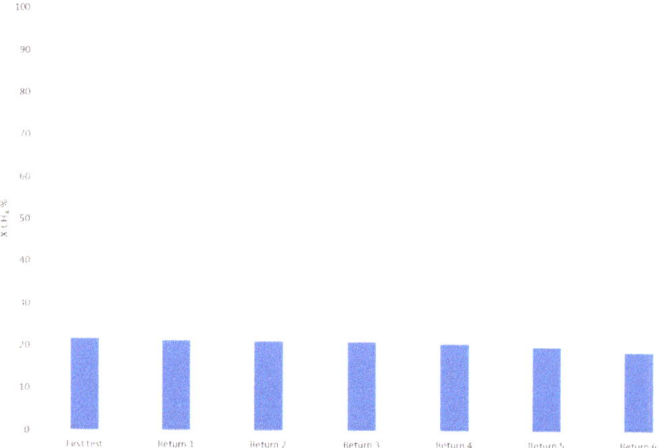

Figure 11. Methane conversion in deactivation tests carried out at 500 °C, 1 atm, 30,000 h^{-1} and S/C 1.5.

This deactivation could be due to the formation of carbon on the active site, owing to Boudouard reaction, which in fact is favored at low temperature and high pressure. The Raman analysis on the spent Rh0.6-CL-R-CZOm catalyst (Figure 12) showed two very small peaks related to the presence of carbon. In particular, the bands detected at 1350 and 1580 cm^{-1} could be attributed to the D-band and G-band of carbon, respectively [37–39]. In general, the G-band derives from the stretching of sp^2 carbon bonds in planes of graphene type [28], while the D-band is caused by the vibrations of carbon atoms in disordered species, such as amorphous carbon [40]. Both bands are rather broad and weak even after a long time on stream and different reaction test conditions, demonstrating the properties of the $Ce_{0.5}Zr_{0.5}O_2$ support synthetized by microemulsion in retarding carbon formation, which was also reported previously [28]. Considering the weakness of the bands, other phenomena, such as the re-oxidation of the catalyst during high-pressure tests, where high water partial pressure is present, need to be taken into account. In those conditions, especially in the first part of the catalytic bed where the presence of hydrogen is still very low, the gas mixture had a higher oxidative potential that could affect the oxidation state of the Rh.

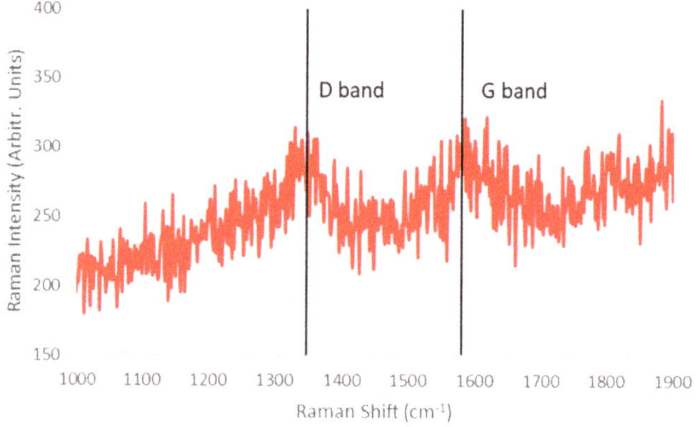

Figure 12. Raman analysis carried out on the used Rh0.6-CL-R-CZOm catalyst. The two bars indicate the presence of carbon.

2.2.2. Catalytic Tests with Low Rh Loading

The catalysts with a lower Rh loading were prepared diluting the 0.6% sample down to 0.05% wt/wt for further tests, to allow a better evaluation of the activity in a kinetic regime and a better understanding of the deactivation phenomena and metal support effects. Furthermore, the study of catalysts with low amounts of active phase is interesting for some specific applications, such as structured catalysts, membrane reactors, and microreactors, where the active phase may be deposited or coated in thin layers or on low-surface-area supports.

Catalysts with a low % of active phase (0.05% wt/wt) on the two different supports (CZOm and ZOm) were prepared and treated in two different ways: either just heated under N_2 or treated firstly under inert gas and then hydrogen. The four produced catalysts were named as follows:

- Rh0.05-CL-N-CZOm (the cluster-based catalyst deposited on CZOm was heated under nitrogen before the tests).
- Rh0.05-CL-R-CZOm (like the previous catalyst, also treated under H_2).
- Rh0.05-CL-N-ZOm (the cluster-based catalyst deposited on ZOm was heated under nitrogen before the tests).
- Rh0.05-CL-R-ZOm (like the previous catalyst, also treated under H_2).

Tests were carried out at fixed experimental conditions (GHSV = 30,000 h^{-1}, P = 1 atm, T = 500 °C, S/C 1.5) to compare the different catalysts' performances.

Before the catalytic tests, a preliminary screening was carried out to compare the Rh-particle size and distribution on the not-reduced (only heated under nitrogen) CZOm- and ZOm-supported catalysts. The TEM (Transmission Electron Microscopy) analyses performed on both supports evidenced the presence of small Rh particles (1-2 nm) with narrow size distribution (Figure 13). However, it is important to note that the identification of particles below 1 nm is difficult and may be strongly underestimated.

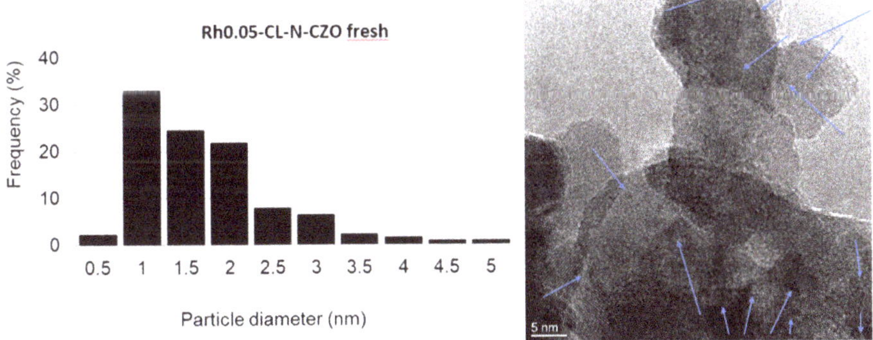

Figure 13. Rh particle-size distribution of the Rh cluster deposited on CZOm support determined by Transmission Electron Microscopy (TEM) analysis.

Figure 14 shows the results obtained with the cluster-based catalysts deposited on CZOm in tests carried out at high temperature. An initial conversion around 35% was observed for the catalyst not treated in hydrogen, with a decreasing trend in the first hour and an almost constant value of 30% during the reaction time. This result is in agreement with the fact that, when using such cluster-derived samples, active species are obtained even without the need of a reduction step (hydrogen treatment). The deactivation of the first part was probably due to the fast sintering of the reactive small particles evidenced by TEM analysis. The sample treated under H_2 was able to produce a higher conversion (40%) with stable results over 300 minutes.

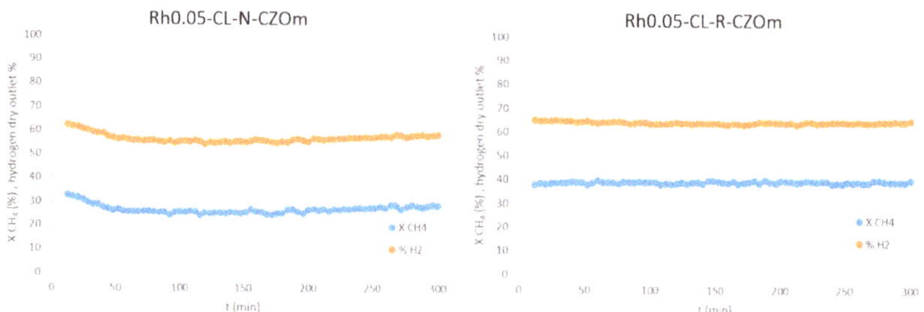

Figure 14. Methane conversion and hydrogen dry outlet percentage obtained on Rh0.05-CL-N-CZOm (**left**) and Rh0.05-CL-R-CZOm (**right**) tested at 750 °C, 30,000 h^{-1}, 1 atm and S/C 1.5.

The same test was conducted on an Rh-impregnated catalyst on the same CeZr oxide and reduced at 500 °C (Rh0.0.5-IWI-R-CZOm), which represents the best catalyst in oxy-reforming reaction [19] (Figure 15). This test showed an initial methane conversion of 36% with a slightly deactivation trend towards 34%. These results were lower than those obtained with Rh0.05-CL-R-CZO, indicating a higher activity for the cluster-derived catalyst and a readily stable conversion. This could be related to a better dispersion of the Rh cluster on the oxide support.

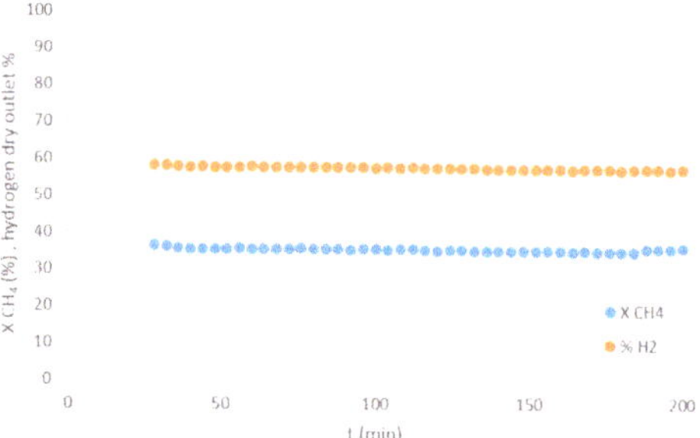

Figure 15. Methane conversion and hydrogen concentration in the dry gas obtained on Rh0.0.5-IWI-R-CZOm tested at 750 °C, 30,000 h^{-1}, 1 atm and S/C 1.5.

The same test was also conducted on the cluster-based catalyst supported on zirconia (Rh0.05-CL-R-ZOm, Figure 16) and treated under H$_2$, which gave a lower methane conversion (around 33%) with respect to the analogous cluster on CZOm. This is probably related to the effect of the Ce(IV)/Ce(III) couple in the sample that acts as steam-dissociating site, and by the higher oxygen mobility in the Ce$_{0.5}$Zr$_{0.5}$O$_2$ phase obtained with the microemulsion synthesis [28,41–44].

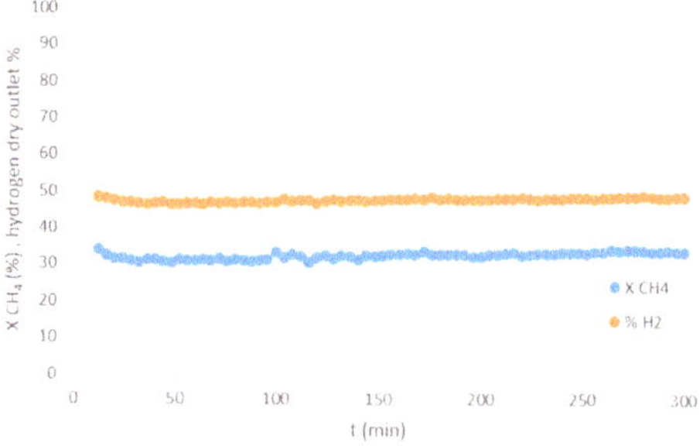

Figure 16. Methane conversion and hydrogen dry outlet percentage obtained on Rh0.05-CL-R-ZOm tested at 750 °C, 30,000 h^{-1}, 1 atm and S/C 1.5.

Having assessed the higher conversion of the cluster-based catalysts at 750 °C, the tests were also conducted at 500 °C. Figure 17 highlights the methane conversion and hydrogen dry outlet concentration versus the reaction time on the two CZOm-supported catalysts.

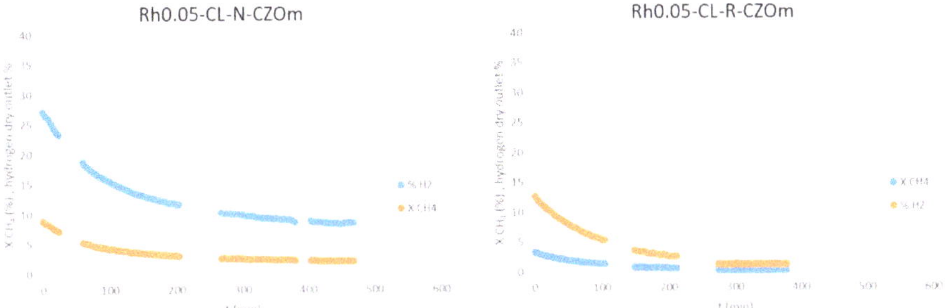

Figure 17. Methane conversion and hydrogen dry outlet percentage obtained on Rh0.05-CL-N-CZOm (**left**) and Rh0.05-CL-R-CZOm (**right**) tested at 500 °C, 30,000 h^{-1}, 1 atm and S/C 1.5.

The unreduced catalyst provided an initial conversion of circa 9%, while the reduced one gave an initial lower conversion (3%). It has been experimentally demonstrated that hydrogen treatment favors the formation of larger particles [26], and this phenomenon could explain the lower activity of the reduced catalyst with respect to the not-reduced one.

In order to explain the deactivation phenomena that took place in both catalysts, it is important to consider that the reaction environment at low temperature is an oxidizing one due to: i) the low amount of produced hydrogen related to lower kinetics of the catalyst, and also related to the low amount of active phase; ii) the high concentration of water. All this could favor the oxidation of Rh, which is faster when small particles are present, i.e., in the unreduced sample. The oxidation is very likely favored by the presence of Ce(IV) and the high oxygen mobility. The possibility to ascribe the deactivation to sintering of the active phase during reaction can be excluded, since small Rh particles were still found on the unreduced spent catalyst, as evidenced by TEM, where it is clear that the Rh metal particles are only slightly larger (centered between 1.5 and 2.5 nm, Figure 18) than those found on the fresh catalyst.

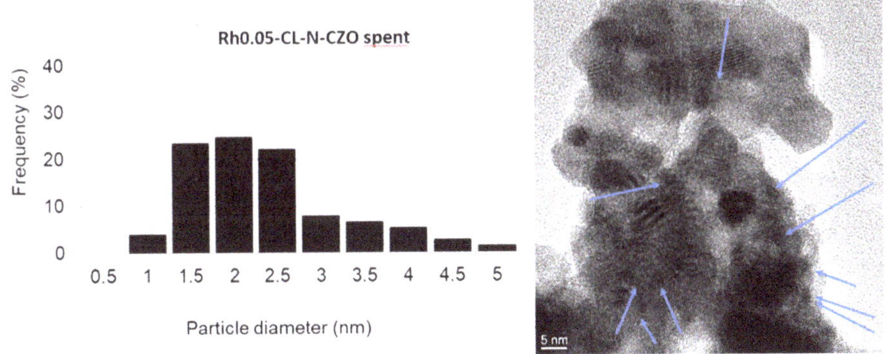

Figure 18. Particle size distribution of the spent Rh0.05-CL-N-CZOm determined via TEM analysis.

The catalytic performances of the Rh cluster deposited on ZOm, both not reduced and reduced (Rh0.05-CL-N-ZOm and Rh0.05-CL-R-ZOm), are shown in Figure 19.

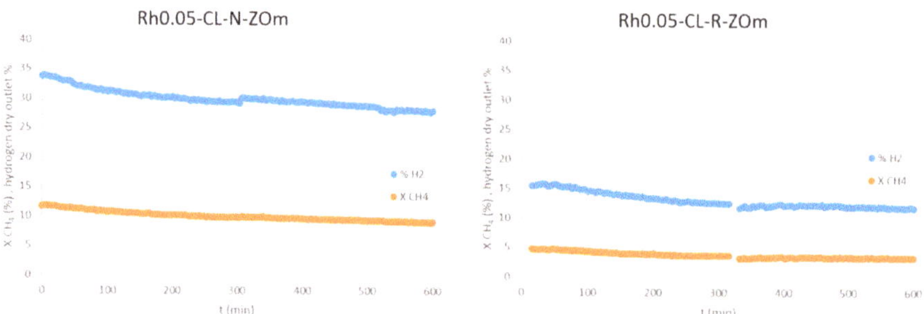

Figure 19. Methane conversion and hydrogen dry outlet percentage obtained on Rh0.05-CL-N-ZOm (**left**) and Rh0.05-CL-R-ZOm (**right**).

The initial conversion of the not-reduced catalyst (10.2%) was higher than the one reached with the CZOm-deposited cluster. Moreover, the fast deactivation that characterized the CZOm-based catalysts was not observed in this case, indicating that the presence of Ce favors the oxidation of the Rh. The reduced catalyst was stable too, but produced a lower conversion than the unreduced one, owing to the larger particles formed during the reduction step [26].

Table 2 reports the turnover frequency (TOF) values of the investigated catalysts at 500 °C. In the case of the CZOm samples, they were calculated on the initial conversion as those catalysts showed deactivation. Their comparison with TOF values of some other Rh-containing catalysts reported in the literature [45–47], however, is not always suggestive, because of the different supports, active-phase loadings, and operative conditions of the latter.

Table 2. Turnover frequency of the catalysts investigated in this work. * Calculated on the initial conversion.

Catalyst	T (°C)	S/C	GHSV (h^{-1})	P (atm)	TOF [molCH$_4$/(molRh*s)]
Rh0.05-CL-N-ZOm	500	1.5	30,000	1	1.47
Rh0.05-CL-R-ZOm	500	1.5	30,000	1	0.70
Rh0.05-CL-N-CZOm	500	1.5	30,000	1	1.32*
Rh0.05-CL-R-CZOm	500	1.5	30,000	1	0.73*

2.2.3. Investigation on Deactivation

It is particularly interesting to elucidate the mechanism of catalyst deactivation, since no such investigation on carbonyl cluster-derived catalyst is reported in literature, as very few low-temperature studies have been carried out. Having observed that the deactivation was related to the presence of Ce and the oxidizing environment, its cause was further investigated by TEM and EDS (Energy Dispersive X-ray Spectrometry) analyses. Having also excluded deactivation by sintering, and assessed the absence of carbon formed over the surface, it was hypothesized that the deactivation was caused by the oxidation of the active particles favored by Ce, which is able to provide oxygen to the active phase thanks to its redox properties, enhanced by the oxygen mobility of the microemulsion catalyst [19].

This phenomenon benefits from the oxidizing environment found in the low-temperature SR conditions, in which a relatively low amount of hydrogen is produced together with high concentrations of unreacted steam. SAED (Selected Area Electron Diffraction) analyses were carried out during TEM on the Rh particles of the unreduced CZOm- and ZOm-supported catalysts, after reaction at low (500 °C) and high (750 °C) temperature with the same Rh loading (0.05% wt/wt). In all samples, the presence of Rh in the analyzed particles was evidenced by the EDS analyses. The spent ZOm-supported samples, both at 500 °C and 750 °C, showed the presence of metallic Rh, as well as the CZOm-supported catalyst after reaction at high temperature (Figure 20). Experimental d-values of 2.25 and 1.84 Å, in fact, could be attributed to the metallic Rh phase, while those at 3.10 and 2.61 Å are related to the support.

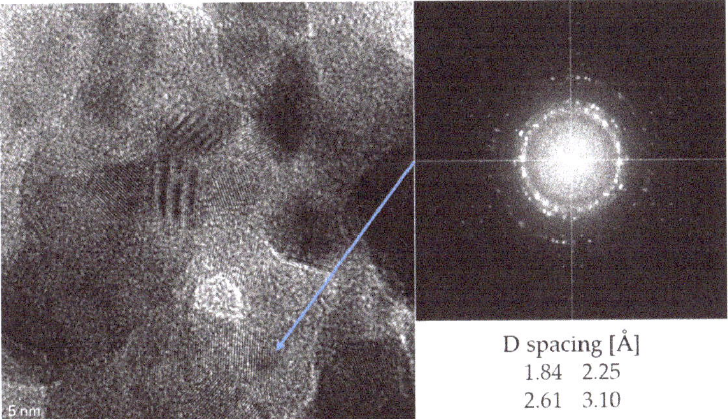

Figure 20. Selected Area Electron Diffraction (SAED) and TEM analyses carried out on a Rh containing area of the spent not-reduced CZOm supported catalyst after reaction at high temperature.

However, no such phase was observed in the spent CZOm-supported catalyst after reaction at low temperature (Figure 21); the measured d-values of 2.65 and 3.05 Å were attributed to the tetragonal phase of $Ce_{0.5}Zr_{0.5}O_2$. It was not possible to detect the presence of Rh oxide because its reflexes overlapped those of the support. Nonetheless, all these evidences can be confidently seen as an indirect confirmation that oxidation of Rh only occurred on the CZOm-supported catalyst after reaction at low temperature, which is the only one that showed deactivation.

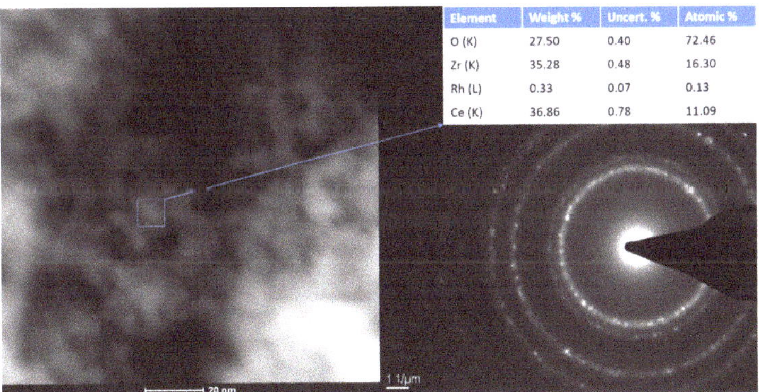

Element	Weight %	Uncert. %	Atomic %
O (K)	27.50	0.40	72.46
Zr (K)	35.28	0.48	16.30
Rh (L)	0.33	0.07	0.13
Ce (K)	36.86	0.78	11.09

Figure 21. SAED and Energy Dispersive X-ray Spectrometry (EDS) analyses carried out on a Rh containing area of the spent not-reduced CZOm supported catalyst after reaction at low temperature.

The hypothesis of the oxidation of Rh is also substantiated by a computational study on metallic Rh clusters on CeO_2, which showed that they are stable in reducing conditions such as those obtained in the high temperature steam reforming, but that the relative oxide is formed in oxidizing ones [36].

3. Materials and Methods

3.1. Catalysts Synthesis

Two supports, namely $Ce_{0.5}Zr_{0.5}O_2$ (CZOm) and ZrO_2 (ZOm), were prepared by *w/o* microemulsion technique as described elsewhere [28] and were calcined at 750 °C for 5 h with a heating ramp of

2 °C/min. In the case of ZrO_2 support, the microemulsion was heated to 70 °C for 30 min at the end of the synthesis in order to destroy the micelles.

The catalysts were prepared by reactive deposition of the $Rh_4(CO)_{12}$ cluster over the selected supports, namely CZOm and ZOm. In a typical experiment, the required amount of carbonyl cluster (0.6% of atomic Rh wt/wt of support) was dissolved in degassed n-hexane (10 mL) and added dropwise to a degassed n-hexane slurry of the desired support kept under CO atmosphere; the obtained slurry was then stirred for 24 h. The completion of the synthesis was evidenced by the discoloration of the solution and its IR analysis, which showed no peaks related to $Rh_4(CO)_{12}$ (νCO at 2075 (vs), 2069 (vs) 2044 (m) and 1885 (s) cm^{-1} in n-hexane).The solvent was thus removed in vacuum at room temperature. The catalyst was then stored under CO atmosphere and removed only to be pelletized and charged in the reactor, then readily kept under nitrogen flow and heated up to 500 °C. In order to provide a catalyst with a lower amount of Rh, the sample was diluted with bare CZOm or ZOm by 12 times, obtaining a catalyst with a loading of Rh of 0.05% (wt/wt of catalyst). In the case of reduced catalysts, the temperature was lowered to 200 °C then hydrogen (10% in nitrogen) was fluxed over the catalysts, while the temperature was again increased to 500 °C. After 15 h the hydrogen flux was stopped, and reactivity tests were carried out.

3.2. Catalyst Characterization

X ray diffraction analyses were carried out using a PW1050/81 diffractometer (Philips/Malvern, Royston, UK) equipped with a graphite monochromator in the diffracted beam and controlled by a PW1710 unit (Cu Kα, λ = 0.15418 nm). A 2θ range from 20° to 80° was investigated at a scanning speed of 40°/h. Nitrogen adsorption-desorption isotherms were determined at liquid nitrogen temperature (−196 °C), using an automatic ASAP 2020 absorptiometer (Micromeritics, Norcross, GA, USA) and analyzed using a software operating standard Brunauer–Emmett–Teller (BET) and BJH methods. Raman analysis was carried out with a micro-spectrometer Raman RM1000 (Renishaw/Thermo Fisher, New Mills, Wotton-under-Edge, Gloucestershire, UK) interfaced to a microscope Leica DMLM (objective 5×, 20×. 50×). The available sources were an Ar^+ laser (λ = 514.5 nm; Pout = 25 mW) and a diode laser (λ = 780.0 nm; Pout = 30mW). In order to eliminate the Rayleigh scattering, the system was equipped with a notch filter for the Ar^+ laser and an edge filter for the diode one. The network was a monochromator with a pass of 1200 lines/mm. The detector was a CCD one (Charge-Coupled Device) with a thermo-electrical cooling (203 K). TEM analyses were carried out using a TEM/STEM TECNAI F20 microscope (FEI, Hillsboro, OR, USA) combined with Energy Dispersive X-Ray Spectrometry (EDS), at 200 keV. The sample preparation was carried out by suspending the powder in ethanol and treating it with ultrasound for 15 min. The suspension was deposited on a "multifoil-carbon film" sustained by a Cu grid. Then the so-prepared system was dried at 100 °C.

3.3. Catalytic Tests

The catalysts were tested in a tubular INCOLOY800HT reactor (Meccanica Padana, San Nicolò (PI), Italy) (length 500 mm; internal diameter 10 mm) placed in a furnace as reported elsewhere [28,48]. The bed temperature was controlled with a thermocouple. 1.0 g of catalyst (30–60 mesh) was loaded into the reactor where the pretreatment and eventual reduction of the active phase were conducted by fluxing a 500 mL flow of N_2 or of an H_2/N_2 (10:90 v/v) gas mixture at 500 °C. Deionized water was fed by a HPLC pump (JASCO, Easton, MD, USA) and vaporized. The outlet gas (H_2, CO, CO_2, non-converted CH_4 and vapor) was condensed in order to eliminate water. The dry gas mixture was analyzed by a 490 micro gas chromatograph (Agilent Technologies, Cernusco sul Naviglio (MI), Italy) with two different columns. Hydrogen was separated through a MS5A 20 m long (carrier: N_2), while CH_4, CO, and CO_2 were separated with a CO_x column 1 m long (carrier: He). Both modules were equipped with a Thermal Conductivity Detector TCD. The CEA-NASA software was used to calculate the outlet composition of the stream at the thermodynamic equilibrium. The software gave the molar gaseous outlet composition (non-converted CH_4, non-converted H_2O, CO, CO_2, H_2, and deposited carbon if present), based on the feed composition in terms of molar percentage, reaction temperature, and pressure. Methane conversion was taken as reference in order to evaluate the catalytic activities of the tested samples and was compared with the one calculated at the thermodynamic equilibrium. Methane conversion accuracy was evaluated by calculating standard deviation. This resulted to be lower than ±0.9% for the tests at 750 °C and lower than ±0.7% in the tests at 500 °C; in a conservative way, we considered accuracy to be ±1% for the test at high temperature and ±0.8% for those at low temperature.

4. Conclusions

This paper describes the synthesis, characterization and catalytic behavior of Rh-based catalysts, obtained by using the $Rh_4(CO)_{12}$ neutral cluster as the active-phase precursor. In particular, the preparation method allowed the deposition of the cluster on the surface of $Ce_{0.5}Zr_{0.5}O_2$ and ZrO_2 supports, which were synthetized by w/o microemulsion technique. The catalysts were found to be active in the low-temperature steam reforming process for syngas production. At high Rh loadings (0.6%) the CZOm-supported catalyst was active at 350 °C and was able to reach the equilibrium conversion, especially at low S/C ratio or at high pressures at 450 and 500 °C. It showed good stability, and this opens the possibility of employing such catalyst in membrane reactors, enhanced reformers or chemical loop based on reforming. At lower concentrations (0.05%) and high temperature, the CZOm-supported cluster sample showed better results with respect to the analogous ZOm-supported one and to a classical Rh-impregnated CeZr catalyst. At low temperature, a deactivation effect was observed for the CZOm-supported catalyst, which could be overcome by employing a ZOm support. A detailed analysis provided evidences that the oxidation of the Rh promoted by Ce and high oxygen mobility was responsible for the fast deactivation. In these conditions, it was also observed that the cluster-based catalyst which had not been treated with hydrogen at 500 °C was more active than the treated one, due to the sintering of the Rh particles. Finally, the unreduced 0.05% Rh cluster deposited on the ZrO_2 support showed significant activity at 500 °C.

Author Contributions: A.F. investigated the catalytic activity and the support characterization and wrote the first draft; S.R. investigated the cluster preparation and deposition, and collaborated to write the first draft; C.F. supervised the catalyst preparation, investigated the IR and cluster analysis, and collaborated in writing, reviewing, and editing; F.B. defined the methodology for support preparation, supervised the project, validated the catalytic data, and took care of the conceptualization and of writing, reviewing, and editing of the paper.

Funding: This research received no external funding.

Acknowledgments: The contribution of Francesca Ospitali of the Department of Industrial Chemistry "Toso Montanari" is acknowledged for the TEM analysis.

Conflicts of Interest: The authors declare no conflict of interest.

References

1. Rostrup-Nielsen, J.R. Catalytic Steam Reforming. In *Catalysis: Science and Technology*; Anderson, J.R., Boudart, M., Eds.; Catalysis; Springer: Berlin/Heidelberg, Germany, 1984; Volume 5, pp. 1–117. ISBN 978-3-642-93247-2.
2. Rostrup-Nielsen, J.R.; Christiansen, L.J.; Bak Hansen, J.-H. Activity of steam reforming catalysts: Role and assessment. *Appl. Catal.* **1988**, *43*, 287–303. [CrossRef]
3. Trimm, D.L. The Steam Reforming of Natural Gas: Problems and Some Solutions. In *Studies in Surface Science and Catalysis*; Bibby, D.M., Chang, C.D., Howe, R.F., Yurchak, S., Eds.; Methane Conversion; Elsevier: Amsterdam, The Netherlands, 1988; Volume 36, pp. 39–50.
4. El-Temtamy, S.A.; El-Salamony, R.A.; Ghoneim, S.A. Review on Innovative Catalytic Reforming of Natural Gas to Syngas. *World J. Eng. Technol.* **2016**, *4*, 720–726.
5. Taji, M.; Farsi, M.; Keshavarz, P. Real time optimization of steam reforming of methane in an industrial hydrogen plant. *Int. J. Hydrog. Energy* **2018**, *43*, 13110–13121. [CrossRef]

6. Azancot, L.; Bobadilla, L.F.; Santos, J.L.; Córdoba, J.M.; Centeno, M.A.; Odriozola, J.A. Influence of the preparation method in the metal-support interaction and reducibility of Ni-Mg-Al based catalysts for methane steam reforming. *Int. J. Hydrog. Energy* **2019**, *44*, 19827–19840. [CrossRef]

7. Berlier, G. Low Temperature Steam Reforming Catalysts for Enriched Methane Production. In *Enriched Methane: The First Step Towards the Hydrogen Economy*; De Falco, M., Basile, A., Eds.; Green Energy and Technology; Springer International Publishing: Cham, Switzerland, 2016; pp. 53–74. ISBN 978-3-319-22192-2.

8. Dincer, I.; Acar, C. Review and evaluation of hydrogen production methods for better sustainability. *Int. J. Hydrog. Energy* **2015**, *40*, 11094–11111. [CrossRef]

9. Yoko, A.; Fukushima, Y.; Shimizu, T.; Kikuchi, Y.; Shimizu, T.; Guzman-Urbina, A.; Ouchi, K.; Hirai, H.; Seong, G.; Tomai, T.; et al. Process assessments for low-temperature methane reforming using oxygen carrier metal oxide nanoparticles. *Chem. Eng. Process. Process Intensif.* **2019**, *142*, 107531. [CrossRef]

10. LeValley, T.L.; Richard, A.R.; Fan, M. The progress in water gas shift and steam reforming hydrogen production technologies–A review. *Int. J. Hydrog. Energy* **2014**, *39*, 16983–17000. [CrossRef]

11. Kim, H.-M.; Jang, W.-J.; Yoo, S.-Y.; Shim, J.-O.; Jeon, K.-W.; Na, H.-S.; Lee, Y.-L.; Jeon, B.-H.; Bae, J.W.; Roh, H.-S. Low temperature steam reforming of methane using metal oxide promoted Ni-Ce0.8Zr0.2O2 catalysts in a compact reformer. *Int. J. Hydrog. Energy* **2018**, *43*, 262–270. [CrossRef]

12. Stefanidis, G.D.; Vlachos, D.G. High vs. low temperature reforming for hydrogen production via microtechnology. *Chem. Eng. Sci.* **2009**, *64*, 4856–4865. [CrossRef]

13. Iulianelli, A.; Liguori, S.; Wilcox, J.; Basile, A. Advances on methane steam reforming to produce hydrogen through membrane reactors technology: A review. *Catal. Rev.* **2016**, *58*, 1–35. [CrossRef]

14. Mendes, D.; Mendes, A.; Madeira, L.M.; Iulianelli, A.; Sousa, J.M.; Basile, A. The water-gas shift reaction: From conventional catalytic systems to Pd-based membrane reactors—A review. *Asia-Pac. J. Chem. Eng.* **2010**, *5*, 111–137. [CrossRef]

15. Basile, F.; Fasolini, A.; Lombardi, E. CHAPTER 7 Membrane Processes for Pure Hydrogen Production from Biomass. In *Membrane Engineering for the Treatment of Gases: Volume 2: Gas-Separation Issues Combined with Membrane Reactors (2)*; The Royal Society of Chemistry: Cambridge, UK, 2018; Volume 2, pp. 212–246. ISBN 978-1-78262-875-0.

16. Basile, F.; Fornasari, G.; Gazzano, M.; Vaccari, A. Rh, Ru and Ir catalysts obtained by HT precursors: Effect of the thermal evolution and composition on the material structure and use. *J. Mater. Chem.* **2002**, *12*, 3296–3303. [CrossRef]

17. Wang, S.; Lu, G.Q.; Millar, G.J. Carbon Dioxide Reforming of Methane to Produce Synthesis Gas over Metal-Supported Catalysts: State of the Art. *Energy Fuels* **1996**, *10*, 896–904. [CrossRef]

18. Hernandez, A.D.; Kaisalo, N.; Simell, P.; Scarsella, M. Effect of H2S and thiophene on the steam reforming activity of nickel and rhodium catalysts in a simulated coke oven gas stream. *Appl. Catal. B Environ.* **2019**, *258*, 117977. [CrossRef]

19. Qi, A.; Wang, S.; Ni, C.; Wu, D. Autothermal reforming of gasoline on Rh-based monolithic catalysts. *Int. J. Hydrog. Energy* **2007**, *32*, 981–991. [CrossRef]

20. Wang, Y.; Chin, Y.H.; Rozmiarek, R.T.; Johnson, B.R.; Gao, Y.; Watson, J.; Tonkovich, A.Y.L.; Vander Wiel, D.P. Highly active and stable Rh/MgOAl2O3 catalysts for methane steam reforming. *Catal. Today* **2004**, *98*, 575–581. [CrossRef]

21. Kikuchi, E.; Tanaka, S.; Yamazaki, Y.; Morita, Y. Steam Reforming of Hydrocarbons on Noble Metal Catalysts (Part 1). *Bull. Jpn. Pet. Inst.* **1974**, *16*, 95–98. [CrossRef]

22. Martinengo, S.; Chini, P.; Giordano, G. Improved synthesis of dodecacarbonyltetrarhodium at atmospheric pressure. *J. Organomet. Chem.* **1971**, *27*, 389–391. [CrossRef]

23. Ojima, I.; Donovan, R.J.; Clos, N. Rhodium and cobalt carbonyl clusters Rh4 (CO) 12, Co2Rh2 (CO) 12, and Co3Rh (CO) 12 as effective catalysts for hydrosilylation of isoprene, cyclohexanone, and cyclohexenone. *Organometallics* **1991**, *10*, 2606–2610. [CrossRef]

24. Li, C.; Widjaja, E.; Garland, M. Rh4 (CO) 12-catalyzed hydroformylation of cyclopentene promoted with HMn (CO) 5. Another example of Rh4 (CO) 12/HMn (CO) 5 bimetallic catalytic binuclear elimination. *Organometallics* **2004**, *23*, 4131–4138. [CrossRef]

25. Basini, L.; Guarinoni, A.; Aragno, A. Molecular and Temperature Aspects in Catalytic Partial Oxidation of Methane. *J. Catal.* **2000**, *190*, 284–295. [CrossRef]

26. Grunwaldt, J.-D.; Basini, L.; Clausen, B.S. In Situ EXAFS Study of Rh/Al2O3 Catalysts for Catalytic Partial Oxidation of Methane. *J. Catal.* **2001**, *200*, 321–329. [CrossRef]

27. Basini, L.; Marchionna, M.; Aragno, A. Drift and mass spectroscopic studies on the reactivity of rhodium clusters at the surface of polycrystalline oxides. *J. Phys. Chem.* **1992**, *96*, 9431–9441. [CrossRef]

28. Basile, F.; Mafessanti, R.; Fasolini, A.; Fornasari, G.; Lombardi, E.; Vaccari, A. Effect of synthetic method on CeZr support and catalytic activity of related Rh catalyst in the oxidative reforming reaction. *J. Eur. Ceram. Soc.* **2019**, *39*, 41–52. [CrossRef]

29. Xu, Y.; Harimoto, T.; Wang, L.; Hirano, T.; Kunieda, H.; Hara, Y.; Miyata, Y. Effect of steam and hydrogen treatments on the catalytic activity of pure Ni honeycomb for methane steam reforming. *Chem. Eng. Process. -Process Intensif.* **2018**, *129*, 63–70. [CrossRef]

30. Iglesias, I.; Baronetti, G.; Alemany, L.; Mariño, F. Insight into Ni/Ce1−xZrxO2−δ support interplay for enhanced methane steam reforming. *Int. J. Hydrog. Energy* **2019**, *44*, 3668–3680. [CrossRef]

31. Goula, M.A.; Charisiou, N.D.; Siakavelas, G.; Tzounis, L.; Tsiaoussis, I.; Panagiotopoulou, P.; Goula, G.; Yentekakis, I.V. Syngas production via the biogas dry reforming reaction over Ni supported on zirconia modified with CeO2 or La2O3 catalysts. *Int. J. Hydrog. Energy* **2017**, *42*, 13724–13740. [CrossRef]

32. Tabanelli, T.; Paone, E.; Blair Vásquez, P.; Pietropaolo, R.; Cavani, F.; Mauriello, F. Transfer Hydrogenation of Methyl and Ethyl Levulinate Promoted by a ZrO2 Catalyst: Comparison of Batch vs Continuous Gas-Flow Conditions. *ACS Sustain. Chem. Eng.* **2019**, *7*, 9937–9947. [CrossRef]

33. Chuah, G.K. An investigation into the preparation of high surface area zirconia. *Catal. Today* **1999**, *49*, 131–139. [CrossRef]

34. Chuah, G.K.; Jaenicke, S.; Cheong, S.A.; Chan, K.S. The influence of preparation conditions on the surface area of zirconia. *Appl. Catal. A Gen.* **1996**, *145*, 267–284. [CrossRef]

35. Srinivasan, R.; Harris, M.B.; Simpson, S.F.; Angelis, R.J.D.; Davis, B.H. Zirconium oxide crystal phase: The role of the pH and time to attain the final pH for precipitation of the hydrous oxide. *J. Mater. Res.* **1988**, *3*, 787–797. [CrossRef]

36. Song, W.; Popa, C.; Jansen, A.P.J.; Hensen, E.J.M. Formation of a Rhodium Surface Oxide Film in Rhn/CeO2(111) Relevant for Catalytic CO Oxidation: A Computational Study. *J. Phys. Chem. C* **2012**, *116*, 22904–22915. [CrossRef]

37. Velasquez, M.; Batiot-Dupeyrat, C.; Gallego, J.; Santamaria, A. Chemical and morphological characterization of multi-walled-carbon nanotubes synthesized by carbon deposition from an ethanol–glycerol blend. *Diam. Relat. Mater.* **2014**, *50*, 38–48. [CrossRef]

38. Charisiou, N.D.; Tzounis, L.; Sebastian, V.; Hinder, S.J.; Baker, M.A.; Polychronopoulou, K.; Goula, M.A. Investigating the correlation between deactivation and the carbon deposited on the surface of Ni/Al2O3 and Ni/La2O3-Al2O3 catalysts during the biogas reforming reaction. *Appl. Surf. Sci.* **2019**, *474*, 42–56. [CrossRef]

39. Carrero, A.; Calles, J.; Vizcaíno, A. Effect of Mg and Ca addition on coke deposition over Cu–Ni/SiO2 catalysts for ethanol steam reforming. *Chem. Eng. J.* **2010**, *163*, 395–402. [CrossRef]

40. Galetti, A.E.; Gomez, M.F.; Arrúa, L.A.; Abello, M.C. Hydrogen production by ethanol reforming over NiZnAl catalysts: Influence of Ce addition on carbon deposition. *Appl. Catal. A Gen.* **2008**, *348*, 94–102. [CrossRef]

41. Fornasiero, P.; Kašpar, J.; Sergo, V.; Graziani, M. Redox Behavior of High-Surface-Area Rh-, Pt-, and Pd-Loaded Ce0.5Zr0.5O2Mixed Oxide. *J. Catal.* **1999**, *182*, 56–69. [CrossRef]

42. Mamontov, E.; Brezny, R.; Koranne, M.; Egami, T. Nanoscale Heterogeneities and Oxygen Storage Capacity of Ce0.5Zr0.5O2. *J. Phys. Chem. B* **2003**, *107*, 13007–13014. [CrossRef]

43. Lemaux, S.; Bensaddik, A.; van der Eerden, A.M.J.; Bitter, J.H.; Koningsberger, D.C. Understanding of Enhanced Oxygen Storage Capacity in Ce0.5Zr0.5O2: The Presence of an Anharmonic Pair Distribution Function in the Zr–O2 Subshell as Analyzed by XAFS Spectroscopy. *J. Phys. Chem. B* **2001**, *105*, 4810–4815. [CrossRef]

44. Luo, M.-F.; Zheng, X.-M. Redox behaviour and catalytic properties of Ce0.5Zr0.5O2-supported palladium catalysts. *Appl. Catal. A Gen.* **1999**, *189*, 15–21. [CrossRef]

45. Ligthart, D.; Van Santen, R.; Hensen, E. Influence of particle size on the activity and stability in steam methane reforming of supported Rh nanoparticles. *J. Catal.* **2011**, *280*, 206–220. [CrossRef]

46. Kusakabe, K.; Sotowa, K.-I.; Eda, T.; Iwamoto, Y. Methane steam reforming over Ce–ZrO2-supported noble metal catalysts at low temperature. *Fuel Process. Technol.* **2004**, *86*, 319–326. [CrossRef]

47. Angeli, S.D.; Turchetti, L.; Monteleone, G.; Lemonidou, A.A. Catalyst development for steam reforming of methane and model biogas at low temperature. *Appl. Catal. B Environ.* **2016**, *181*, 34–46. [CrossRef]
48. Fasolini, A.; Abate, S.; Barbera, D.; Centi, G.; Basile, F. Pure H2 production by methane oxy-reforming over Rh-Mg-Al hydrotalcite-derived catalysts coupled with a Pd membrane. *Appl. Catal. A Gen.* **2019**, *581*, 91–102. [CrossRef]

Article

Methane and Ethane Steam Reforming over MgAl₂O₄-Supported Rh and Ir Catalysts: Catalytic Implications for Natural Gas Reforming Application

Johnny Saavedra Lopez, Vanessa Lebarbier Dagle, Chinmay A. Deshmane, Libor Kovarik, Robert S. Wegeng and Robert A. Dagle *

Institute for Integrated Catalysis, Pacific Northwest National Laboratory, 902 Battelle Blvd, Richland, WA 99352, USA

* Correspondence: Robert.Dagle@pnnl.gov

Received: 13 August 2019; Accepted: 19 September 2019; Published: 25 September 2019

Abstract: Solar concentrators employed in conjunction with highly efficient micro- and meso-channel reactors offer the potential for cost-effective upgrading of the energy content of natural gas, providing a near-term path towards a future solar-fuel economy with reduced carbon dioxide emissions. To fully exploit the heat and mass transfer advantages offered by micro- and meso-channel reactors, highly active and stable natural gas steam reforming catalysts are required. In this paper, we report the catalytic performance of MgAl₂O₄-supported Rh (5 wt.%), Ir (5 wt.%), and Ni (15 wt.%) catalysts used for steam reforming of natural gas. Both Rh- and Ir-based catalysts are known to be more active and durable than conventional Ni-based formulations, and recently Ir has been reported to be more active than Rh for methane steam reforming on a turnover basis. Thus, the effectiveness of all three metals to perform natural gas steam reforming was evaluated in this study. Here, the Rh- and Ir-supported catalysts both exhibited higher activity than Ni for steam methane reforming. However, using simulated natural gas feedstock (94.5% methane, 4.0% ethane, 1.0% propane, and 0.5% butane), the Ir catalyst was the least active (on a turnover basis) for steam reforming of higher hydrocarbons (C₂₊) contained in the feedstock when operated at <750 °C. To further investigate the role of higher hydrocarbons, we used an ethane feed and found that hydrogenolysis precedes the steam reforming reaction and that C–C bond scission over Ir is kinetically slow compared to Rh. Catalyst durability studies revealed the Rh catalyst to be stable under steam methane reforming conditions, as evidenced by two 100-hour duration experiments performed at 850 and 900 °C (steam to carbon [S/C] molar feed ratio = 2.0 mol). However, with the natural gas simulant feed, the Rh catalyst exhibited catalyst deactivation, which we attribute to coking deposits derived from higher hydrocarbons contained in the feedstock. Increasing the S/C molar feed ratio from 1.5 to 2.0 reduced the deactivation rate and stable catalytic performance was demonstrated for 120 h when operated at 850 °C. However, catalytic deactivation was observed when operating at 900 °C. While improvements in steam reforming performance can be achieved through choice of catalyst composition, this study also highlights the importance of considering the effect of higher hydrocarbons contained in natural gas, operating conditions (e.g., temperature, S/C feed ratio), and their effect on catalyst stability. The results of this study conclude that a Rh-supported catalyst was developed that enables very high activities and excellent catalytic stability for both the steam reforming of methane and other higher hydrocarbons contained in natural gas, and under conditions of operation that are amendable to solar thermochemical operations.

Keywords: heterogeneous catalysis; syngas production; solar thermochemical; iridium catalyst; rhodium catalyst

---

1. Introduction

Approximately 95% of industrial hydrogen in the United States is currently produced through natural gas reforming [1,2]. The product of natural gas reforming is a mixture of H_2, CO, and CO_2, (referred to as syngas), which is a versatile and useful feedstock for producing a number of industrially relevant chemical commodities such as methanol, dimethyl ether [3], and Fischer-Tropsch products [4,5]. Natural gas reforming involves a series of reactions:

$$CH_4 + H_2O \leftrightarrow CO + 3H_2 \tag{1}$$

$$C_nH_m + nH_2O \rightarrow nCO + \left(\frac{m + 2n}{2}\right)H_2 \tag{2}$$

$$CO + H_2O \leftrightarrow CO_2 + H_2 \tag{3}$$

Reaction 1 is the steam methane reforming (SMR) reaction, which is endothermic and therefore requires high temperatures (e.g., 600–850 °C) for operation. Reaction 2 is the reaction for steam reforming of higher hydrocarbons (C_{2+}). Both the SMR (Reaction 1) and water-gas-shift (Reaction 3) reactions are subject to thermodynamic equilibrium. Supported metal catalysts (e.g., Rh, Ru, and Ni metals supported on alumina or alumina spinel type formulations) are typically employed for steam reforming reactions [6,7].

Solar concentrators are a prospective method to provide the thermal requirements necessary to carry out the reforming process, with the added advantage of reduced carbon dioxide emissions since fossil fuel is not used to generate process heat [8–10]. Pacific Northwest National Laboratory has demonstrated an integrated solar thermochemical reaction system that combines solar concentrators with micro- and meso-channel reactors and heat exchangers that accomplish more than 20% solar augment of methane higher heating value. Further, a solar-to-chemical energy conversion efficiency slightly over 70% has been achieved [8]. At PNNL, the solar concentrators being utilized include those that provide concentration ratios high enough to achieve the temperatures required for methane steam reforming. These include parabolic dish concentrators and central receivers [8]. PNNL solar reformers typically operate in the 700–800 °C temperature range—with some portions of the reactor falling to less than 700 °C near the entry point to the reaction channel. PNNL solar reformers start reacting at modestly high temperatures (>600 °C) and exit the catalyst zone at substantiality at higher temperatures [8]. Typical steam/carbon (S/C) molar feed ratios utilized are in the 2.5–3.0 range. We routinely achieve 90% or greater conversion in on-sun testing under these conditions [8]. Catalysts used in these highly efficient compact reactors must meet the following requirements:

- They must be active over a wide range of operating temperatures, which are largely dictated by efficiency of the solar collector and local solar irradiance.
- To facilitate small reactor footprints, they must operate at high throughput rates.
- They must work under continuous shutdown cycles (i.e., sun/shade).
- To reduce energy requirements for steam vaporization, they must operate with reduced steam concentrations (e.g., low S/C molar ratios).
- They must tolerate temperature and gas composition changes caused by abrupt variations in local conditions.

Conventional Ni-based steam reforming catalysts have drawbacks pertaining to activity and durability that make them unsuitable for solar-driven applications. Noble metal-based catalysts, while generally more expensive than Ni-based catalysts, offer higher catalytic activities that enable faster throughput rates and smaller reactor hardware [11]. Furthermore, catalyst durability is improved, with reduced sintering and deactivation by coke when operating at the high temperatures required for typical operation [9]. Additionally, we note how, unlike with conventional fixed-bed reactor systems, with microreactors the use of more expensive precious metal catalysts may be made economical. Monolith-type substrates utilizing highly active catalysts are integrated in microreactors to minimize heat- and mass-transfer resistances and maximize catalyst efficiency. The use of smaller, more efficient systems may compensate for the higher cost of the catalyst material [12].

SMR catalysts commonly employ Ni or Rh metals and use alumina or alumina spinel (e.g., $MgAl_2O_4$) type supports [13–15]. SMR is a structure-sensitive reaction in which turnover rates increase with decreasing metal particle size [16,17]. The support choice influences both the resulting metal particle size and catalyst stability [17,18]. Facilitating increased metal dispersion and enabling improved stability are among the reasons for using $MgAl_2O_4$ over other common supports (e.g., Al_2O_3, SiO_2, and ZrO_2). Noble metal based catalysts are known to be more resistant to carbon fouling and metal sintering compared to Ni [19–21]. We have previously reported $MgAl_2O_4$-supported Rh and Ir catalysts to have improved activity and stability compared to Ni and other metals (e.g., Pd, Pt) [20]. Catalysts with very well dispersed Rh (2 nm) and Ir (1 nm) metal clusters were obtained with the use of a $MgAl_2O_4$ support. High dispersion was maintained even after high temperature operation (e.g., 850 °C) and with the use of high metal loadings (e.g., 5–10 wt.%). Taken together, well dispersed Rh and Ir catalysts with high metal loadings were previously reported by our group to be highly active and stable for the SMR reaction.

While methane is the predominant compound in natural gas, natural gas also contains up to 20 vol.% higher C_{2+} constituents, with the exact composition depending on the source [22,23]. The activation of methane requires a higher temperature than required for activation of C_{2+} natural gas components [24]. Thus, C_{2+} hydrocarbons are typically converted more readily when compared to methane. However, C_{2+} hydrocarbons facilitate coke formation [25,26]. Regardless of the hydrocarbon used, upon reaction the metal and metal oxide support surfaces are populated by a variety of reactive species (e.g., C*, H*, CH_x*, CO*, O*, OH*) [8,27–33]. While SMR has been studied extensively, relatively few studies have been dedicated to studying its reaction mechanism when higher hydrocarbons are present in the feed, which would be more representative of real feed mixtures [19,21,31,34,35]. The presence of higher hydrocarbons facilitate catalyst deactivation [25,26]. Previously, we investigated steam reforming of biomass gasifier-derived hydrocarbons, which includes tar (polyaromatic hydrocarbons) species. We evaluated steam reforming of benzene, as a model tar species, over $MgAl_2O_4$-supported Rh and Ir catalysts. The Rh catalyst was more active than Ir on a turnover basis due to differences in the C–C bond breaking step (which was found to be rate limiting) [19].

The objective of this study is to assess the catalytic performance of Rh-, Ir-, and Ni-supported catalysts for steam reforming of natural gas. Both Rh- and Ir-based catalysts are known to be more active and durable than conventional Ni-based formulations, and recently Ir has been reported to be more active than Rh for methane steam reforming on a turnover basis [20,36]. Thus, the effectiveness of all three metals to perform natural gas steam reforming was evaluated in this study. Steam reforming kinetic and mechanistic comparisons were elucidated using both discrete methane and ethane feeds. While the catalytic performance of various precious metal based catalysts (i.e., Rh, Ir, and Ni) have been widely studied for the methane steam reforming reaction, we are not aware of any studies focused on their comparison when under ethane steam reforming conditions. Finally, we comparatively assess both activity and stability when using a simulant natural gas feedstock that was a mixture of methane and C_{2+} hydrocarbons. We note that in order to investigate the kinetics of the catalysts we are operating at higher throughputs and lower conversions than what are typically utilized in solar thermochemical application.

2. Results and Discussion

The performance of $MgAl_2O_4$-supported Rh, Ir, and Ni catalysts was investigated for steam reforming of natural gas under industrially relevant conditions. The methods explored included the use of (1) simulated natural gas feedstock, (2) separate methane and ethane model feeds to elucidate contributions from each, (3) catalyst durability tests, and (4) evaluation over a wide temperature range. These conditions are all of particular interest for solar thermochemical applications where highly active and durable catalysts are required for a range of conditions.

2.1. Catalyst Characterization

Catalyst physical characterization details are presented in Table 1. These MgAl$_2$O$_4$-supported metal catalysts have been characterized in our previous publications, and used as catalysts for steam reforming of methane [20], gasifier-derived hydrocarbons including benzene [19] and complex mixtures including tar [21], and biomass-derived ethylene glycol [37] and aqueous products produced via fast pyrolysis [38]. It was reported that the metals form small and stable clusters when supported on MgAl$_2$O$_4$ (approximately 1, 2, and 7 nm Ir, Rh, and Ni median cluster sizes, respectively) even under steam reforming conditions at 850 °C [20,21]. In this study, we report additional catalyst characterization include TPR as well as the hydrogen adsorption over the temperature range of interest for natural gas steam reforming (i.e., 600–850 °C).

Table 1. Catalyst physiochemical characterization. Scanning transmission electron microscopy (STEM) images for the Ir and Rh catalysts are included in the supplementary information (Figure S1).

Catalyst	Metal Loading (wt.%)	Metal Particle Size TEM (nm)	Catalyst Surface Area BET [a] (m^2/g)
5Ir/MgAl$_2$O$_4$ [20]	5.0	1.0	133
5Rh/MgAl$_2$O$_4$ [20]	5.0	2.0	117
15Ni/MgAl$_2$O$_4$ [21]	15.0	6.6	86

[a] Brunauer–Emmett–Teller method.

Figure 1 shows TPR profiles for the Rh-, Ir-, and Ni-supported catalysts used in this study. Results indicate an easier reduction for Rh with a single reduction peak at 112 °C. Ni requires the highest temperature to reach a similar reduction level (peak centered at 725 °C) and shows evidence of at least two reduction stages (shown by overlapping peaks). The Ir catalyst shows several reduction stages at 168, 220, 253, and 613 °C. For all the three catalysts (Rh, Ir, and Ni) after 800 °C was reached, no further hydrogen was chemisorbed, indicating a total (or near total) reduction of the metal clusters. However, the temperature needed to produce reduced metal clusters (measured by hydrogen consumption) varied significantly. Results in the literature suggest Rh disperses very well on a wide variety of supports, with low temperature single peaks being characteristic for Rh [39,40], thus corroborating our results. The several reduction stages required for Ir can be ascribed to a broader particle size distribution or a broad distribution of oxidation states. Reports in the literature indicate that smaller Ir particles require a higher reduction temperature compared to those for larger particles [41,42]. Nevertheless, the reduction of Ir-based catalysts is a relatively complex phenomenon that largely depends on the processing conditions and thermal history of each sample (Ir reduction is an autocatalytic process and IrO$_x$ are volatile species detrimental to metal dispersion) [43]. The Ni catalyst TPR profile also is consistent with literature results for catalysts with a similar particle size (6 nm) [44,45]. Before reaction catalysts are reduced at 850 °C for 16 h in order to ensure a degree of reduction for the metal clusters. We note that in our prior reports we have shown how the MgAl$_2$O$_4$-supported Ir, Rh, and Ni clusters are fully reduced when the catalyst was reduced under hydrogen at 850 °C. However, we note that the total reduction cannot be determined, especially for Ni (because of its lower dispersion), when under operando conditions.

H$_2$ activation, adsorption strength relative to carbon, and coverage on metal surfaces are important metrics for steam reforming relevant reactions [33,41,46]. However, obtaining accurate characterization under operando conditions presents technical and experimental challenges. In an attempt to study H$_2$ adsorption capacity relative to metal active area, we combined volumetric H$_2$ adsorption at 600 °C (as representative of reforming conditions) with metal dispersion, as determined from metal particle sizes revealed by the STEM imaging (2 nm, 1 nm, and 7 nm for the Rh, Ir, and Ni, respectively) [20,21]. The H$_2$ adsorption data presented in Figure 2 was calculated as H-coverage evolution (mol$_H$/ mol$_{Msurf}$, M= Rh, Ir, Ni) for accumulated H$_2$ added (by pulses) into the adsorption cell (μmol). These adsorption results indicate that bare Rh and Ir surfaces adsorb H$_2$ completely until saturation, which is indicated by the inflexion point and the asymptotical part of the plots (after adding ~40 and ~20 mmol of H_2, respectively). Saturation is reached at coverages of H-M_{surf} of 1 and 0.7 for Rh, and Ir, respectively, which indicates a surface stoichiometry of 1:1 (H:Rhs_{urf}) and 0.7:1 (H:Ir_{Surf}). In the case of Ni, the adsorption pattern is dissimilar in three main aspects: (1) H-uptake is low (0.2 mol_H/mol_{Nisurf}) compared to Rh or Ir, (2) there is no clear inflexion point for the adsorption curve, and (3) the calculated stoichiometry H-Ni_{surf} is low (only 20% of surface nickel atoms adsorb hydrogen). H_2-uptake at 800 °C is lower compared to the uptake at 600 °C (data shown Figure S2). At 800 °C, hydrogen saturates the metal surfaces at 1:1 (Rh), 0.5:1 (Ir), and 0.08:1 (Ni).

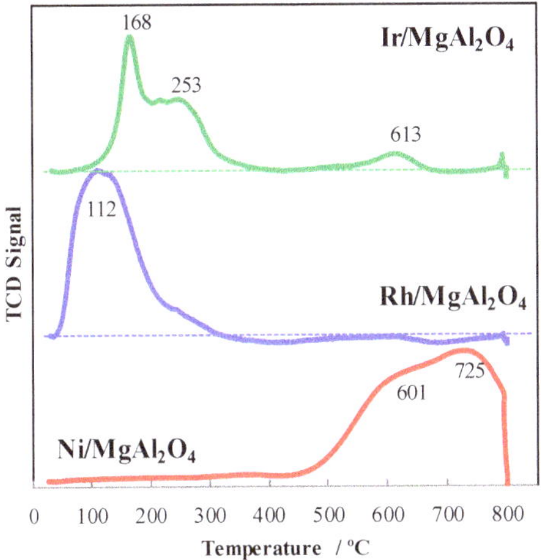

Figure 1. Temperature programmed reduction (TPR) profiles for $MgAl_2O_4$-supported Ir, Rh, and Ni catalysts. Prior to TPR the catalysts were calcined at 500 °C for 4 h. TPR conditions were 50 mg of catalyst, temperature ramping rate 5 °C/min, feed rate of 50 mL/min of 5% H_2/Ar gas.

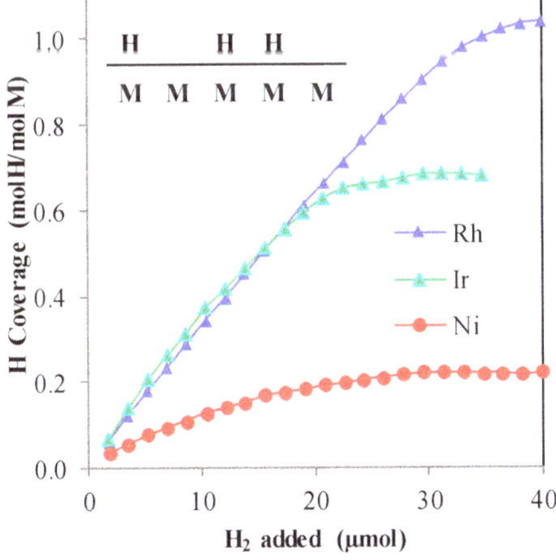

Figure 2. Hydrogen surface saturation curves over $MgAl_2O_4$-supported Rh (blue), Ir (green), and Ni (red) catalysts at 600 °C. Coverage was calculated using accumulated H_2-uptake and metal dispersion data obtained from STEM imaging (Table 1). 50 mg of catalyst was calcined before analysis and reduced at the conditions used in the catalytic experiments (16 h at 850 °C, 100 mL/min, 10% H_2 in N_2). H_2 was pulsed using a 100 µL gas loop.

The physicochemical characterizations of the catalysts shown in this section are significant for the following reasons:

- STEM characterization shows that high dispersion that can be achieved with Rh and Ir catalysts supported on $MgAl_2O_4$ compared to Ni.
- TPR characterization showed that, associated with a greater dispersion, reducibility is enhanced, particularly for Rh where TPR profiles indicate a narrower particle size distribution and full reduction at low temperatures (broad single peak centered at 112 °C).
- H_2 adsorption demonstrated that surface stoichiometry for H_2 on metal surfaces can vary greatly under reforming conditions.

We found that while full surface coverage with a stoichiometry 1:1 H-Rh$_{surf}$ can be achieved at reforming temperatures in the 600–800 °C range for the Rh catalyst, this coverage is significantly lower for Ir (0.7:1 H-Ir$_{surf}$) and even lower for Ni (0.2:1 H-Ni$_{surf}$).

2.2. Steam Reforming Activity Comparison Using Natural Gas Simulant

Steam reforming using a simulant gas mixture representative of natural gas (94.5% methane, 4% ethane, 1% propane, 0.5% butane) was performed over $MgAl_2O_4$-supported 5% Ir and 5% Rh catalysts. Conversion of methane and ethane is plotted separately and shown in Figure 3 (propane conversion in Figure S3). Short contact times (τ = 4.5 ms) favor low conversion levels. In the 600–875 °C temperature range, methane conversion follows a linear trend for both catalysts (Figure 3a). In spite of similar methane conversion, conversion of ethane (Figure 3b) and propane (Figure S3) is significantly lower for the Ir catalyst in the lower temperature range (<750 °C). These differences in activity are important for two main reasons: (1) gas feeds with higher concentration of C_{2+} hydrocarbons lower the overall performance of Ir-based catalysts (more hydrocarbons remain unreacted) and (2) hampered hydrocarbon hydrogenolysis/reforming activity is an indication that the underlying mechanism is different for Ir compared to Rh in spite of similar methane reforming activity. We note that the methane conversions presented in Figure 3a are below those of equilibrium (with the equilibrium conversions indicated with the dotted line). We also note that the equilibrium ethane conversion shown in Figure 3b is near completion for the entire range of conditions investigated.

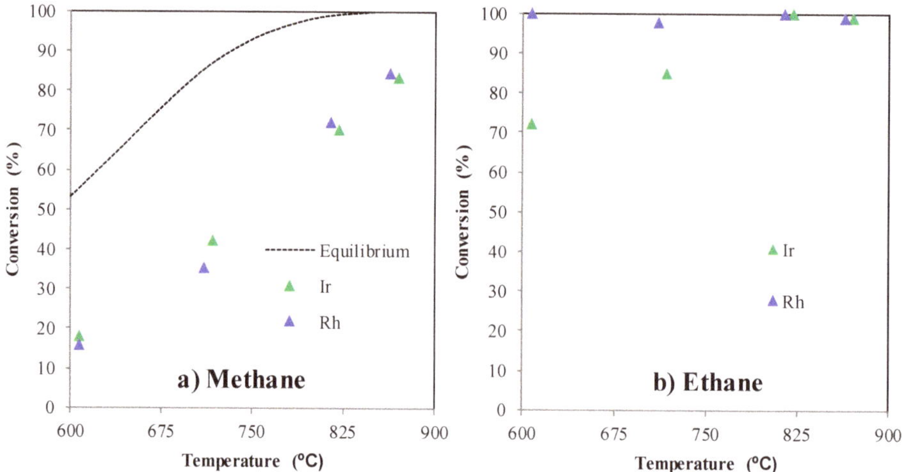

Figure 3. Comparison of (**a**) methane conversion and (**b**) ethane conversion for the Rh and Ir catalysts when steam reforming using natural gas simulant feedstock (S/C = 1.5 mol, τ = 4.5 ms).

We note that H_2/CO ratio is an important metric for reforming catalysts when the steam reforming reaction is integrated to downstream processes. In practice, several strategies can be used to tune the H_2/CO ratio of the reforming product (e.g., decreased S/C ratio during SMR, CO_2 recirculation, incorporating an additional WGS step) towards reactions of interest (e.g., Fischer–Trospch). The H_2/CO ratio for $RhMgAl_2O_4$ was evaluated over a wide range of conversions (20%–100%) at the range of operative temperatures of SMR in solar thermochemical applications (600–800 °C) by changing the contact time of the methane stream (for τ values from 1.2 to 10 ms) at a constant S/C ratio (S/C = 3). Results for conversion vs. temperature, and H_2/CO ratio vs. conversion are presented in the supplementary information as Figure S8. Interestingly, Figure S8 shows that independently on the temperature, H_2/CO ratio decreases exponentially with methane conversion reaching a minimal value of 5 at the highest conversion (ca. 100%). This will lead to the conclusion that WGS equilibrium determines the resulting H_2/CO ratio. As the H_2 partial pressure increases as an effect of higher reforming conversion, more CO is formed through the reverse water gas shift reaction (RWGS, $CO_2 + H_2 \rightarrow CO + H_2O$).

2.3. Methane and Ethane Steam Reforming Activity Comparison

Based on the results described above, the Ir catalyst shows lower activity than Rh towards the steam reforming of higher hydrocarbons. With the aim of gaining knowledge on fundamental differences between the different hydrocarbon constituents, we performed a set of separate experiments using either methane or ethane feed. Figure 4 shows the metal surface normalized catalytic rate ($moles_{CH4}/mol_{metal}*s$) for the steam reforming of methane and ethane for the Rh, Ir, and Ni supported catalysts as a function of temperature. These results show that for the Ir catalyst, the ethane steam reforming rates (Figure 4b) are significantly lower compared to methane steam reforming rates (Figure 4a). For the Rh and Ni catalysts, ethane conversion is more facile than methane conversion.

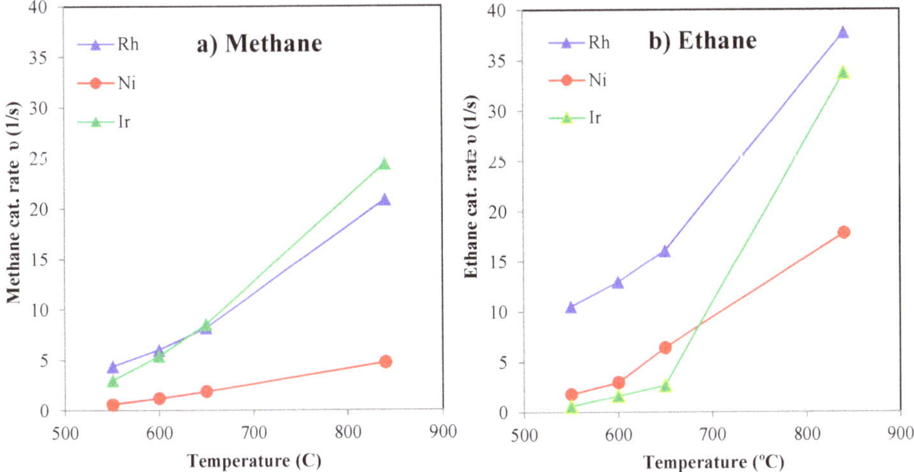

Figure 4. Comparison of turnover steam reforming rates for the Rh, Ir, and Ni catalysts using (**a**) methane, and (**b**) ethane feeds (S/C = 3, τ = 30 ms; methane feed = 15.9 vol%, ethane feed = 14.8 vol%).

Over typical metal supported catalysts activation of the initial C–H bond of methane is more difficult than for ethane [47]. Dehydrogenation of the initial C–H and the creation of C radical + H for ethane is slightly more favorable than for methane (C–H bond strength 101 vs. 105 kcal/mol, respectively). Further, the C–C bond of ethane is much weaker (90.1 kcal/mol) [48]. These differences between methane and ethane typically make ethane more reactive than methane. Few literature studies can be used to make a direct comparison or to determine trends with respect to reactivity of methane vs. ethane reforming because most studies use methane as the only model reactant. Specific studies on ethane reforming have yielded contradictory results depending on the metal used as the catalyst. On one hand, Schädel et al. [35] showed that over Rh, ethane reacts faster than methane, which is consistent with our findings for Rh and Ni catalysts. On the other hand, Graf et al. [34] reported that over Pt, methane reforming occurs the fastest, which is similar to our findings for Ir. Similar trends, where ethane reactivity is higher compared to methane, have been found in other metal-catalyzed reactions such as ethane combustion [49] and ethane hydrogenolysis over Pt and Pd [50]. Interestingly, Figure 3b also shows that the Ir catalyst is relatively inactive at low temperatures but then the catalytic rates quickly accelerate at higher temperatures (>750 °C). This rapid transition from 'inactive' to 'active' prompted our study to pursue a more detailed understanding for ethane reactivity over Ir catalyst. The following section is dedicated to hypothesizing the possible reason(s) behind the low reactivity of ethane over Ir catalysts at temperatures lower than 750 °C.

2.4. Ethane Steam Reforming over Rh vs. Ir: Mechanistic Insights

Results shown in Figure 4 indicate that over Ir catalyst at 600 °C, surface normalized ethane reforming rates (0.6 s^{-1}) are much slower compared to methane reforming rates (3 s^{-1}). This difference in catalytic rate could be attributed to several factors: (1) potential formation of carbon deposits over Ir surface when ethane is fed, (2) decreased number of active sites due to competitive adsorption (e.g., ethyl, O, OH, H, C_1, and C_2 species), and the (3) chemical identity of the metal and its bonding characteristics. To compare hydrocarbon reactivity over metal surfaces, catalytic measurements should be performed under the same coverage [51]; however, this comparison is not always easy to achieve. As a simplified approach to compare ethane reforming activity over Ir vs. Rh, we performed catalytic measurements at 600 °C using the same concentration of ethane in the gas over similar amounts of metal exposed (Table 2). The average Rh cluster size (2 nm) was twice that of the Ir cluster (1 nm); however, the Ir molecular weight (192 g/mol) is twice that of Rh (103 g/mol). Thus, the amount of exposed metal is similar for both catalysts on a molar basis. With similar number of exposed metal sites, the observed product selectivity and associated reaction rates should be determined by the underlying chemical reaction mechanism.

Table 2. Catalytic results for ethane steam reforming over $MgAl_2O_4$-supported Rh and Ir catalysts at 600 °C (S/C = 2.75 mol, τ = 28.3 ms).

Condition	5%Rh/MgAl$_2$O$_4$	5%Ir/MgAl$_2$O$_4$
Metal on catalyst surface (μmol) [a]	23.5	25.2
Conversion after 1 h	60.0	8.3
Conversion after 2 h	58.9	8.1
Deactivation rate (%)	1.8	2.4
ppm ethylene formed	50	2000
Selectivity towards methane (mol C %)	32	9.6
Carbon on spent catalyst (wt.%)	bdl [b]	2.8 [c]

[a] Based on TEM particle size measurements. [b] Below limit of detection of CNHS elemental analysis instrument (0.3%). [c] 0.28% measured for the dilution of 1:10 catalyst:Al$_2$O$_3$.

Shown in Table 2 are ethane steam reforming results using the Rh- and Ir-supported catalysts. For both catalysts steady state is reached after 15 min, and catalytic activity is reported after 1 h and 2 hours' time-on-stream (see Figure S4). The main reaction products detected by gas chromatography analysis include H_2, CO_2, CO, methane, and ethylene. Reactivity of ethane over Rh is different than Ir in four main aspects: (1) conversion is significantly higher (60% vs. 8%), over a similar exposed metal surface area, (2) ethylene formation is markedly lower (50 ppm vs. 2000 ppm), (3) post mortem analysis showed significant less accumulated carbon, and (4) methane is the main reaction product (methane selectivity of 32% vs. 9.6%).

Methane is the primary hydrocarbon product over Rh when under ethane steam reforming conditions. Methane can be formed as a hydrogenation product of CO or CO_2 [33] or as a product of ethane hydrogenolysis (according to the Sinfelt–Taylor mechanism) [52,53]. These two reactions are related and depend on the metal's carbon/hydrogen adsorption characteristics and its ability to breakdown adsorbed C_2 radicals [54]. In Section 3.1 we reported marked differences in H_2 adsorption and coverage for the Rh catalyst vs. Ir or Ni. Although the H_2 adsorption experiment does not allow comparison for H/C adsorption, they demonstrate improved H retention for the Rh catalyst. This is a particularly useful property that favors the aforementioned methane formation mechanisms (hydrogenation of CO/CO_2 species and hydrogenolysis of ethane) under reforming conditions. Parallel to methane formation, the presence of ethylene product is an indicator of formation (and slow decomposition) of ethyl radicals on the catalyst surface. Higher formation of ethylene over the Ir catalyst is a strong indicator that the overall catalytic ethane reforming rate is hampered by a slow C–C bond scission capacity compared to Rh.

The marked difference in conversion for the Ir and Rh catalysts shown in Table 2 makes it difficult to fairly compare product selectivity. Conversion for the Ir catalyst was increased, by increasing the contact time (from 28 to 166 ms), to compare product selectivity at similar conversion levels. As shown in Figure 5, under these conditions methane is the main reaction product for both catalysts (60% vs. 30% methane selectivity for Ir and Rh, respectively), with CO and CO_2 also being produced as main products, and ethylene being produced via incomplete decomposition of ethane. Figure 5c shows changes in selectivity as a function of change in contact time over the Ir catalyst. Methane selectivity increases as ethane conversion increases (selectivity towards methane increases from 8 to 31% for ethane conversion increasing from 8 to 50%). Interestingly, selectivity towards ethylene formation decreases as the conversion increases (2 to 0.6%), which is significantly higher than selectivity towards ethylene over the Rh catalyst at a similar conversion (0.02%, Insert in Figure 5c). These changes in product selectivity—increased methane and decreased ethylene—at higher ethane conversion levels can be explained in a scenario where C–C bond scission forming methane is the main and primary route for ethane decomposition before reforming reactions (CO and CO_2 formation) take place, and is similar for both metals (Figure S5).

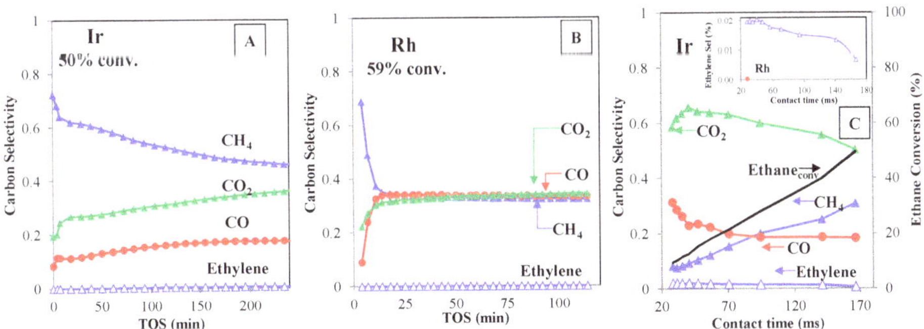

Figure 5. Product selectivity for ethane steam reforming at 600 °C (S/C = 2.75 mol, τ = 28 ms [Rh]; τ = 167 ms [Ir]; ethane feed = 16 vol %) over Ir/MgAl$_2$O$_4$ catalyst at 50% ethane conversion (**A**), over Rh/MgAl$_2$O$_4$ catalyst at 59% ethane conversion (**B**). Variation of carbon selectivity with ethane contact time over Ir/MgAl$_2$O$_4$ catalyst (**C**).

These observations for ethane reforming over Ir compared to Rh (higher ethylene formation, lower selectivity towards methane) can be combined to suggest that hydrogenolysis of ethane is relatively slow over Ir and facile over Rh. This hypothesis is additionally supported with several control experiments performed over Ir catalyst and summarized into the following four main points:

(1) Ethane steam reforming rate does not increase with higher S/C ratios. A wide range of S/C ratios were used to eliminate oxidation (reforming steps) as being responsible for the lower reaction rate on Ir (S/C up to 10). This is illustrated in Figure 6. This result is consistent with kinetic literature studies reporting that the reforming reaction is independent of the partial water pressure (zero order with respect to water) [36].

(2) Low activity over Ir catalyst is not caused by coke formation. In spite of higher carbon content on Ir spent catalyst (Table 2), coke (or at least hard coke) is not formed. A series of control experiments showed that high activity is consistently low for ethane reforming after repeated methane/ethane cycles (see Table S1).

(3) Lower ethane partial pressure increases catalytic conversion. This is consistent with hydrocarbon hydrogenolysis literature reporting that hydrocarbons equilibrate with metal surfaces by sequential dehydrogenation steps. As a consequence, a lower ethane partial pressure (from 14 to 8 mol %) causes a lower coverage that increases the number of active sites (ethane conversion increases from 10 to 47%).

(4) Ethane reforming over Ir supported on Al_2O_3 shows similar trends. Ir supported by Al_2O_3 is active for methane reforming; however, ethane steam reforming is severely hampered. Ir supported on alumina has a bigger particle size (4 nm) [20].

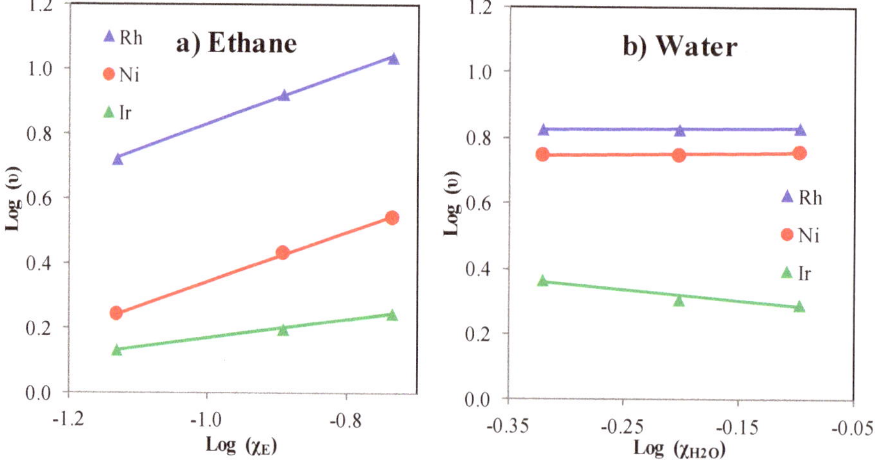

Figure 6. Logarithmic plot for catalytic rate vs. partial pressure of (**a**) ethane and (**b**) water.

2.5. Ethane Reforming Kinetic Measurements

Methane reforming kinetics have been extensively reported in the literature. Results indicate that methane decomposition and reforming rates are first order in methane; C–H activation is the rate limiting step and occurs on metal surfaces similar to conventional hydrocarbon chain reaction mechanism [32,55]. Ethane activation over metal surfaces is more complex, involving a series of partially dehydrogenated intermediates that lead to C–C bond breaking [41,56]. Strongly adsorbed H is considered to be competitive with hydrocarbon adsorption and limits its reactivity (in general, and over metal surfaces, the ethane hydrogenolysis rate is inhibited by excess H_2) [50,57,58]. One might think that interaction of highly dehydrogenated species in the absence of active hydrogen could lead to a higher coke formation; however, there is limited experimental evidence that shows that co-feeding hydrogen prevents coke formation better than water does [59]. Regarding the activation of the oxidant (e.g., steam), it is been largely reported that methane has zero order dependence on water (or CO_2). This indicates that after C–H activation, all the catalytic steps, including dehydrogenation of CH_x species, CO/CO_2 formation, and WGS reaction, should have no kinetic relevance.

To study the influence that the concentrations of reactants have in catalytic rate, we varied ethane and water concentrations in the gas feed at 600 °C. Kinetic results for catalytic rate vs. partial pressure are shown in Figure 6. Calculated reaction orders for both ethane and water and activation energies are shown in Table 3. The reation order to ethane is less than one and zero or near zero to water. For methane steam reforming, the reaction is first order to methane, and the rate limiting steps have been found to be adsorption and activation of the C–H bond [36]. Thus, there is less dependance of hydrocarbon partial pressure on reaction rate for ethane steam reforming than for methane steam reforming.

Table 3. Kinetic parameters calculated for ethane steam reforming reaction at 600 °C.

Catalyst	n_{Ethane}	n_{water}	E_a (kJ/mol) [a]
$Rh/MgAl_2O_4$	0.79	0	26.8
$Ni/MgAl_2O_4$	0.76	0	79.1
$Ir/MgAl_2O_4$	0.29	−0.3	95.7

[a] Measured at 550, 600, and 650 °C.

The results presented in Table 2 indicate that the Ir catalyst accumulated more carbon on the surface than Rh after the ethane reforming reaction at 600 °C; this presumably is due to a slower C–C scission rate. Analysis of the hydrogenolysis mechanism suggests that if a surface step is hampering the catalytic cycle, the initiation step (adsorption/activation of C–H bond) loses kinetic importance (resulting in a lower reaction order). Our low reaction orders for the Ir catalyst (n_{ethane} = 0.29 and n_{water} = −0.3) are consistant with high carbon coverage of the catalyst surface; in this catalyst, the reaction of adsorbed H with C_2 species occur before the C–C bond breaking step, causing formation of ethylene gas. In the ethane hydrogenolysis reaction, the hydrogen adsorption capacity of the metal (M–H coordination), its adsorption strength, and the coverage relative to other species (C_2, C_1) are important factors in overall catalytic rate and coke formation. The initiation and termination reactions in hydrocarbon chain mechanisms on metals require a high affinity of the metals for H. Our previous analysis in this section indicate that the slowest step is related to the creation (and accumulation) of adsorbed carbon species (C_1 and C_2); these results might be related to a poor capacity to activate and retain H. The low reaction order found for ethane reforming (0.29), a higher ethylene formation level, and higher carbon accumulation (Table 2) support the conclusion that surface ethyl decomposition (part of ethane hydrogenolysis reaction) is superseding the global reaction order.

Ethane steam reforming studies at 600 °C on Rh performed by Graf et al [59] indicated that in the presence of H_2, ethane reacts to form methane at a much higher rate than towards the formation of reforming products; additionally, their analysis of reaction products suggests that ethane reforming and hydrogenolysis proceed in a parallel fashion. Our analysis of reforming products differs in two important key aspects from the conclusions by Graf et al. First, by changing the contact time, we found that a higher ethane conversion is accompanied by a higher selectivity towards methane and a lower formation of ethylene (Figure 5c, Figure S5). This indicates that hydrogenolysis reaction proceeds reforming reactions; in other words, reforming steps are sequential to ethane hydrogenolysis. Second, the amount of H available and its adsorption strength to hydrogenolysis reaction becomes critical, more than what it is the presence of oxidant species [41,58]. In this scenario, surface active hydrogen is necessary to hydrogenate C_1 species on catalyst surface that will cause excessive carbon buildup if they don't undergo reforming reactions. Several key factors for improved catalysis are (1) an optimal amount of steam, (2) a fast C–C bond scission rate, and (3) a high H retention capacity by the catalyst. These key factors will avoid the accumulation of highly dehydrogenated carbon species (especially C and CH [80 kcal/mol]) that promote coke formation.

2.6. Catalyst Stability

Catalyst longevity is one of the cornerstones in the study of reforming catalysts. Sintering of metal particles, loss of support surface area, and extensive coke formation can cause catalyst fouling at the harsh conditions necessary to perform reforming reactions. Spinel supported metals (especially Ir and Rh) have been shown to maintain physical stability under these harsh reaction conditions with minimal coke formation [20,21]. In this section, we analyze the effect that the use of a simulant natural gas has in catalyst performance, focusing on longer term stability. In the prior sections we demonstrated that C_{2+} steam reforming activity is limited over Ir-based catalysts, presumably due to a reduced capacity of breaking ethane's C–C bond becoming rate limiting at low reaction temperatures. In the following section, we present results for the Rh catalyst (chosen as benchmark) and evaluate the influence that reaction conditions have on catalyst performance. The aim is to find reaction conditions that allow stable catalytic operation even with fast throughputs and relatively low S/C ratios.

For the reforming experiments shown in Figure 3, methane was fed at a rate of 5.3 mmol/min and the reaction was carried out on similar metal surfaces (23.5 and 25.2 µmol of Rh and Ir respectively). The catalytic rate was reported for each temperature after 1 h where steady state was reached. Using these reaction conditions, especially with short reaction times (1 h time on stream [TOS]), no deactivation was seen; however, if the reaction was allowed to proceed for longer times, deactivation was evident. The results shown in Figure S6 (for the reaction at 850 °C) show how conversion starts to decline slightly after 8 h of reaction. Under ethane reforming conditions, a more pronounced deactivation was found after a few hours on stream for the reaction at 600 °C (see Table 1). It is clearly evident that the presence of C_{2+} hydrocarbons, and especially at a high carbon throughput, has a negative effect on catalyst life, which largely depends on the reaction temperature and feed rate.

Catalyst durability experiments for methane steam reforming over the Rh catalyst are presented in Figure 7a. After over 100 h at both 850 and 900 °C (τ = 2.3 ms, S/C = 2), no deactivation was observed. Figure 7b depicts results when using a simulated natural gas feed mixture containing methane, ethane, propane, and butane. For simplicity, only methane conversion results are shown in Figures 7b and 8. After 100 h, catalysts showed continued deactivation and retention of only part of the initial activity (91%, 78%, and 53% activity retained for the Ir, Rh, and Ni catalysts, respectively). These results illustrate how the presence of C_{2+} hydrocarbons are detrimental to catalyst durability. However, we note that these results were obtained at a very low molar S/C ratio of 1.5. We attribute catalyst deactivation to increased carbon fouling (coke buildup) of the higher hydrocarbons on the catalyst surface. As shown in Figure 7c when the molar S/C feed ratio was increased to 2.0 the catalyst was quite stable at 850 and even 900 °C for 120 h duration. Although we note that prolonged duration testing at 900 °C revealed some deactivation, at 850 °C the reaction was quite stable.

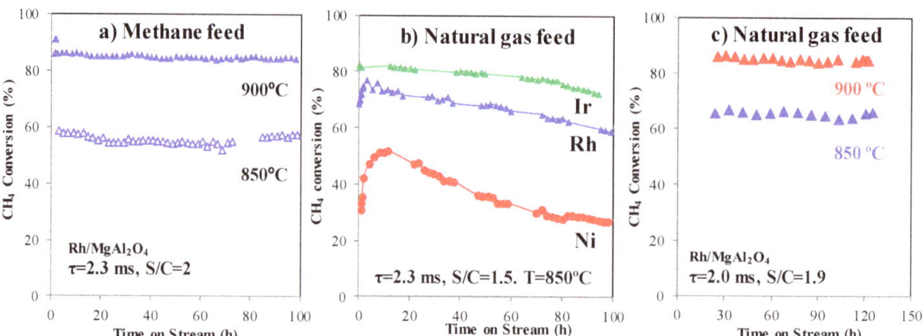

Figure 7. Methane conversion vs. time on stream (TOS) using (**a**) pure methane feed over Rh at 850 and 900 °C (S/C = 2.3 ms, τ = 2.3 ms), and using natural gas simulant feed over (**b**) Rh, Ir, and Ni at 850 °C (S/C = 1.5 ms, τ = 2.3 ms), and (**c**) Rh at 850 and 950 °C (S/C = 1.9 ms, τ = 2.0 ms).

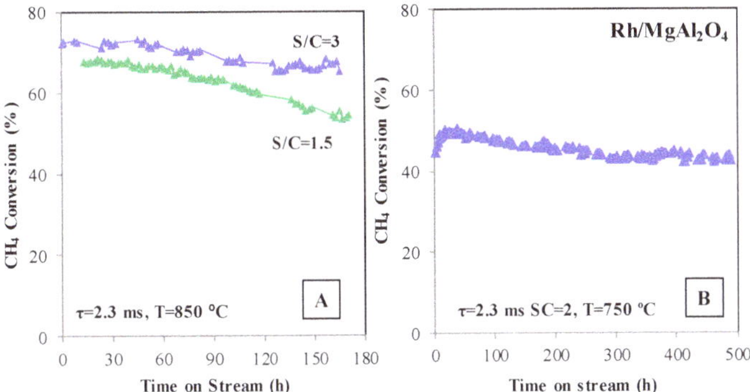

Figure 8. Methane conversion vs. TOS over Rh/MgAl$_2$O$_4$ catalyst as a function of steam/carbon molar feed ratio at 850 °C (**A**), and at 750 °C for 500 h TOS (**B**).

There are two common strategies for avoiding/suppressing coking on reforming catalysts: (1) increase the amount of steam fed and (2) conduct the reaction at lower temperatures. On one hand, increasing the amount of steam increases the formation of oxidation products (CO, and CO$_2$), while hydrocarbon adsorption/activation rates remain unaffected. On the other hand, the main effect of a lower reaction temperature is a decrease in the rates of hydrocarbon activation; as a result, reforming reaction rates supersede the coke rate formation, finally leading to a more stable long-term operation. Results for these two situations are presented in Figure 8 for the Rh catalyst. At 850 °C, the S/C molar feed ratio was varied. Increasing the S/C ratio enhanced catalyst durability. Figure 8a shows how running with an S/C molar feed ratio of 3 catalytic activity was largely maintained after 100 h of testing (some initial deactivation appeared to have occurred). By comparison, only 78% of catalytic activity was maintained with a lower S/C molar feed ratio of 1.5. Thus, increasing the S/C ratio improves catalyst durability. Figure 8b shows that a reduced operation temperature has a beneficial effect on catalyst life. Even after a 500 h test duration, the Rh catalyst retained its activity when operated at 750 °C.

3. Experimental Methods

3.1. Catalyst Preparation

As reported elsewhere [20], a series of catalysts were prepared by incipient wetness impregnation of MgAl$_2$O$_4$ (Puralox 30/140 from Sasol) with a solution of Rh nitrate (10 wt.% Rh in nitric acid), Ir nitrate (19.3 wt.% Ir in nitric acid), and Ni nitrate hexahydrate salt (Sigma-Aldrich, St. Louis, MO, USA). The resulting metal loadings were 5 wt.% Rh, 5 wt.% Ir, and 15 wt.% Ni. After impregnation, the catalysts were dried at 120 °C for 8 h and calcined at 500 °C for 4 h under static air.

3.2. Catalyst Characterization

Scanning transmission electron microscopy (STEM) measurements over reduced samples (850 °C, 16 h, 10% H$_2$/N$_2$) were conducted with a FEI Titan 80–300 microscope (Hillsboro, OR, USA) operated at 300 kV. The FEI Titan is equipped with a CEOS GmbH double-hexapole aberration corrector (Heidelberg, Germany) for the probe-forming lens, which allows imaging at ~0.1 nm resolution in STEM mode. The STEM images were acquired on high angle annular dark field with an inner collection angle of 52 mrad. In general, sample preparation involved mounting powder pre-reduced samples on copper grids covered with lacey carbon support films and then loading them immediately into the instrument airlock to minimize an exposure to atmospheric O$_2$.

Nitrogen adsorption was measured at 77 K with an automatic adsorptiometer (Micromeritics ASAP 2000, Norcross, GA, USA). The samples were pre-treated at 383 K for 12 h under vacuum.

The surface areas were determined from adsorption values for five relative pressures (P/P_0) ranging from 0.05 to 0.2 using the Brunauer–Emmett–Teller method.

Temperature programmed reduction (TPR) and H_2 chemisorption experiments were performed using an AutoChem II 2920 automated chemisorption analyzer (Micromeritics, Norcross, GA, USA). Metal reducibility was determined by TPR using changes in a thermal conductivity signal of the effluent gas. Samples (100 mg) were heated to 800 °C at 5 °C/min in a gas stream of 5% H_2 in Ar (50 sccm). Volumetric pulse H_2 adsorption measurements were carried out at 600 °C and 800 °C. First, 50 mg of the sample was reduced at 850 °C for 16 h using H_2 (10% in N_2, 100 mL/min) and purged for 4 h in pure N_2. After ramping to the adsorption temperature, 5% H_2/Ar was pulsed using a 100 µL loop with 1 min intervals between injections.

The amount of solid carbon deposited on the spent catalysts was measured by a Shimadzu Total Carbon Analyzer (TOC-5000A with a SSM-5000A Solid Sample Module, Shimadzu, Kyoto, Japan). A 1% O_2/Ar mixture was blown through the sample starting from 40 °C until the temperature reached 900 °C. The temperature ramping rate was 10 °C/min.

3.3. Activity Measurements

Catalytic activity tests were conducted in an 8 mm inner diameter Inconel fixed-bed reactor. For each test, the catalyst (9 mg, 60–100 mesh), diluted with α-Al_2O_3 (90 mg, 60–100 mesh), was loaded between two layers of quartz wool inside the reactor. Temperature was monitored with a thermocouple placed in the middle of the catalyst bed. Before reaction, the catalyst was reduced at 850 °C for 16 h using H_2 (10% in N_2, 100 mL/min).

Methane, ethane and natural gas simulant (94.5% methane, 4% ethane, 1% propane, 0.5% butane) gases were supplied by Matheson (Longview, WA, USA). Deionized water (18.6 mΩ) was introduced using a high-performance liquid chromatography pump (ChromTech series 1500, Bad Camberg, Germany) through $^1/_{16}$ inch stainless steel tubing into a vaporizer where the temperature was set at 250 °C. The catalysts were tested at atmospheric pressure at temperatures ranging from 550–900 °C, over a range of gas-hour-space-velocities (22,000–356,000 h^{-1}). Note that we commonly refer to throughput in terms of contact time (τ) which ranges from 10–170 ms. Flow rates of dry gas products in the effluent gases were monitored by a digital flow meter (DryCal) (Mesa Labs, Butler, NJ, USA). Gas products were analyzed online using a two-channel Agilent Micro GC (3000A series) (Santa Clara, CA, USA) equipped with thermal conductivity detector.

In a typical experiment, a mass of 9 mg of catalyst was loaded to the reactor, 220 sccm of gas was fed down flow to the catalytic bed (80 sccm of N_2, 35 sccm of methane, 105.8 sccm of H_2O [0.085 mL/min]). Under those conditions, the calculated space velocity (GHSV) was 119,223 h^{-1}, which corresponds to a τ of 30.2 ms and an S/C molar ratio of 3.0. For ethane reforming experiments, changes in space velocities were achieved with changes in gas flow (3.5–35 sccm of ethane for the Ir catalyst case, 35–100 sccm of ethane for Rh catalyst case) while concentrations (14.8 mol% ethane) and S/C ratios were maintained constant (S/C = 2.75).

4. Conclusions

This study compares $MgAl_2O_4$-supported Rh and Ir catalysts for the reforming reaction of methane, ethane, and natural gas simulant (a C_1-C_4 mixture) under industrially relevant conditions. In reforming of a natural gas simulant mixture, the Ir catalyst showed a lower capacity for converting C_{2+} hydrocarbons compared to Rh in the lower 600–700 °C range of reforming temperatures. At higher temperatures (700–900 °C), there is no real distinction in the activity for the two metals. In a more detailed study of the reaction at 600 °C, the Ir-based catalyst showed a limited capacity for ethane reforming compared to methane. Additionally, the formation of ethylene as a byproduct was accompanied by a higher amount of carbon deposited on the catalyst when ethane is used as a reactant. On the contrary, the reforming rate of ethane was faster over the Rh catalyst. Less carbon was deposited and a negligible amount of ethylene was formed. A lower reaction rate of ethane over Ir is likely related to a lower C–C scission rate. Literature reports for ethane hydrogenolysis explain how this decreased rate depends on the nature of the metal. The kinetic study found there to be a lower reaction order for ethane reforming over Ir compared to Rh and Ni (0.29, 0.79, 0.76, respectively), which is consistent with a higher coverage of the catalysts by carbon deposits. Calculated activation energies also were found to be higher for Ir compared to Rh and Ni (95, 26, 79 kJ/mol, respectively) in the lower temperature reforming range.

Adsorbed hydrogen assists hydrocarbon reactions on metal supported catalysts by hydrogenating C_1 species to form methane. In an attempt to extend the analysis that H adsorption has on Ir and Rh reforming catalysts, we performed H_2 adsorption measurements on Ir, Rh, and Ni at reforming reaction temperatures (600 °C). Results indicate that hydrogen saturates the Rh surface in a ratio 1:1 with respect to metal atoms on the surface. Coverages over Ir and Ni were found lower (0.7:1 for Ir, and 0.2:1 for Ni). A summary of main findings is presented in Table 4.

Table 4. Summary of key findings.

Catalyst	Metal Loading (wt.%)	Metal Dispersion (%)	Metal Particle Size (nm)	H Stoichiometry H/M_{surf} (mol/mol)	n_{Ethane}	n_{water}	E_a (kJ/mol)
$5Rh/MgAl_2O_4$	5	50	2.0	1:1	0.79	0	26.8
$5Ir/MgAl_2O_4$	5	100	1.0	0.7:1	0.29	−0.3	95.7
$15Ni/MgAl_2O_4$	15	15.4	6.5	0.2:1	0.76	0	79.1

Given the low capacity of Ir to reform ethane and higher hydrocarbons, Rh was chosen for additional catalyst durability studies. Long-term stability tests revealed the Rh catalyst to be very stable under SMR conditions, and under relatively harsh conditions (up to 900 °C and with S/C molar feed ratios up to 1.5). However, when a more complex mixture of hydrocarbons was added to the methane feed, catalyst stability was adversely affected. Here, the S/C molar feed ratio and operating temperature can be adjusted to extend catalyst life. Stable catalyst operation was observed for the Rh catalyst when using a natural gas simulant and operating under a relatively low S/C ratio of 2.0 and at 850 °C. Thus, we note that improvements in catalyst life can be achieved through both proper choice of the catalyst material and operational conditions. In addition, temperature and S/C molar feed ratio are critical processing variables to consider when optimizing catalyst performance. The results of this study conclude that a Rh-supported catalyst was developed that enables very high activities and excellent catalytic stability for both the steam reforming of methane and other higher hydrocarbons contained in natural gas, and under conditions of operation that are amendable to solar thermochemical operations.

Supplementary Materials: The following are available online at http://www.mdpi.com/2073-4344/9/10/801/s1, Figure S1. (A) TEM analysis of $5Ir/MgAl_2O_4$ (top) and $5Rh/MgAl_2O_4$ (bottom) catalysts used in this study. (B) Effect of ageing on $Rh/MgAl_2O_4$ catalyst. (950 °C) under H_2. Figure S2. Hydrogen uptake at 600 and 800 °C. Evolution of hydrogen uptake is calculated as H coverage (per mole of metal surface) vs H_2 pulsed. Volumetric pulse hydrogen adsorption measurements were carried out at 600 °C and 800 °C. First, 50 mg of the sample was reduced at 850 °C for 16 h using H_2 (flow 10% in N_2, 100 mL/min) and purged for 4 in pure N_2. After ramping at the adsorption temperature, 5% Hydrogen/Ar is pulsed using a 100 µL loop with 1 minute intervals between injections. Figure S3. Conversion of the individual components of a simulant natural gas mixture. Butane conversion was complete. Reaction conditions: S/C:1.5, t = 4.5 ms, 1h TOS, Simulant gas feed (94.5%$_v$ methane, 4% ethane, 1% propane, 0.5% butane) was supplied by Matheson. Figure S4. Ethane reforming conversion vs time on stream at 600 °C. Reaction conditions: S/C = 2.75, 9 mg of catalyst. τ = 28.3 ms, 35 sccm ethane, 80 sccm N_2, 9 mg of catalyst. Table S1. Deactivation check experiments for the Ir catalyst under ethane reforming experiments shown in Figure S4. Fresh catalyst was tested for methane activity before ("initial") and after reaction with ethane to check for deactivation ("final"). Figure S5. Ethane reforming conversion and carbon selectivity vs contact time (ms) for the reforming of ethane at 600 °C over $MgAl_2O_4$-supported Ir and Rh catalysts (A-top), ethylene selectivity (mol %) vs contact time (ms) (B-middle), and linear correlation for ethane conversion vs contact time. Reaction conditions: S/C = 2.75, 9 mg of catalyst. Changes in contact time were achieved by changing gas flow over the same mass of catalyst in a continuous experiment. Each point corresponds to a steady state measurement after stabilizing for 1h. Ethane over iridium catalyst was changed from 35 to 3.5 sccm. For the case of Rh, ethane flow was varied from 35 to 100 sccm. Figure S6. Ethane reforming conversion over 5Ir/MgAl$_2$O$_4$ catalyst at 600 °C. (A) Ethane conversion vs time on stream at increase carbon/steam ratio (10 for empty symbols) at two different concentration of ethane in the gas (8 and 14 vol %). (B) Ethane conversion vs time on stream comparing Ir activity over two different supports, MgAl$_2$O$_4$ (red) and Al$_2$O$_3$ (blue). Figure S7. (A) Product selectivity for ethane steam reforming over Ir at 600 °C (8.2% conversion). (B) Ethane conversion over Rh and Ir catalysts at 600°C (S/C = 3 mol, τ = 28 ms (Rh), τ = 167 ms (Ir)). Figure S8. Conversion vs time on stream for methane steam reforming over the Rh and Ir supported catalysts at 850 °C (S/C = 3 mol, τ = 12.4 ms; Methane feed = 22.6 vol. %). Figure S9. Methane conversion (A) and H$_2$/CO ratio for methane steam reforming products (B) over benchmark 5% Rh/MgAl$_2$O$_4$ catalyst. CH4, S/C = 3, 9 mg of catalyst. 5% Rh MgAl$_2$O$_4$ catalyst reduced in-situ at 850 °C for 16 h under flowing 10% H$_2$ in N$_2$.

Author Contributions: Conceptualization, J.S.L., R.A.D. and V.L.D.; Methodology, R.A.D., V.L.D., L.K., and J.S.L.; Investigation, J.S.L., V.L.D., L.K. and C.A.D.; Data Curation and Formal Analysis, J.S.L. and R.A.D.; Writing—Original Draft Preparation, J.S.L.; Writing—Review & Editing and Supervision, R.A.D.; Project Administration and Funding Acquisition, R.A.D. and R.S.W.

Funding: This work was financially supported by the U.S. Department of Energy's (DOE) Solar Energy Technology Office (SETO) and was performed at the Pacific Northwest National Laboratory (PNNL) under Contract DE-AC05-76RL01830.

Acknowledgments: Catalyst characterization equipment use was granted by a user proposal at the William R. Wiley Environmental Molecular Sciences Laboratory, which is a national scientific user facility sponsored by the DOE Office of Biological and Environmental Research and located at PNNL. The authors would like to thank David L. King for offering a technical review and Cary A. Counts for providing technical editing support of this manuscript. The views and opinions of the authors expressed herein do not necessarily state or reflect those of the United States Government or any agency thereof. Neither the United States Government nor any agency thereof, nor any of their employees, makes any warranty, expressed or implied, or assumes any legal liability or responsibility for the accuracy, completeness, or usefulness of any information, apparatus, product, or process disclosed, or represents that its use would not infringe privately owned rights.

Conflicts of Interest: The authors declare no conflict of interest.

References

1. Nikolaidis, P.; Poullikkas, A. A comparative overview of hydrogen production processes. *Renew. Sustain. Energy Rev.* **2017**, *67*, 597–611. [CrossRef]

2. DOE. Hydrogen Production: Natural Gas Reforming, Washington, DC, USA. Available online: https://www.energy.gov/eere/fuelcells/hydrogen-production-natural-gas-reforming (accessed on 23 April 2019).

3. Azizi, Z.; Rezaeimanesh, M.; Tohidian, M.R. Rahimpour, Dimethyl ether: A review of technologies and production challenges. *Chem. Eng. Process. Process Intensif.* **2014**, *82*, 150–172. [CrossRef]

4. Wood, D.A.; Nwaoha, C.; Towler, B.F. Gas-to-liquids (GTL): A review of an industry offering several routes for monetizing natural gas. *J. Nat. Gas Sci. Eng.* **2012**, *9*, 196–208. [CrossRef]

5. Glasser, D.; Hildebrandt, D.; Liu, X.; Lu, X.; Masuku, C.M. Recent advances in understanding the Fischer–Tropsch synthesis (FTS) reaction. *Curr. Opin. Chem. Eng.* **2012**, *1*, 296–302. [CrossRef]

6. Pakhare, D.; Spivey, J. A review of dry (CO$_2$) reforming of methane over noble metal catalysts. *Chem. Soc. Rev.* **2014**, *43*, 7813–7837. [CrossRef] [PubMed]

7. Jones, G.; Jakobsen, J.; Shim, S.; Kleis, J.; Andersson, M.; Rossmeisl, J.; Abildpedersen, F.; Bligaard, T.; Helveg, S.; Hinnemann, B.; et al. First principles calculations and experimental insight into methane steam reforming over transition metal catalysts. *J. Catal.* **2008**, *259*, 147–160. [CrossRef]

8. Zheng, R.; Diver, R.; Caldwell, D.; Fritz, B.; Cameron, R.; Humble, P.; TeGrotenhuis, W.; Dagle, R.; Wegeng, R. Integrated Solar Thermochemical Reaction System for Steam Methane Reforming. *Energy Procedia* **2015**, *69*, 1192–1200. [CrossRef]

9. Simakov, D.S.A.; Wright, M.M.; Ahmed, S.; Mokheimer, E.M.A.; Román-Leshkov, Y. Solar thermal catalytic reforming of natural gas: A review on chemistry, catalysis and system design. *Catal. Sci. Technol.* **2015**, *5*, 1991–2016. [CrossRef]

10. Steinfel, A. Solar thermochemical production of hydrogen—A review. *Sol. Energy* **2005**, *78*, 603–615. [CrossRef]

11. Wang, Y.; Chin, Y.H.; Rozmiarek, R.T.; Johnson, B.R.; Gao, Y.; Watson, J.; Tonkovich, A.Y.L.; Wiel, D.P.V. Highly active and stable Rh/MgOAl2O3 catalysts for methane steam reforming. *Catal. Today* **2004**, *98*, 575–581. [CrossRef]

12. Palo, D.R.; Stenkamp, V.S.; Dagle, R.A.; Jovanovic, G.N. Industrial Applications of Microchannel Process Technology in the United States. In *Micro Process Engineering*; Kockmann, N., Ed.; WILEY-VCH Verlag GmbH & Co. KGaA: Weinheim, Germany, 2008; pp. 387–414.

13. Kehres, J.; Andreasen, J.W.; Fløystad, J.B.; Liu, H.; Molenbroek, A.; Jakobsen, J.G.; Chorkendorff, I.; Nielsen, J.H.; Høydalsvik, K.; Breiby, D.W.; et al. Reduction of a Ni/Spinel Catalyst for Methane Reforming. *J. Phys. Chem. C* **2015**, *119*, 1424–1432. [CrossRef]

14. Salhi, N.; Boulahouache, A.; Petit, C.; Kiennemann, A.; Rabia, C. Steam reforming of methane to syngas over NiAl2O4 spinel catalysts. *Int. J. Hydrog. Energy* **2011**, *36*, 11433–11439. [CrossRef]

15. Boukha, Z.; Jiménez-González, C.; de Rivas, B.; González-Velasco, J.R.; Gutiérrez-Ortiz, J.I.; López-Fonseca, R. Synthesis, characterisation and performance evaluation of spinel-derived Ni/Al2O3 catalysts for various methane reforming reactions. *Appl. Catal. B Environ.* **2014**, *158–159*, 190–201. [CrossRef]

16. Aramouni, N.A.K.; Touma, J.G.; Tarboush, B.A.; Zeaiter, J.; Ahmad, M.N. Catalyst design for dry reforming of methane: Analysis review. *Renew. Sustain. Energy Rev.* **2018**, *82*, 2570–2585. [CrossRef]

17. Ligthart, D.A.J.M.; van Santen, R.A.; Hensen, E.J.M. Influence of particle size on the activity and stability in steam methane reforming of supported Rh nanoparticles. *J. Catal.* **2011**, *280*, 206–220. [CrossRef]

18. Duarte, R.B.; Krumeich, F.; van Bokhoven, J.A. Structure, Activity, and Stability of Atomically Dispersed Rh in Methane Steam Reforming. *ACS Catal.* **2014**, *4*, 1279–1286. [CrossRef]

19. Mei, D.; Lebarbier, V.M.; Rousseau, R.; Glezakou, V.-A.; Albrecht, K.O.; Kovarik, L.; Flake, M.; Dagle, R.A. Comparative Investigation of Benzene Steam Reforming over Spinel Supported Rh and Ir Catalysts. *ACS Catal.* **2013**, *3*, 1133–1143. [CrossRef]

20. Mei, D.; Glezakou, V.-A.; Lebarbier, V.; Kovarik, L.; Wan, H.; Albrecht, K.O.; Gerber, M.; Rousseau, R.; Dagle, R.A. Highly active and stable MgAl2O4-supported Rh and Ir catalysts for methane steam reforming: A combined experimental and theoretical study. *J. Catal.* **2014**, *316*, 11–23. [CrossRef]

21. Dagle, V.L.; Dagle, R.; Kovarik, L.; Genc, A.; Wang, Y.-G.; Bowden, M.; Wan, H.; Flake, M.; Glezakou, V.-A.; King, D.L.; et al. Steam reforming of hydrocarbons from biomass-derived syngas over MgAl2O4-supported transition metals and bimetallic IrNi catalysts. *Appl. Catal. B Environ.* **2016**, *184*, 142–152. [CrossRef]

22. Speight, J.G. *Composition and Properties, Natural Gas—A Basic Handbook*; Gulf Publishing Company: Houston, TX, USA, 2007.

23. U.S.E.I. Administration. *Ethane Production Growth Led to Record U.S. Natural Gas Plant Liquids Production in 2017, 2018*; US Energy Initiatives Corporation: Santa Clarita, CA, USA, 2018.

24. Adesina, A.A.; Trimm, D.L.; Cant, N.W. Kinetic study of iso-octane steam reforming over a nickel-based catalyst. *Chem. Eng. J.* **2004**, *99*, 131–136.

25. Sperle, T.; Chen, D.; Lødeng, R.; Holmen, A. Pre-reforming of natural gas on a Ni catalyst. *Appl. Catal. A Gen.* **2005**, *282*, 195–204. [CrossRef]

26. Takeguchi, T.; Kani, Y.; Yano, T.; Kikuchi, R.; Eguchi, K.; Tsujimoto, K.; Uchida, Y.; Ueno, A.; Omoshiki, K.; Aizawa, M. Study on steam reforming of CH4 and C2 hydrocarbons and carbon deposition on Ni-YSZ cermets. *J. Power Sources* **2002**, *112*, 588–595. [CrossRef]

27. Jeong, H.; Kang, M. Hydrogen production from butane steam reforming over Ni/Ag loaded MgAl2O4 catalyst. *Appl. Catal. B Environ.* **2010**, *95*, 446–455. [CrossRef]

28. Baek, B.; Aboiralor, A.; Wang, S.; Kharidehal, P.; Grabow, L.C.; Massa, J.D. Strategy to improve catalytic trend predictions for methane oxidation and reforming. *AIChE J.* **2017**, *63*, 66–77. [CrossRef]

29. Angeli, S.D.; Monteleone, G.; Giaconia, A.; Lemonidou, A.A. State-of-the-art catalysts for CH4 steam reforming at low temperature. *Int. J. Hydrog. Energy* **2014**, *39*, 1979–1997. [CrossRef]

30. Halabi, M.H.; de Croon, M.H.J.M.; van der Schaaf, J.; Cobden, P.D.; Schouten, J.C. Low temperature catalytic methane steam reforming over ceria–zirconia supported rhodium. *Appl. Catal. A Gen.* **2010**, *389*, 68–79. [CrossRef]

31. Morlanés, N. Reaction mechanism of naphtha steam reforming on nickel-based catalysts, and FTIR spectroscopy with CO adsorption to elucidate real active sites. *Int. J. Hydrog. Energy* **2013**, *38*, 3588–3596. [CrossRef]

32. Wei, J. Structural requirements and reaction pathways in methane activation and chemical conversion catalyzed by rhodium. *J. Catal.* **2004**, *225*, 116–127. [CrossRef]

33. Kneale, B.; Ross, J.R.H. The steam reforming of ethane over nickel/alumina catalysts. *Faraday Discuss. Chem. Soc.* **1981**, *72*, 157–171. [CrossRef]

34. Graf, P.O.; Mojet, B.L.; van Ommen, J.G.; Lefferts, L. Comparative study of steam reforming of methane, ethane and ethylene on Pt, Rh and Pd supported on yttrium-stabilized zirconia. *Appl. Catal. A Gen.* **2007**, *332*, 310–317. [CrossRef]

35. Schädel, B.T.; Duisberg, M.; Deutschmann, O. Steam reforming of methane, ethane, propane, butane, and natural gas over a rhodium-based catalyst. *Catal. Today* **2009**, *142*, 42–51. [CrossRef]

36. Wei, J.; Iglesia, E. Mechanism and Site Requirements for Activation and Chemical Conversion of Methane on Supported Pt Clusters and Turnover Rate Comparisons among Noble Metals. *J. Phys. Chem. B* **2004**, *108*, 4094–4103. [CrossRef]

37. Mei, D.; Dagle, V.L.; Xing, R.; Albrecht, K.O.; Dagle, R.A. Steam Reforming of Ethylene Glycol over $MgAl_2O_4$ Supported Rh, Ni, and Co Catalysts. *ACS Catal.* **2016**, *6*, 315–325. [CrossRef]

38. Xing, R.; Dagle, V.L.; Flake, M.; Kovarik, L.; Albrecht, K.O.; Deshmane, C.; Dagle, R.A. Steam reforming of fast pyrolysis-derived aqueous phase oxygenates over Co, Ni, and Rh metals supported on $MgAl_2O_4$. *Catal. Today* **2016**, *269*, 166–174. [CrossRef]

39. Huang, C.; Ma, Z.; Miao, C.; Yue, Y.; Hua, W.; Gao, Z. Catalytic decomposition of N2O over Rh/Zn–Al2O3 catalysts. *RSC Adv.* **2017**, *7*, 4243–4252. [CrossRef]

40. Mizuno, T.; Matsumura, Y.; Nakajima, T.; Mishima, S. Effect of support on catalytic properties of Rh catalysts for steam reforming of 2-propanol. *Int. J. Hydrog. Energy* **2003**, *28*, 1393–1399. [CrossRef]

41. Flaherty, D.W.; Hibbitts, D.D.; Gürbüz, E.I.; Iglesia, E. Theoretical and kinetic assessment of the mechanism of ethane hydrogenolysis on metal surfaces saturated with chemisorbed hydrogen. *J. Catal.* **2014**, *311*, 350–356. [CrossRef]

42. Hernández-Cristóbal, O.; Díaz, G.; Gómez-Cortés, A. Effect of the Reduction Temperature on the Activity and Selectivity of Titania-Supported Iridium Nanoparticles for Methylcyclopentane Reaction. *Ind. Eng. Chem. Res.* **2014**, *53*, 10097–10104. [CrossRef]

43. Kip, B.J.; van Grondelle, J.; Martens, J.H.A.; Prins, R. Preparation and characterization of very highly dispersed iridium on Al2O3 and SiO2. *Appl. Catal.* **1986**, *26*, 353–373. [CrossRef]

44. Nuernberg, G.D.B.; Fajardo, H.V.; Foletto, E.L.; Hickel-Probst, S.M.; Carreño, N.L.V.; Probst, L.F.D.; Barrault, J. Methane conversion to hydrogen and nanotubes on Pt/Ni catalysts supported over spinel $MgAl_2O_4$. *Catal. Today* **2011**, *176*, 465–469. [CrossRef]

45. Horváth, A.; Guczi, L.; Kocsonya, A.; Sáfrán, G.; la Parola, V.; Liotta, L.F.; Pantaleo, G.; Venezia, A.M. Sol-derived $AuNi/MgAl_2O_4$ catalysts: Formation, structure and activity in dry reforming of methane. *Appl. Catal. A Gen.* **2013**, *468*, 250–259. [CrossRef]

46. Gillan, C.; Fowles, M.; French, S.; Jackson, S.D. Ethane Steam Reforming over a Platinum/Alumina Catalyst: Effect of Sulfur Poisoning. *Ind. Eng. Chem. Res.* **2013**, *52*, 13350–13356. [CrossRef]

47. Latimer, A.A.; Kulkarni, A.R.; Aljama, H.; Montoya, J.H.; Yoo, J.S.; Tsai, C.; Abild-Pedersen, F.; Studt, F.; Nørskov, J.K. Understanding trends in C–H bond activation in heterogeneous catalysis. *Nat. Mater.* **2016**, *16*, 225. [CrossRef] [PubMed]

48. Blanksby, S.J.; Ellison, G.B. Bond Dissociation Energies of Organic Molecules. *Acc. Chem. Res.* **2003**, *36*, 255–263. [CrossRef]

49. Burch, R.; Loader, P.K.; Urbano, F.J. Some aspects of hydrocarbon activation on platinum group metal combustion catalysts. *Catal. Today* **1996**, *27*, 243–248. [CrossRef]

50. Sinfelt, J.H.; Yates, D.J.C. Catalytic hydrogenolysis of ethane over the noble metals of Group VIII. *J. Catal.* **1967**, *8*, 82–90. [CrossRef]

51. Bond, G.C. Kinetics of alkane reactions on metal catalysts: Activation energies and the compensation effect. *Catal. Today* **1999**, *49*, 41–48. [CrossRef]

52. Kuz'min, I.V.; Zeigarnik, A.V. Microkinetic Modeling of Ethane Hydrogenolysis on Metals. *Kinet. Catal.* **2004**, *45*, 561–569. [CrossRef]

53. Vincent, R.S.; Lindstedt, R.P.; Malik, N.A.; Reid, I.A.B.; Messenger, B.E. The chemistry of ethane dehydrogenation over a supported platinum catalyst. *J. Catal.* **2008**, *260*, 37–64. [CrossRef]

54. Goddard, S.A.; Amiridis, M.D.; Rekoske, J.E.; Cardona-Martinez, N.; Dumesic, J.A. Kinetic simulation of heterogeneous catalytic processes: Ethane hydrogenolysis over supported group VIII metals. *J. Catal.* **1989**, *117*, 155–169. [CrossRef]

55. Mark, M.F.; Maier, W.F. CO2-Reforming of Methane on Supported Rh and Ir Catalysts. *J. Catal.* **1996**, *164*, 122–130. [CrossRef]

56. Sinfelt, J.H. Kinetics of ethane hydrogenolysis. *J. Catal.* **1972**, *27*, 468–471. [CrossRef]
57. Sinfelt, J.H. Specificity in Catalytic Hydrogenolysis by Metals. In *Advances in Catalysis*; Eley, D.D., Pines, H., Weisz, P.B., Eds.; Academic Press: Cambridge, MA, USA, 1973; pp. 91–119.
58. Guczi, L.; Gudkov, B.S.; Tétényi, P. The mechanism of catalytic hydrogenolysis of ethane over nickel. *J. Catal.* **1972**, *24*, 187–196. [CrossRef]
59. Kahle, L.C.S.; Roussière, T.; Maier, L.; Delgado, K.H.; Wasserschaff, G.; Schunk, S.A.; Deutschmann, O. Methane Dry Reforming at High Temperature and Elevated Pressure: Impact of Gas-Phase Reactions. *Ind. Eng. Chem. Res.* **2013**, *52*, 11920–11930. [CrossRef]

Article

Catalytic Dry Reforming and Cracking of Ethylene for Carbon Nanofilaments and Hydrogen Production Using a Catalyst Derived from a Mining Residue

Abir Azara [1,2], El-Hadi Benyoussef [2], Faroudja Mohellebi [2], Mostafa Chamoumi [1], François Gitzhofer [1] and Nicolas Abatzoglou [1,*]

1   Department of Chemical & Biotechnological Engineering, Université de Sherbrooke, 2500 Boulevard de l'Université, Sherbrooke, QC J1K 2R1, Canada; abir.azara@usherbrooke.ca (A.A.); Mostafa.Chamoumi@USherbrooke.ca (M.C.); francois.gitzhofer@usherbrooke.ca (F.G.)
2   Laboratoire de Valorisation des Énergies Fossiles, École National Polytechnique, 10 Avenue Hassen Badi El Harrach BP182, Alger 16200, Algeria; el-hadi.benyoussef@enp.edu.dz (E.-H.B.); mohellebifaroudja@yahoo.fr (F.M.)
*   Correspondence: Nicolas.Abatzoglou@USherbrooke.ca

Received: 7 November 2019; Accepted: 9 December 2019; Published: 14 December 2019

Abstract: In this study, iron-rich mining residue (UGSO) was used as a support to prepare a new Ni-based catalyst via a solid-state reaction protocol. Ni-UGSO with different Ni weight percentages wt.% (5, 10, and 13) were tested for C_2H_4 dry reforming (DR) and catalytic cracking (CC) after activation with H_2. The reactions were conducted in a differential fixed-bed reactor at 550–750 °C and standard atmospheric pressure, using 0.5 g of catalyst. Pure gases were fed at a molar ratio of $C_2H_4/CO_2 = 3$ for the DR reaction and $C_2H_4/Ar = 3$ for the CC reaction. The flow rate is defined by a $GHSV = 4800\ mL_{STP}/h.g_{cat}$. The catalyst performance is evaluated by calculating the C_2H_4 conversion as well as carbon and H_2 yields. All fresh, activated, and spent catalysts, as well as deposited carbon, were characterized by Brunauer–Emmett–Teller (BET), X-ray diffraction (XRD), scanning electron microscopy (SEM), energy dispersive X-ray spectrometry (EDX), transmission electron microscopy (TEM), temperature programmed reduction (TPR), and thermogravimetric analysis (TGA). The results so far show that the highest carbon and H_2 yields are obtained with Ni-UGSO 13% at 750 °C for the CC reaction and at 650 °C for the DR reaction. The deposited carbon was found to be filamentous and of various sizes (i.e., diameters and lengths). The analyses of the results show that iron is responsible for the growth of carbon nanofilaments (CNF) and nickel is responsible for the split of C–C bonds. In terms of conversion and yield efficiencies, the performance of the catalytic formulations tested is proven at least equivalent to other Ni-based catalyst performances described by the literature.

Keywords: dry reforming; catalytic cracking; ethylene; carbon nanofilaments; hydrogen

1. Introduction

Hydrogen is an energy vector and is mainly used in the synthesis of several chemicals such as methanol, ammonia, and liquid hydrocarbons via the Fischer–Tropsch process. Concerning carbon nanofilaments (CNF), several studies have shown that they have noteworthy properties, including high surface area, high mechanical resistance, and high electrical and thermal conductivities [1]. This is why research efforts have focused on optimizing and controlling the formation of these types of carbon structures, instead of inhibiting their growth [2]. Although, usually, carbon formation on catalysts causes the deactivation of the catalyst [3], CNF are shown to grow in such a way that the catalytically active sites maintain their activity [4]. CNF properties make them a good substitute to high-cost materials used currently in various applications such as reinforcement of composites [5], manufacturing of double-layer condensers [6], fabrication of anodes in lithium batteries [7,8], adsorption [9], support for catalysts [10], or catalysts themselves [11].

H_2 is generated mainly from hydrocarbons via thermocatalytic processes such as steam reforming (SR), autothermal reforming (ATR), partial oxidation (POX), dry reforming (DR), and catalytic decomposition or cracking (CC) [12]. However, methane SR is the only industrial production technology used so far [13]. SR is an endothermic reaction and requires a high energy input. Temperatures as high as 950 °C and relatively high steam/C ratios are required to reach high H_2 yields and avoid carbon formation and, consequently, premature catalyst deactivation. In the last decade, many researches have focused on DR that uses CO_2 instead of H_2O to produce not only H_2 but also CNF. Hence, DR reaction has not only economic interests but also an environmental interest, which is the contribution on the sequestration of CO_2—a greenhouse gas [14–17].

The feedstock composition, the choice of catalysts including the support and active metals, as well as the operating conditions, especially the temperature, are the main elements that have been largely studied to optimize H_2 and CNF production [18]. Methane [19–21], n-octane [22], ethanol [15], and biogas [23] are the main reactants used. The use of pyrolytically-produced gases has rarely been cited in the current literature. Arena et al. [24] have developed an innovative process for mass production of multiwall carbon nanotubes (MWCNT) by pyrolysis of virgin or recycled polyolefins. Regarding CNF production, the literature is rather scarce. Svinterekos et al. [25] used lignin (a natural polymer found in plants) combined with recycled polyethylene terephthalate (PET) to make precursor fibers that are used for the electrospinning of CNF. The work presented here is part of a larger research endeavor aimed at the conversion of waste plastic streams into added plus-value products such as CNF. Since the gases produced by plastic pyrolysis are composed mainly of unsaturated hydrocarbons, the first step of this study is focused on using C_2H_4 as a surrogate molecule. The dry reforming reaction of ethylene is not yet well reported in the literature; the products of this type of reaction are considerably dependent on the nature of the catalyst used. In the presence of a transition metal catalyst, carbon and synthesis gas are the products obtained from ethylene DR [26]. However, the Mn and Cr oxides convert ethylene to butadiene and propylene. For example, with a MnO/SiO_2 catalyst at 850 °C, the products obtained are C_4H_6 with a selectivity of 25%, C_3H_6 with a selectivity of 18%, and traces of CH_4, C_3H_8, and C_4H_8 [26].

The theoretical reaction of ethylene DR is given by Equation (1) below [26]:

$$C_2H_4 + 2CO_2 \rightarrow 4CO + 2H_2 \ (\Delta H^{\circ}_{298} = 292.5 \ \text{MJ/kmol}) \tag{1}$$

Other known reactions that take place during ethylene DR are Equations (2)–(5):

$$\text{Ethylene decomposition: } C_2H_4 \rightarrow 2C + 2H_2 \ (\Delta H^{\circ}_{298} = -52.5 \ \text{MJ/kmol}) \tag{2}$$

$$\text{Reverse water gas shift reaction (RWGS): } CO_2 + H_2 \rightleftharpoons CO + H_2O \ (\Delta H^{\circ}_{298} = 41.0 \ \text{MJ/kmol}) \tag{3}$$

$$\text{Boudouard reaction: } 2CO \rightleftharpoons CO_2 + C \ (\Delta H^{\circ}_{298} = -172.0 \ \text{MJ/kmol}) \tag{4}$$

$$\text{Carbon gasification: } C + H_2O \rightarrow CO + H_2 \ (\Delta H^{\circ}_{298} = 131.3 \ \text{MJ/kmol}) \tag{5}$$

In general, at the temperatures used for these reactions and in the presence of the chosen catalysts, hydrocarbon molecules (HC) are converted into free radicals in the gas phase or at the catalyst's surface (intermediates). The reforming agent, CO_2 in the case of DR, is also dissociated into oxygen intermediates (O *) and CO. Oxygen-containing intermediates oxidize HC intermediates to produce CO and subsequently produce CO_2 and carbon via the Boudouard reaction (Equation (4)). The atomic carbon formed during the Boudouard reaction first diffuses and dissolves into the metal particles until saturation is reached and then the graphitic carbon starts precipitating to form CNF [27]. For the CC reaction, HC intermediates self-decompose to produce H_2 and carbon [28] and the atomic carbon formed follows the same process of diffusion, saturation, and finally precipitation [29].

Transition metal-based catalysts, particularly iron and nickel, are recognized for their ability to decompose carbonaceous gases into filamentous carbon and hydrogen. This capacity for carbon formation is due to the high diffusion rate of carbon in these metals at high temperatures.

The coefficients of diffusion of carbon into transition metals at 550 °C are 1.2×10^{-7}, 0.8×10^{-7}, and 0.2×10^{-7} cm^2/s, for Ni, Fe, and Co, respectively [29]. Consequently, the carbon yield would increase as follows: Ni-based catalyst > Fe-based catalyst > Co-based catalyst. This order, however, was not confirmed by Romero et al. [29], who studied the influence of these active metals (Co, Ni, Fe) and the influence of the zeolite type support on the synthesis of highly graphitized carbon nanofibers produced from the catalytic decomposition of ethylene. They found that the order is rather Ni > Co > Fe. They affirmed that this difference is due to the zeolite support that has a different synergistic effect, which explains the important role played by the support, on the activity of the catalyst.

Recently, our research group (GRTP-C & P) collaborated with Rio Tinto Iron and Titanium (RTIT) for the valorization of a mining residue (upgraded slag oxide (UGSO)) of the upgraded slag (UGS) process to produce titanium slag from ilmenite. Since UGSO is largely composed of iron oxides, in addition to Mg and Al oxides, it has been used to produce an effective Ni-functionalized spinel catalyst tested in methane DR, methane mixed reforming [30], and pyrolytic oils SR [31]. In this work, we investigate the efficiency of this new catalyst in ethylene DR and CC reactions to produce H_2 and CNF.

2. Results and Discussion

2.1. Fresh Catalyst Characterization

Table 1 illustrates the BET surface area, average pore volume, and average pore diameter for UGSO and Ni-UGSO with different Ni wt.%. We can observe that Ni-UGSO has a smaller BET surface area, smaller pore volume, and smaller pore diameter than UGSO; this is due to the formation of other phases (spinels) as shown on the XRD pattern (Figure 1) that cause a rearrangement of UGSO structure (the signification of each symbol in XRD patterns are presented in Table 2). Regarding the effect of Ni wt.%, no statistically significant change was found. Generally, Ni addition leads to the reduction of specific surface area, pore volume, and pore size of the catalyst.

Table 1. Textural properties of Ni-UGSO with different Ni contents (5, 10, and 13 wt.%).

Catalyst	BET Surface Area (m^2·g^{-1})	Average Pore Volume (cm^3·g^{-1}) [a]	Average Pore Diameter (nm) [b]	FWHM (nm)	Ni spinel Crystal Size (nm) [c]
UGSO	4.96	0.0256	20.1	–	–
Ni-UGSO 5%	2.91	0.0134	18.2	0.429	3.38
Ni-UGSO 10%	2.66	0.0126	19.2	0.426	3.41
Ni-UGSO 13%	2.87	0.0132	17.7	0.429	3.38

[a] Pore volume was obtained from $P/P_0 = 0.97$. [b] Pore diameter was obtained from Barret–Joyner–Halenda (BJH) desorption method. [c] Ni crystallite size was calculated from Scherrer Equation.

Table 2. XRD phase legend.

Symbol	Phase	Symbol	Phase
◆	MgFeAlO$_4$, MgFe$_2$O$_4$, Fe$_3$O$_4$, AlFe$_2$O$_4$,	☾	(FeNi)O
■	NiFe$_2$O$_4$, FeNiAlO$_4$	○	Fe
▲	NiO	◇	Ni
●	MgO	△	FeNi$_3$
⬠	Carbon	★	Fe$_2$O$_3$
✖	Fe$_3$C	▫	FeO

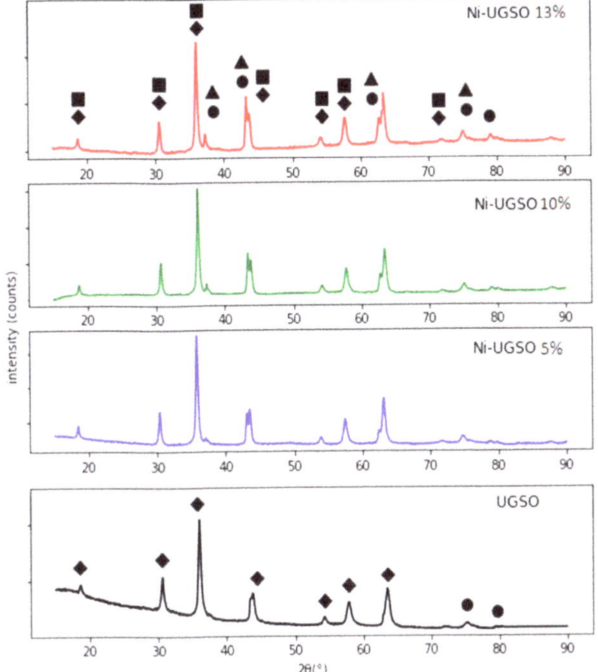

Figure 1. XRD analysis of Ni-UGSO with different Ni contents (0, 5, 10, and 13 wt.%).

We can also observe that the $NiFe_2O_4$ crystal sizes are nanometric and there is no difference in the crystal size in function of wt.% of Ni on the catalyst.

XRD patterns of fresh catalysts with different wt.% of Ni are shown in Figure 1, which shows that the patterns of the three catalysts are identical. The same family phases have been detected whatever the Ni percentage. In summary, the catalysts are mainly composed of two phases: spinels, in the most probable order of formation (figure of merit (FOM) smallest); $MgFeAlO_4$, $MgFe_2O_4$, Fe_3O_4, $NiFeAlO_4$, $AlFe_2O_4$, $NiFe_2O_4$; and monoxides (NiO, MgO), which coexist in their solid solution. When comparing to the pattern of fresh UGSO, the new crystalline phases are a clear indication that the Ni has been well integrated into the structure of the UGSO.

According to the TEM images (Figure 2a,c), the catalyst particles are faceted and have a size distribution ranging between 70 nm and 355 nm.

SAED patterns (Figure 2b,d) indicate that the catalyst is composed of the spinels $NiFe_2O_4$ and Fe_3O_4, and oxides NiO and (MgFe)O (Table 3). These results corroborate the XRD analysis results.

Table 3. Indexation of d-spacing measured by SAED.

Measured D-Spacing (A°)	Indexation	Theoretical D-Spacing (A°)
1.48 and 1.46	(4 4 0) (MgFe)O/NiO	1.47
1.67	(4 2 2) $NiFe_2O_4$/Fe_3O_4	1.7
2.03 and 2.06	(4 0 0) (MgFe)O/NiO	2.08
2.47 and 2.41	(3 1 1) $NiFe_2O_4$/Fe_3O_4	2.51
2.9	(2 2 0) $NiFe_2O_4$/Fe_3O_4	2.94
4.61	(1 1 1) $NiFe_2O_4$/Fe_3O_4	4.8

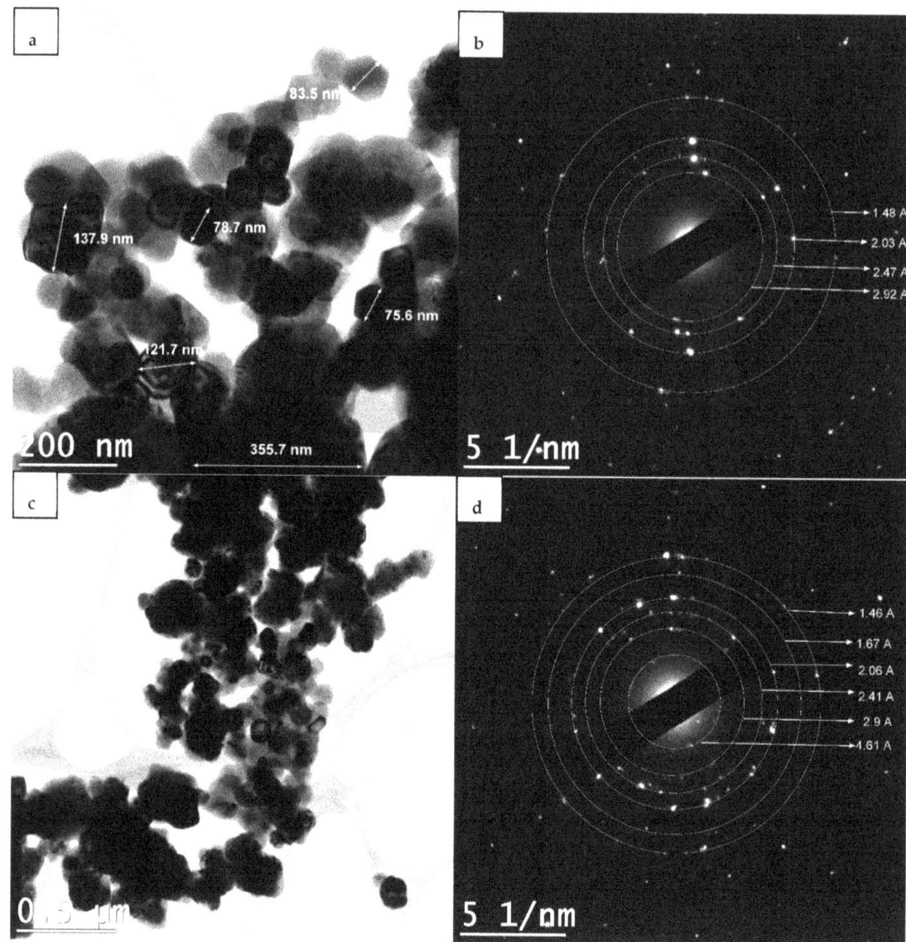

Figure 2. TEM analysis of Ni-UGSO 13% (**a,c**) and its corresponding selected area electron diffraction (SAED) (**b,d**).

2.2. Catalyst Activation and Characterization before DR and CC Reactions

Before the DR and CC reactions, Ni-UGSO was activated by H_2. Concerning structural properties, we notice that the activation has increased the BET surface area, pore volume, and pore diameter (Table 4). The effect of the activation is the reduction of metal oxides into metal particles as we can see in XRD pattern (Figure 3), especially into Ni and Fe metal and their alloys, which leads to a pore enlargement and a BET increase due to the nanometric size of the metallic species proved by TEM (Figure 4).

Table 4. Textural properties of Ni-UGSO 13% before and after activation.

Catalyst	BET Surface Area $(m^2 \cdot g^{-1})$	Average Pore Volume $(cm^3 \cdot g^{-1})$ [a]	Average Pore Diameter (nm) [b]	FWMH	Ni Crystallite Size (nm) [c]
Fresh Ni-UGSO 13%	2.87	0.0132	17.7	0.43	3.4
Activated Ni-UGSO 13%	4.81	0.0238	19.1	1.39	1.2

[a] Pore volume was obtained from $P/P_0 = 0.97$. [b] Pore diameter was obtained from Barret–Joyner–Halenda (BJH) desorption method. [c] Ni crystallite size was calculated from Scherrer Equation.

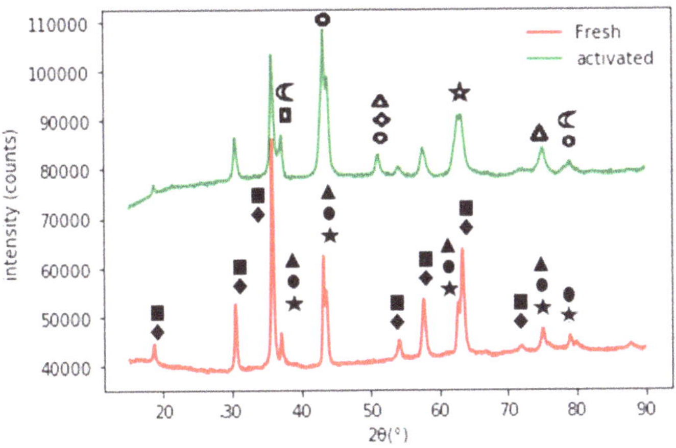

Figure 3. XRD analysis of Ni-UGSO 13% before and after activation.

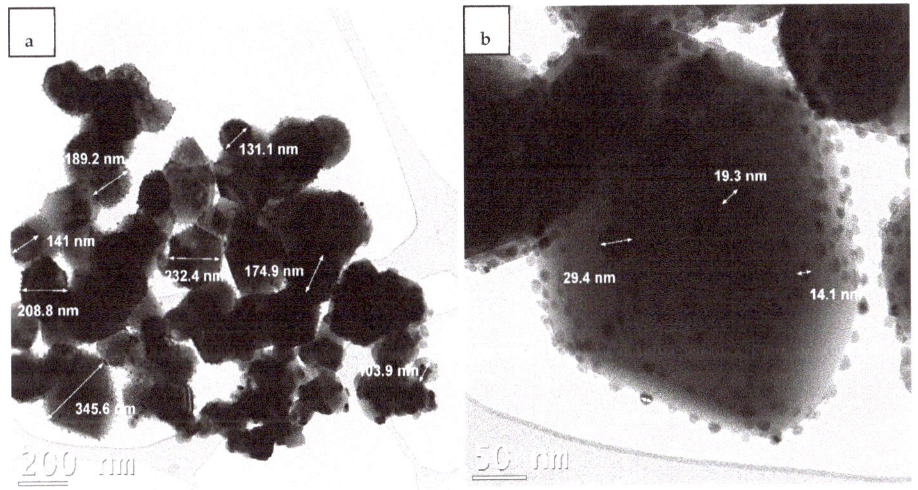

Figure 4. TEM analysis of activated Ni-UGSO 13%, (**a**) support particle size and (**b**) crystallite size.

When comparing the XRD patterns of the catalyst structure before and after activation by H_2, we can observe the appearance of peaks attributed to the metallic phases Ni, Fe, and their alloys. Yu et al. [32] have shown that the reduction of catalysts containing Ni and Fe leads to the formation of their alloys such as tarnite and kamacite, and the proportion of Ni:Fe on the alloy after reduction depends on their initial mass ratio. We can also observe the presence of FeO, which means that the magnetite has been reduced partially into wüstite and iron.

TPR analysis was used to determine the reduction temperatures of the different metal species present in the catalyst. The TPR profiles for the three different catalysts with different wt.% (Figure 5) have the same shape with three distinctive peaks. The difference is in the amount of H_2 consumed, which, as expected, increases with the wt.% of Ni in the catalyst.

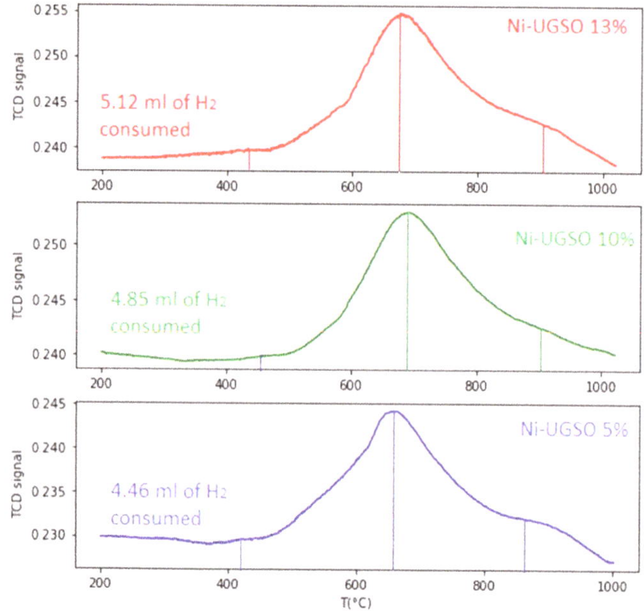

Figure 5. TPR analysis for Ni-UGSO with different Ni contents (5, 10, and 13 wt.%).

In fact, the reduction temperature of metals depends on their interaction with the support and their location, as well as the structure to which it belongs (oxide or spinel). The reduction of Ni-UGSO by H_2 essentially leads to the formation of metallic Fe and Ni particles in addition to their solid solution. Al and Mg are resistant to reduction and remain in their oxidized state.

The analysis of the Ni-UGSO 13% TPR pattern depicts the main reduction peak (the one in the middle) and two others. The first peak can be attributed to the reduction of free NiO (not in interaction with all other phases) and the reduction of Fe^{3+} to Fe^{2+}. The second peak can be assigned to the reduction of both Fe^{3+} species to Fe^{2+} and Fe, and Ni^{2+} to Ni (NiO moderately interacting with other phases). The third peak can be attributed to the reduction of NiO strongly interacting with MgO or having a strong interaction with spinel $MgFeAlO_4$ [30].

As shown in Figure 4, the reduction of the catalyst by H_2 led to the formation of metal crystallites with different sizes on the surface of crystals that have not been reduced (Al and Mg oxides). Similar results were found by Romero et al. [29], who studied the reduction of zeolite-supported Ni- and Fe-based catalysts. They observed Fe and Ni crystallites with different distributions formed on the surface of the zeolite. In fact, the activation of catalysts by H_2 led to the reduction of oxides into small metallic particles, which are the active phase for the growth of CNF.

As depicted in Figure 4b, the crystallite sizes are in the range of 10–30 nm. Yu et al. [32] have found that the reduction of the Ni:Fe (6:1) catalyst has an average crystallite size of 6 nm with a Gaussian-like distribution. The difference observed when comparing our results with those in the literature might be due to a different Ni:Fe ratio and/or to the reduction conditions (nature of the substrate and rate of heat and mass transfer). Some sintering seems to have taken place due to reduction because, if we compare Figures 4a and 2, we notice that the support particle size has increased (100–400 nm).

The EDX pattern (Figure 6) shows that these crystallites are composed mainly of Ni and Fe and no O has been detected. This proves that these crystallites are metallic and they are Ni, Fe, and/or Ni–Fe alloys, thus corroborating the already presented XRD results.

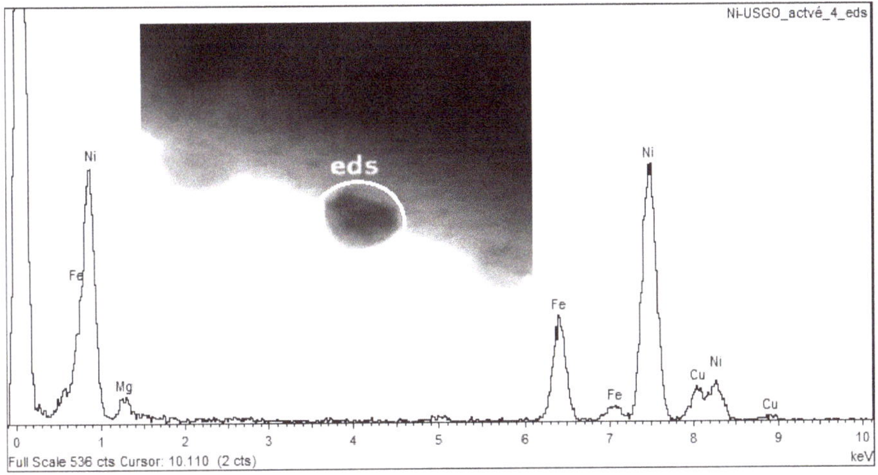

Figure 6. EDX analysis of activated Ni-UGSO 13%.

2.3. Ni-UGSO Catalyst Performance

2.3.1. Thermodynamic Investigation

FactSage software was used to study the thermodynamic equilibrium of the C_2H_4 CC and DR reactions at different conditions of temperature (450–850 °C) and molar ratios C_2H_4/CO_2 (1/1-3/1) at atmospheric pressure. Equilibrium composition, heat, and enthalpy of the reaction, as well as the amount of deposited carbon, were studied during this research. This investigation allowed us to choose the experimental conditions. The results are shown in Figures 7–9.

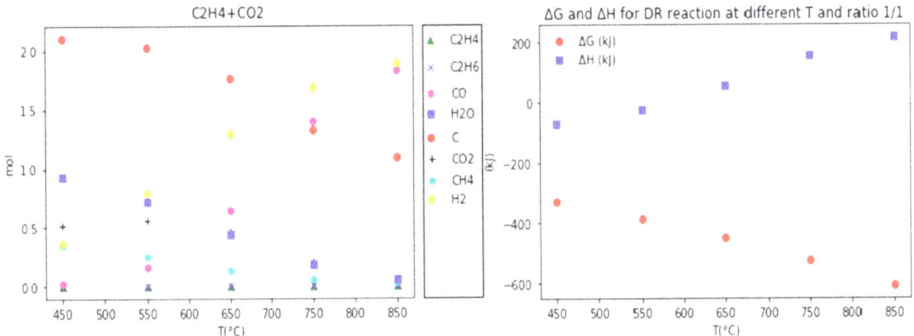

Figure 7. Thermodynamic study of DR reaction at different temperature at ratio 1/1.

ΔH is negative for the CC reaction, which means that the reaction is exothermic, and it increases with the increase of temperature (Figure 9). The ΔH of the DR reaction is negative for temperatures below 600 °C for both ratios. This means that at temperatures higher than 600 °C the reaction is endothermic. The heat of this reaction increases with the increase of the amount of ethylene in the feedstock.

The decline of ΔG with temperature illustrates that the equilibrium is displaced toward the products. For the CC reaction, at temperatures below 700 °C, ΔG is positive and, therefore, the reaction is not taking place. For the DR reaction, ΔG is higher at higher C_2H_4/CO_2 ratios, which means that conversion is favored.

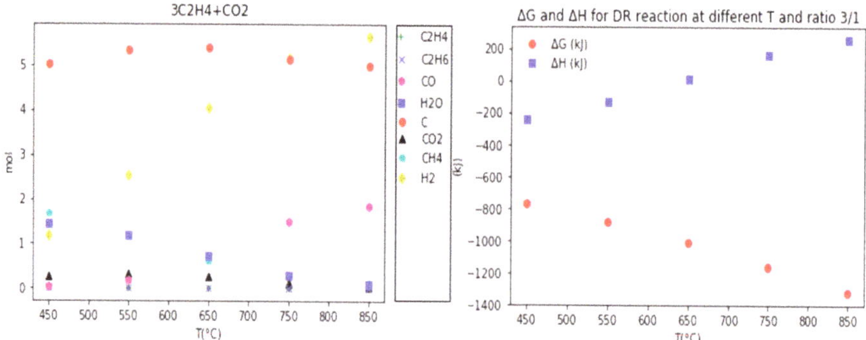

Figure 8. Thermodynamic study of DR reaction at different temperature at ratio 3/1.

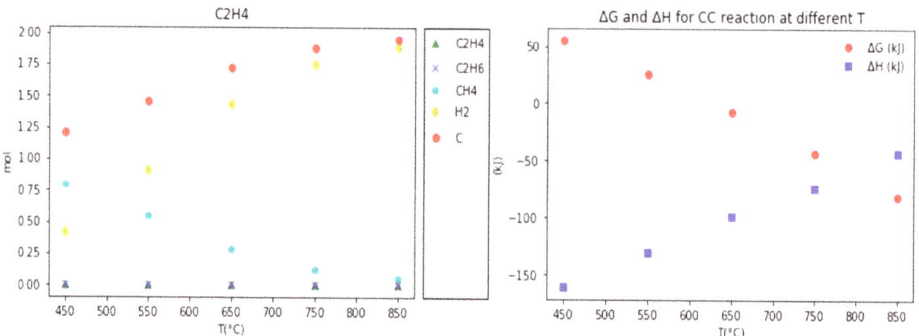

Figure 9. Thermodynamic study of CC reaction at different temperature.

For the CC reaction, H_2 and carbon amounts at the equilibrium increase with temperature (Figure 9). However, for the DR reaction, the H_2 amount increases with the temperature for both ratios, while the C amount decreases with the temperature at ratio 1/1 and reaches its maximum at 650 °C at a ratio of 3/1.

Based on these results, the three following temperatures have been chosen to be studied experimentally: 550 °C, 650 °C, and 750 °C. Since we want to maximize carbon and H_2 production, a 3/1 ratio of C_2H_4/CO_2 was chosen.

2.3.2. Study of CC Reaction

The effect of temperature and Ni percentage on CNF and H_2 yields, as well as carbon growth rate, are presented in Figure 10 and Table 5, respectively.

Table 5. Carbon growth rate for the CC reaction using Ni-UGSO with different Ni wt.% (5, 10, and 13) at T = 550 °C, 650 °C, and 750 °C for 2 h TOS.

Catalyst	Carbon Growth Rate ($g_C \cdot g_{cat}^{-1} \cdot h^{-1}$) At		
	550 °C	650 °C	750 °C
Ni-UGSO 5%	1.42	1.7	1.78
Ni-UGSO 10%	1.78	1.82	1.94
Ni-UGSO 13%	2.1	2.2	2.8

Yu et al. [32], who used a bimetallic catalyst of Ni–Fe with different Ni loading levels, found that productivity is higher for higher T and higher Ni loading. This work shows that C and H_2 yields increase with temperature (Figure 10). This is in accordance with the previously reported literature [32]. Moreover, the carbon growth rate increases with T because the solubility and diffusion of the carbon in the solid metallic phases also increases with T [33]. This effect is even more pronounced at higher percentages of Ni in the catalyst.

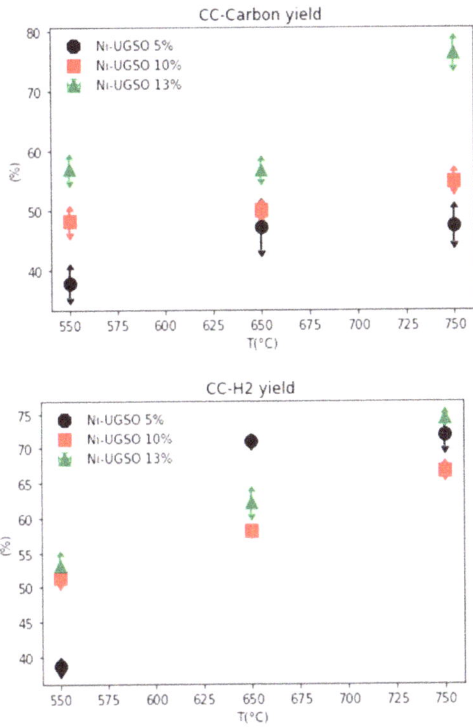

Figure 10. Carbon and H_2 yields for the CC reaction using Ni-UGSO with different Ni wt.% (5, 10, and 13) at T = 550 °C, 650 °C, and 750 °C for 2 h time-on-stream (TOS).

Since the highest carbon and H_2 yields were observed at T = 650 °C and wt.% of Ni = 13%, the reaction results at these conditions are presented in detail in Table 6 and Figure 11.

Table 6. General experimental results for CC reaction using Ni-UGSO 13% at 750 °C for 2 h TOS.

C_2H_4 (mL/min)	30
Ar (mL/min)	10
Catalyst weight (g)	0.5
TOS (min)	120
$GHSV_{STP}$ (mL·h^{-1}. g)	4800
C_2H_4/Ar	3
Ar/C_2H_4	0.33
Carbon (g)	2.8
Carbon production rate (g$_C$·g$_{cat}^{-1}$·min^{-1})	0.047
Carbon production rate (g$_C$·g$_{cat}^{-1}$·h^{-1})	2.8
Carbon yield (%)	76.25
Total H_2 yield (%)	74.46
Total C_2H_4 conversion (%)	92.24
Mass balance error for C (%)	3.31
Mass balance error for H (%)	0.08

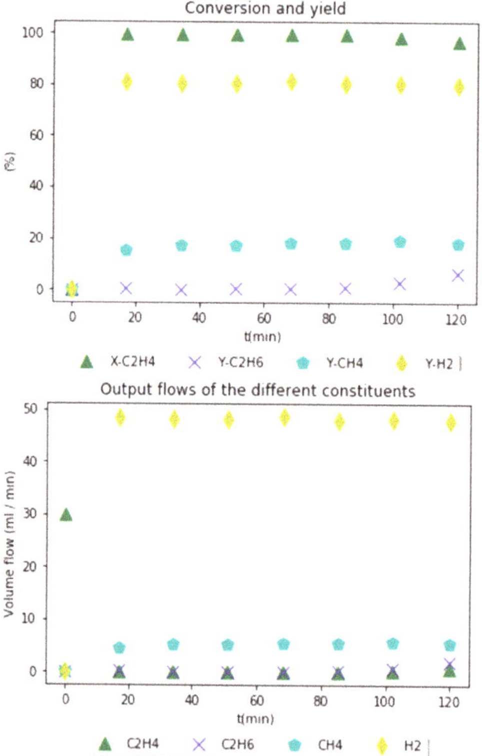

Figure 11. Experimental results for CC reaction using Ni-UGSO 13% at 750 °C.

Overall, the highest carbon and H_2 yields, Y_{H2} = 74.46% and Y_C = 76.25%, respectively, are observed at T = 750 °C and 13% of Ni. Thus, the carbon turnover frequency (TOF) expressed per mass of catalysts was 2.8 $g_C \cdot g_{cat}^{-1} \cdot h^{-1}$ at a flow rate of 30 mL/min (Table 6). When compared to the catalysts used in the literature, this catalyst has shown similar to better performance. Yu et al. [32] have produced 3 $g_C \cdot g_{cat}^{-1} \cdot h^{-1}$ and 2.55 $g_C \cdot g_{cat}^{-1} \cdot h^{-1}$ using bimetallic catalysts Ni–Fe(6-1) and Ni–Fe(5-5), respectively, with a feed of $C_2H_4/CO/H_2$ (30/10/10). Diaz et al. [34] studied Ni-SiO$_2$ catalyst for the catalytic decomposition of ethylene to produce carbon, between 600 °C and 700 °C. They obtained the maximum of carbon at 600 °C with 2 $g_C \cdot g_{cat}^{-1} \cdot h^{-1}$ for 60 mL/min of C_2H_4.

Figure 11 shows that the steady-state has been reached very fast during the first 20 min of TOS. The conversion of C_2H_4 is nearly 100% and starts slightly decreasing in the last 20 min. The hydrogen yield is also constant around 80% for 120 min while the rate of carbon formation is also high and equal to 2.82 $g_C \cdot g_{ca}^{-1} . h^{-1}$. The observed high and constant rates of carbon and H_2 formation are due to the high activity of the catalyst at the beginning of the reaction; moreover, even though carbon was formed, the catalyst did not show any deactivation during the TOS of operation. The latter can be explained by the type of carbon formed. Indeed, the carbon formed was analyzed by SEM and it has been proven that it was under the form of CNF (Figure 12), which was not affecting considerably the access of the reactants at the surface of the catalyst.

Figure 12. SEM analysis of CNF produced at 750 °C using Ni-UGSO 13% for CC reaction.

2.3.3. Study of DR Reaction

The effect of temperature and Ni percentage on CNF and H_2 yields, as well as carbon growth rate, are illustrated in Figure 13 and Table 7, respectively.

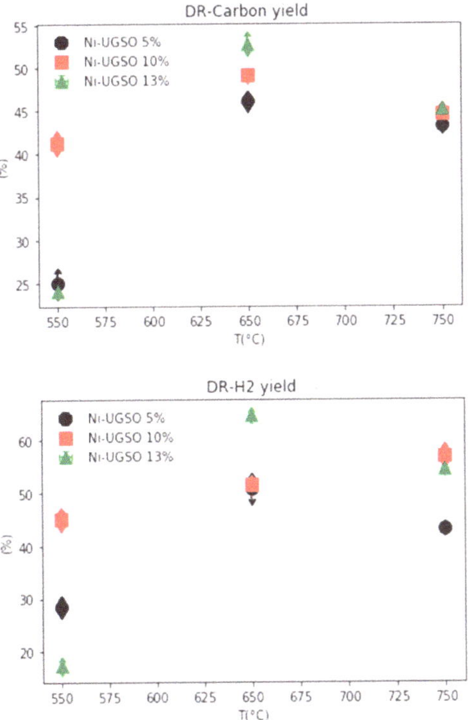

Figure 13. Carbon and H_2 yield for the DR reaction using Ni-UGSO with different Ni wt.% (5, 10, and 13) at T = 550 °C, 650 °C, and 750 °C for 2 h TOS.

Table 7. Carbon growth rate for the DR reaction using Ni-UGSO with different Ni wt.% (5, 10, and 13) at T = 550 °C, 650 °C, and 750 °C for 2 h TOS.

Catalyst	Carbon Growth Rate ($g_C \cdot g_{cat}^{-1} \cdot h^{-1}$) At		
	550 °C	650 °C	750 °C
Ni-UGSO 5%	1.1	1.96	1.83
Ni-UGSO 10%	1.7	2.05	1.9
Ni-UGSO 13%	0.9	2.25	1.96

Figure 13 shows that the carbon yield for all Ni contents has a maximum at 650 °C. Although it seems that the same applies to the H_2 yield, the latter keeps increasing in the case of the 10% Ni content catalyst. The amount of Ni active sites is a parameter that plays a significant role in terms of catalytic activity. The BET results in Table 1 show that the specific surface and the average pore volume is not a function of the Ni content in the range between 5 and 13 wt.%. The small difference observed in the case of 10 wt.% Ni catalyst is within the experimental error and cannot be used as a differentiation argument. In light of the above, the difference in H_2 yield observed in the case of the 10 wt.% Ni catalysts might be explained in the following way:

(a)　As expected by thermodynamic calculations, the temperature around 600 °C is optimal for carbon production and this is clearly shown by the experiments in Figure 13;

(b)　At higher temperatures (i.e., 750 °C), carbon production decreases, but the H_2 yield must increase. Although this is the trend observed with the 10 wt.% Ni catalyst, in the case of 5 wt.% Ni, this increase is nil. In the case of 13 wt.% Ni, we observe the opposite. The most plausible speculation is that, at 5 wt.% Ni, the active catalytic sites are low, while, at 13 wt.% Ni, the Ni distribution is less than optimal. It is well known that in almost all heterogeneous catalysts there is an optimal active metal content below which and above which the catalytic activity decreases.

Since the target of the manuscript is the comparison of two regimes with a number of Ni-UGSO formulations, there are no available surface data to support further discussion. Our continuous efforts are now focusing namely on these aspects.

Both carbon and hydrogen yields are maximal at 650 °C. This behavior can be explained by carbon and H_2 formation and consumption reactions. In fact, carbon is produced from C_2H_4 decomposition and Boudouard reaction and consumed by the gasification reaction, while H_2 is produced by C_2H_4 decomposition and consumed by RWGS reaction. C_2H_4 decomposition and RWGS reactions are favored by high temperatures. Boudouard has a thermodynamic maximum of carbon formation around 550 °C. At T lower than 650 °C, the formation rate exceeds the consumption rate

Concerning the effects of Ni, the yield of carbon and H_2 increases with the increase of the Ni weight percentage in the catalyst, and this is attributed to the higher catalytic activity at higher Ni loading levels and consequently faster reaction rates. We can notice an exception for Ni-UGSO 13% at T = 550 °C where the yields are very low. This could be explained by the fact that the catalyst at such a low T with such a high load of Ni has not reached its highest activity within 2 h.

Since the highest carbon and H_2 yields were observed at T = 650 °C and wt.% of Ni = 13%, the reaction results at these conditions are presented in detail in Table 8 and Figure 14.

Ni-based catalysts have been used in the past for DR reactions, especially for methane and ethanol, but there are few studies on ethylene dry reforming. Jankhah et al. [15] examined in detail the dry reforming reaction of ethanol using activated stainless-steel strips as a catalyst (strip surface of 0.04 m^2). Experiments have shown that the results that give the best yields of carbon and H_2 are obtained at a temperature of 550 °C. They have obtained a carbon rate equal to 3.6 $g \cdot h^{-1}$ and an H_2 yield of 76.33%. Since this catalyst is 2D and not 3D, the equivalent carbon TOF is related to the catalyst surface and not to the weight and is equal to 90 $g \cdot h^{-1} \cdot m^{-2}$.

Table 8. General experimental results for the DR reaction at 650 °C and Ni-UGSO 13% for 2 h TOS.

C_2H_4 (mL/min)	30
CO_2 (mL/min)	10
Catalyst weight (g)	0.5
TOS (min)	120
$GHSV_{STP}$ (mL·h^{-1}·g^{-1})	4800
C_2H_4/CO_2	3
CO_2/C_2H_4	0.33
Carbon (g)	2.25
Carbon production rate (g$_C$·g$_{cat}$$^{-1}$·min^{-1})	0.0375
Carbon production rate (g$_C$·g$_{cat}$$^{-1}$·h^{-1})	2.25
Carbon yield (%)	53.57
Total H_2 yield (%)	67.47
Total C_2H_4 conversion (%)	91.29
Total CO_2 conversion (%)	88.48
Mass balance error for C (%)	9.16
Mass balance error for H (%)	1.46
Mass balance error for O (%)	8.81

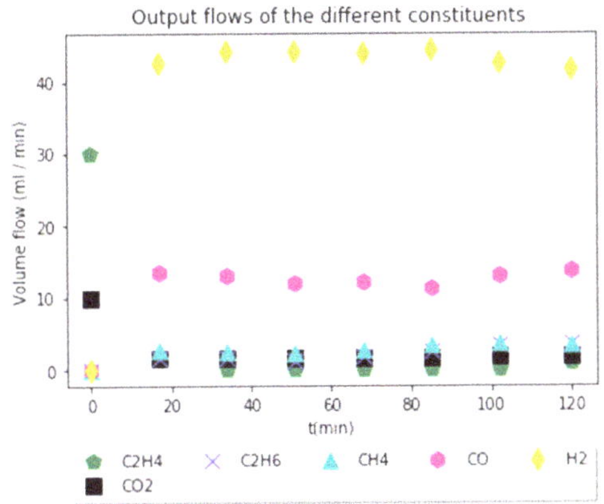

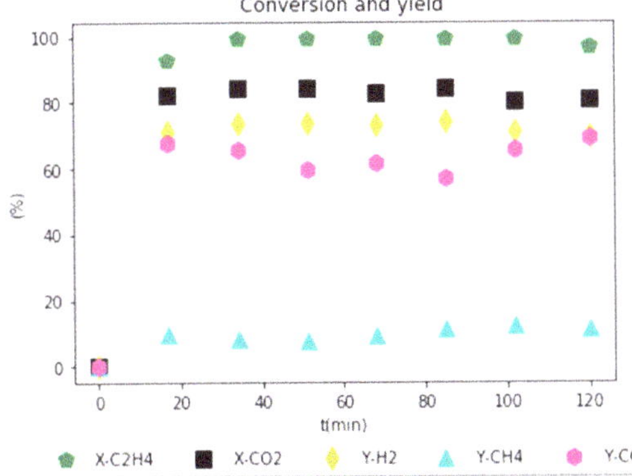

Figure 14. Experimental results for DR reaction using Ni-UGSO 13% at 650 °C for 2 h TOS.

We observed that, during the first 100 min, the conversion of C_2H_4 is near 100% and starts slightly decreasing during the last 20 min. Hydrogen yield is constant for 120 min and equal to 65% (Figure 14), and the rate of carbon formation is high and equal to 2.25 g$_C$·g$_{cat}$$^{-1}$·h^{-1}. These high and constant rates of carbon and H_2 formation are due to the high activity of the catalyst and can be explained by the following: the H_2 formed contributes to the additional activation of the catalyst through the reduction of iron oxides. This is proven by the presence of Fe and Ni metal peaks and the disappearance of iron oxide peaks on used catalyst XRD (Figure 15). It has been demonstrated that the carbon under the form of catalytically induced CNF itself has catalytic properties [4]. Although the activity measured through carbon TOF per mass of CNF is lower, if the TOF is calculated per mass of carbides content of the CNF, it is shown to be higher. This explains, at least partially, why the catalytic activity remains high even when the catalyst surface is covered by CNF.

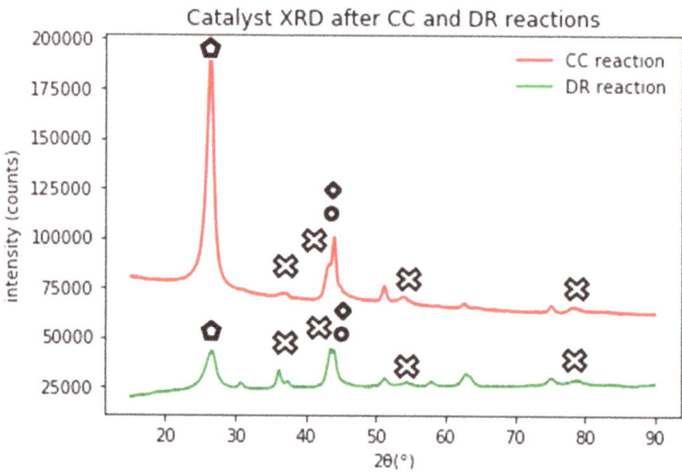

Figure 15. XRD analysis of Ni-UGSO 13% after the CC reaction at 750 °C and after the DR reaction at 650 °C for 2 h TOS.

The carbon formed was analyzed by SEM and it has been proven that it is mainly under the form of CNF (Figure 16).

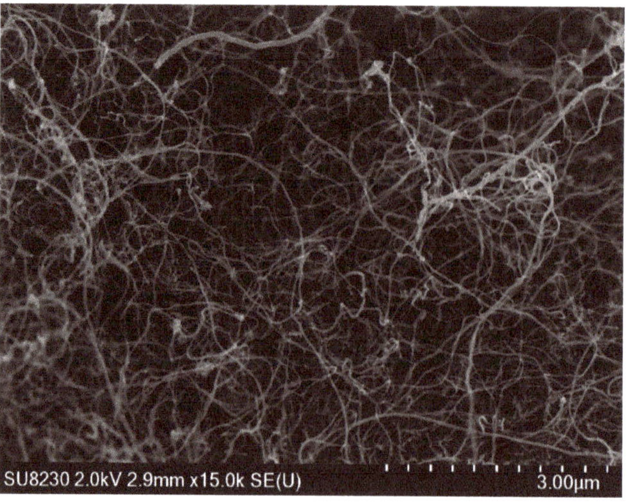

Figure 16. SEM analysis of CNF produced at 650 °C using Ni-UGSO 13% for DR reaction for 2 h TOS.

2.4. Characterization of CNF and Spent Catalyst

2.4.1. XRD Analysis

Figure 15 shows the XRD of deposited carbon on the used catalyst after the DR and CC reactions. The peak at $2\theta = 26°$ confirms that the carbon formed is graphitic, no peaks of oxides have been detected, and only Ni and Fe were present in the patters, which proves that their oxides were reduced during the reactions. Carbide formation was expected because carbides are known to be the precursor of CNF especially with iron-based catalysts. Fe_3C is metastable under the reaction conditions so it is decomposed to CNF and α-Fe [35]. Nickel plays a key role in the formation of CNF because, when UGSO was used alone, catalyst activity was low [30]. When nickel was added, the quantity and quality of CNF were found to have improved. Nickel catalyzes the C–C bond cleavage, thus producing carbon species radicals and atomic carbon that diffuses and dissolves in the iron to form a solid solution of iron carbides [36].

2.4.2. TGA

Figure 17 shows the results of TGA analysis. For the DR sample, at T up to 450 °C, a mass gain of 0.25% was detected. Between 450 °C and 765 °C, there was a mass loss of about 60%; finally, between 765 °C and 890 °C, a mass gain of 0.05% was observed.

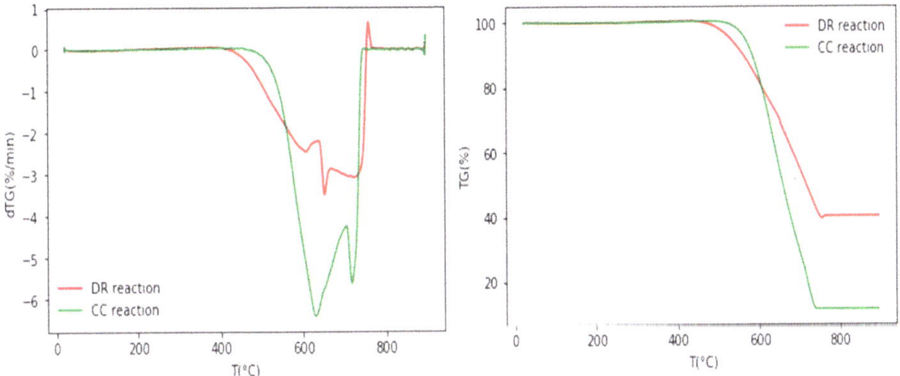

Figure 17. TGA analysis of Ni-UGSO 13% after CC reaction at 750 °C and after DR reaction at 650 °C for 2 h TOS.

For the CC sample, up until 500 °C, a mass gain of 0.23% was measured. From 500 °C to 750 °C, there was a mass loss of 88.62% and, finally, between 750 °C and 900 °C, a mass gain of 0.05% was observed.

The higher the temperature of oxidation, the higher the degree of structural order. Thus, as can be seen in Figure 17, the oxidation of CNF produced by DR (CNF-DR) begins at a temperature lower than that in the case of CNF formed by CC (CNF-CC) (450 °C vs. 500 °C). In the literature, it has been reported that the oxidation of graphite and C_{60} in TGA occurs at 645 °C and 420 °C, respectively [37]. The oxidation temperature of CNF-CC is similar to that reported for CNT [38] and higher than that reported by Sui et al. [39]. They are all lower than the graphite oxidation temperature. Serp et al. [40] have confirmed that CNT and CNF are more reactive than graphite. They have shown that CNF samples with 10% of remaining metal (produced from ethylene on Fe/SiO_2 catalysts) present a maximum gasification rate at 650 °C. The single-wall carbon nanotube (SWCNT), which is the carbon nanostructure that has the least remaining metal percentage (less than 1% of metal) and the least defects on its surface, presents a maximum gasification rate at 800 °C. MWCNT, with 3% and 7.5% of residual metal, presents a maximum rate at 650 °C and 550 °C, respectively. According to these findings, the presence of defects on the CNF surface and the presence of residual metal within the carbon nanostructures that can catalyze carbon gasification cause a shift to lower temperatures. While the oxidation resistance of the DR carbon and the CC carbon is different, we can say that either CNF-DR carbon is more structured than CNF-CC, or that it contains more metal, or even that their surfaces are not the same, which means that they are two distinct types of CNF. The TEM analyses reported below help us to identify the type of CNF produced. It is well known that there are different types of CNF depending on the arrangement of the graphene plans. Accordingly, CNF are classified into three categories: platelets, fishbone, and stacked-cup CNF [2].

2.4.3. SEM-EDX Analysis

SEM images have shown that, in this experiment, the carbon is under the form of filaments of varying diameters. For CNF-DR, the diameter range is 15–50 nm (Figure 18) and for CNF-CC it is 25–75 nm (Figure 19). Using backscattered electron imaging (Figure 20), we can see that the metal particles are located on the top of the nanofilaments.

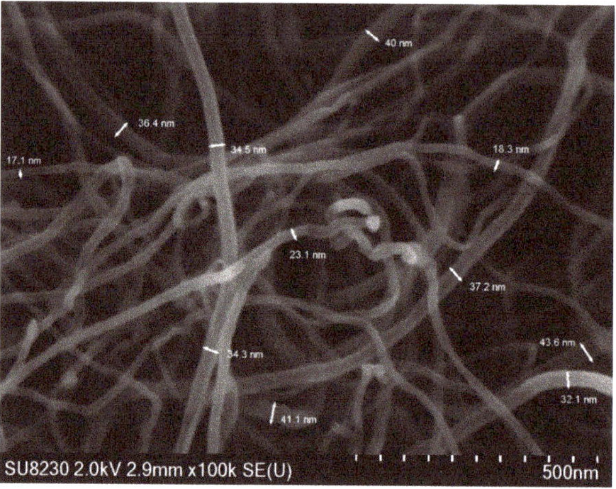

Figure 18. SEM analysis of carbon deposited on Ni-UGSO 13% after DR reaction at 650 °C for 2 h TOS.

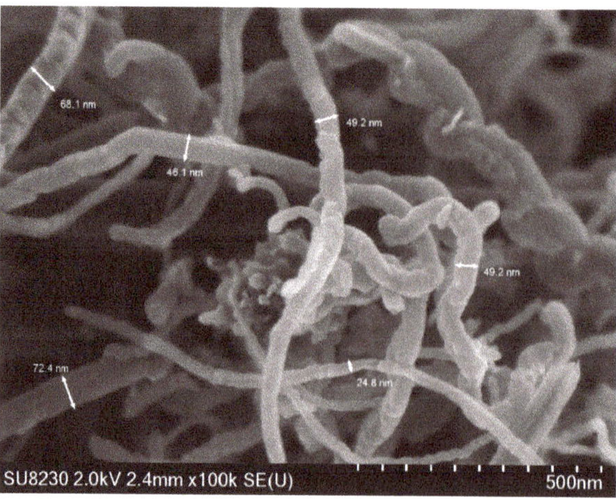

Figure 19. SEM analysis of carbon deposited on Ni-UGSO 13% after the CC reaction at 750 °C for 2 h TOS.

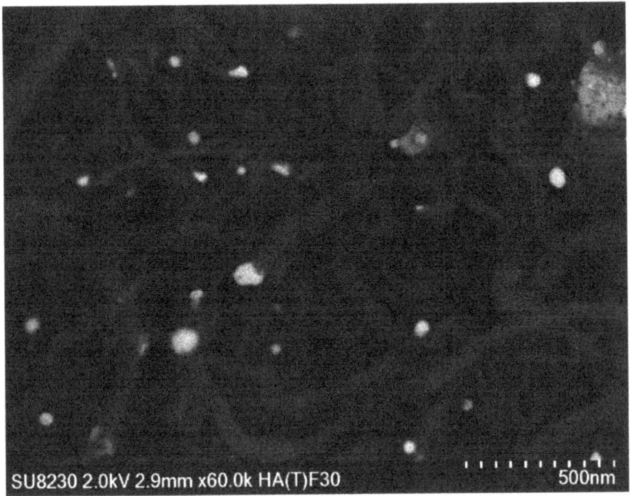

Figure 20. SEM analysis (using backscattered electron imaging) of carbon deposited on Ni-UGSO 13% after DR reaction at 650 °C for 2 h TOS.

The EDX images presented in Figures 21 and 22 give a chemical analysis of the spent catalyst and carbon deposited. We notice on the spectrum that the peaks of carbon are intense for both reactions, which proves the existence of carbon corresponding to CNF, as we can see on the SEM images.

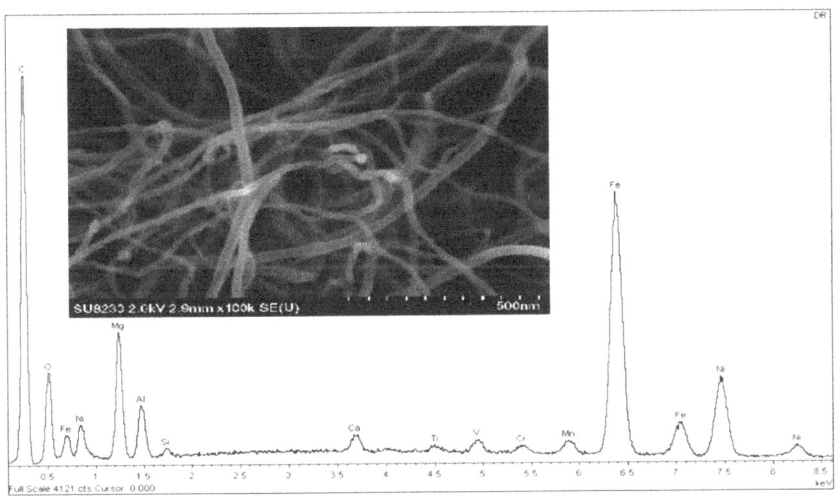

Figure 21. EDX analysis of carbon deposited on Ni-UGSO 13% after the DR reaction at 650 °C for 2 h TOS.

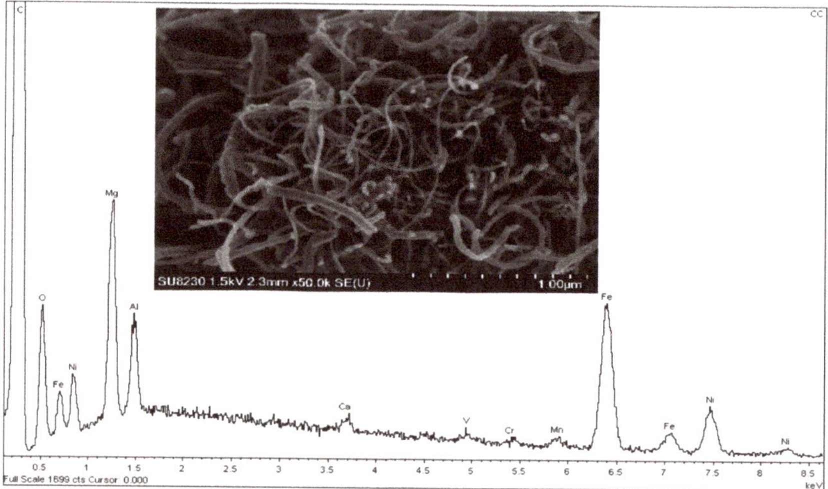

Figure 22. EDX analysis of carbon deposited on Ni-UGSO 13% after the CC reaction at 750 °C for 2 h TOS.

2.4.4. TEM-EDX Analysis

- DR Reaction Sample

Examination of the carbon formed in TEM demonstrated that the carbon formed consists entirely of cylindrical and straight CNF. However, all CNF produced are of the fishbone type (graphene sheets at a certain angle relative to the hollow core fiber main axis). Romero et al. [29] have indicated in their article that Fe-based catalysts are responsible for the growth of two types of CNF, tubular (sheets parallel to fiber axis), and platelets; while Ni-based catalysts are responsible for the growth of fishbone type CNF only. However, in an earlier work of our research group [15] where steel was used as a dry reforming catalyst to produce CNF, it has been found that different forms of CNF were formed during the reactions, including fishbone ones. Yu et al. [32] have done a study in conditions similar to ours, which consists of decomposing a $C_2H_4/H_2/CO$ mixture with a Ni–Fe bimetallic catalyst and have found that the CNF formed during their study were of the fishbone type. They have reported that the fishbone is probably formed when a Boudourd reaction took place. In light of these results, we can deduce that the catalyst is not the only factor that influences the type of carbon formed; other factors, such as the type of gas supplied and the temperature, also have an influence as discussed further in Section 2.5.

When the graphitic sheets stack with one another, the angle formed between the sheets and the fiber axis was not always the same for all of the produced CNF. Different angles (11°, 17°, 23°) are observed. It also seems that the diameter of the hollow core depends on this angle: the bigger the angle, the smaller the diameter (Figure 23). Since the catalyst's structure changes over TOS, it is rather impossible to control the width of the CNF as it is also discussed further on (Section 2.5).

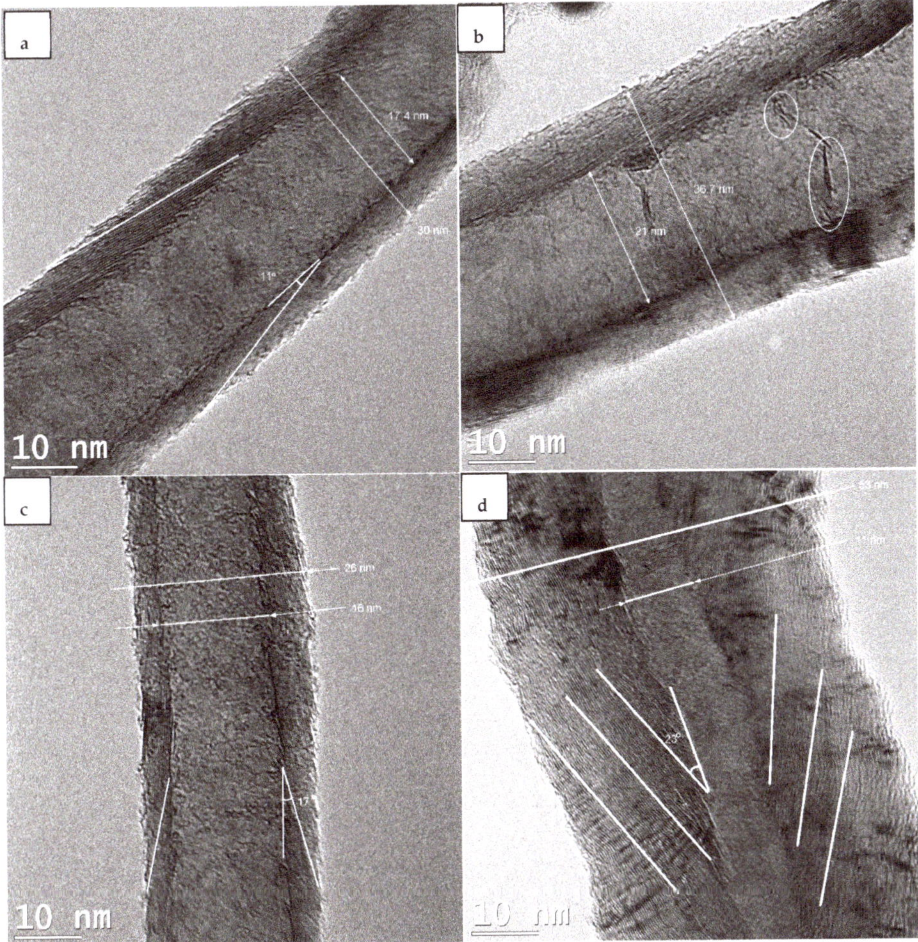

Figure 23. TEM analysis of carbon deposited on Ni-UGSO 13% after the DR reaction at 650 °C for 2 h TOS, (**a**) CNF with d = 30 nm, (**b**) CNF with d = 36.7 nm, (**c**) CNF with d = 26 nm, (**d**) CNF with d = 53 nm.

The distance between the graphitic sheets is 0.340 nm (Figure 24), which is higher than the corresponding distance of graphite (0.335 nm). This means that CNF have structural defects and their structure is only ordered locally, not globally [29]. These defects are shown in as waved lines. The fact that the CNF are less ordered and have defects has been verified by TGA profiles (Figure 17), where the latter demonstrates that lower temperatures are necessary to oxidize CNF. Some zones that are darker than others can also be observed; they are due to layers not being stacked identically, and whose local density is different. The degree of graphitization (g) is calculated using this Equation: $d_{hkl} = 3.354 + 0.086 (1 - g)$, where d_{hkl} is the interplanar distance [41]. Thus, the g -value of the CNF produced in this study is g = 46.5%. Romero et al. [29] have found that CNF produced from ethylene decomposition over a Fe-Ni-based catalyst have an interplanar distance of 3.42 Å, therefore g = 23%, which means that they are less graphitized than those obtained in this work.

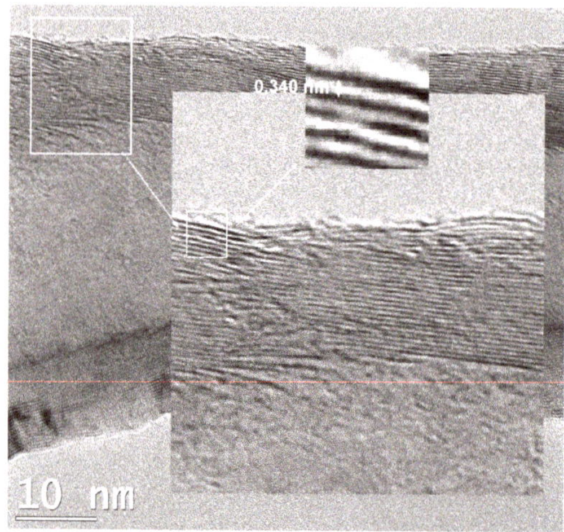

Figure 24. Interplanar distance of graphene sheets.

SAED analysis confirmed that the interplanar distance of CNF is 0.340 nm. The second d_{hkl} can be attributed to planes (102), (220), and (031) of Fe_3C according to JCPDF File # 35-0772, or to planes (111) and (110) of Ni and Fe, respectively. From the presence of Fe_3C, it can be deduced that CNF have grown on Fe. The existence of the Ni atoms in the metallic particles at the bottom of the CNF, as it is proven by SAED (Figure 25) and confirmed by the EDX analysis (Figure 21), confirms that the Ni has mainly participated in one of the stages of the growth of the CNF, which is the decomposition of the HC, while the iron is the main contributor in the second and third stages (dissolution and precipitation). In fact, it is known that Ni and Fe differ in their ability to decompose HC and solubilize carbon. Ni rapidly dehydrogenates the adsorbed HC while Fe is slower, and iron solubilizes carbon better than Ni. Indeed, the solubility of the carbon in the Ni in the range of temperature at which we worked is very low. Lander et al. [42] have experimentally developed an Equation that gives the solubility of carbon in nickel between 700 °C and 1300 °C, which is as follows: $lnS = 2.48 - \frac{4.880}{T}$, where S is the solubility in grams of carbon per 100 gr of nickel and the temperature is in °C. The S value at 700 °C is relatively low (2.5%). This will also be discussed further in Section 2.5.

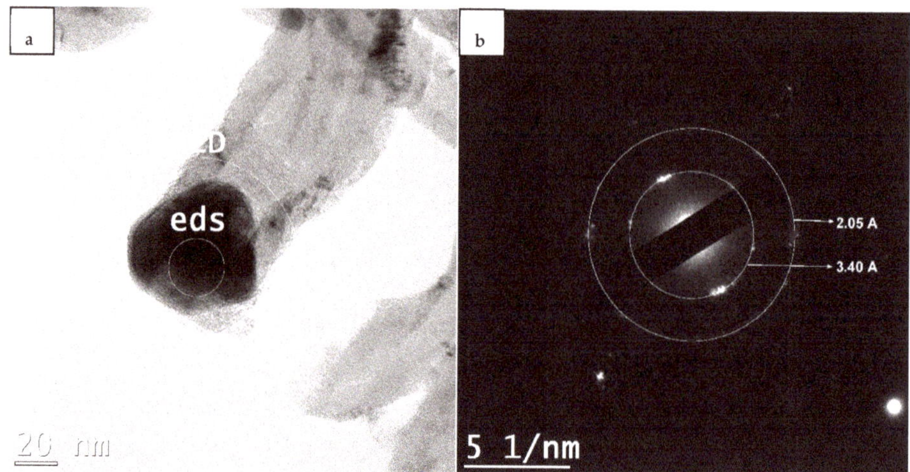

Figure 25. (**a**) Metallic particle at the tip of CNF. (**b**) SAED of this particle.

- CC Reaction Sample

For the CC reaction, we observed that there are different types of CNF, such as the tubular shape (layers are parallel to fiber axis like MWCNT) with a hollow core (Figure 26b,d), where we can also observe that some layers have torn ends. It seems that the sheets tended to connect to fill the inside of the CNF structure. Another type of structure is the bamboo type (Figure 26c). We were also able to observe that CNF formed with irregular stacking of graphene planes (Figure 26a), and we can observe that the graphene planes started out parallel to one another and that the angle of inclination with the fiber axis subsequently changed. The appearance of these CNF is quite similar to those formed by the decomposition of C_2H_4/H_2 over Fe:Ni catalyst studied by Park and Baker [43].

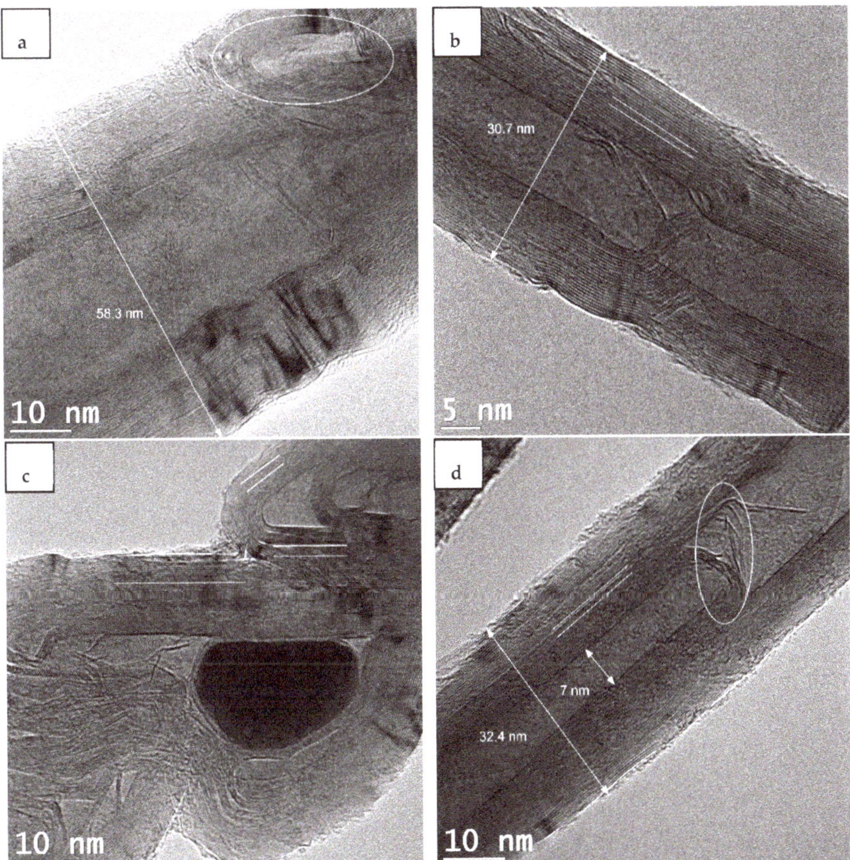

Figure 26. TEM analysis of carbon deposited on Ni-UGSO 13% after CC reaction at 750 °C for 2 h TOS, (**a**) CNF formed with irregular stacked graphene planes, (**b**) and (**d**) tubular CNF with hollow core, (**c**) bamboo CNF.

In Figure 27, we observe that the metal particle is not on the tip of the filament contrary to what was found in DR, but it is encapsulated inside the filament. This could be explained by the fragmentation of the main particle, as its fragments could have been entrained within the body structure of the filament during the growth phase. The same behavior was found by Park and Baker [43] who worked in conditions similar to those used for this work (decomposition of C_2H_4 over Ni–Fe catalyst). The metal particle that was encapsulated by the CNF during the CC reaction appears to have a smooth globular morphology, in contrast to the structure of the catalyst particles at the end of the CNF produced in the DR reaction, which have more angular forms.

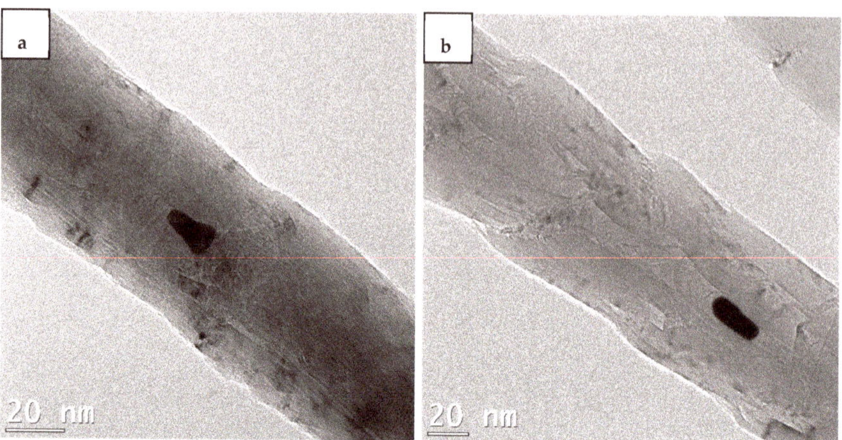

Figure 27. TEM analysis showing a catalyst particle inserted in two different nanofilaments.

The EDX analysis presented in Figure 28 shows Fe and Ni peaks in the pattern as well as C peak, which proves that the metallic particle is composed of both metals and encapsulated by carbon.

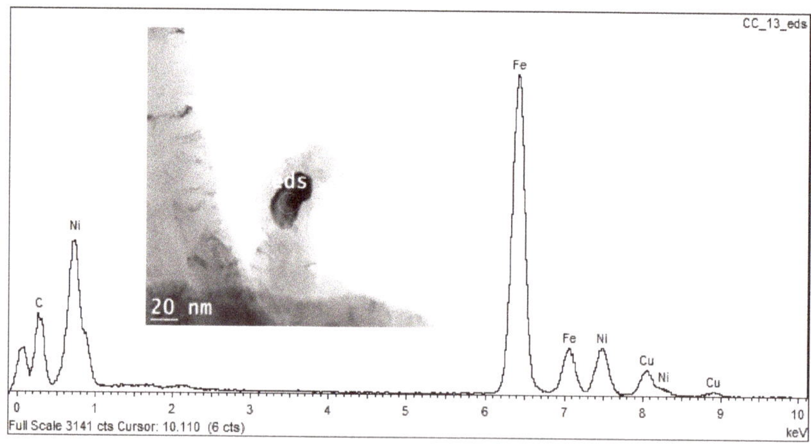

Figure 28. EDX analysis of carbon deposited on Ni-UGSO 13% after CC reaction at 750 °C for 2 h TOS.

2.5. Mechanistic Understanding for the Growth of CNF

The results of this study, which aimed to test the catalytic performance of a new catalyst derived from a mining residue (Ni-UGSO), have shown that Ni-UGSO is also a good catalyst for CNF production from ethylene cracking and dry reforming.

The influence of the catalyst composition as well as of the precursor gas composition and reaction conditions are discussed below.

2.5.1. Influence of the Catalyst on the Growth of CNF

The effects of Ni and Fe on the growth of CNF is different depending on whether it is a CC reaction (decomposition of C_2H_4) or a DR reaction (decomposition of C_2H_4 and disproportionation of CO).

Park et al. [43] have mentioned in their work that Ni-based catalysts are good for the decomposition of ethylene but were not as potent for catalyzing the Boudouard reaction, whereas Fe-based catalysts exhibited the opposite behavior. This fact was verified by our work, where we found that Ni is responsible for ethylene decomposition while Fe is responsible for the growth of CNF. They studied a bimetallic Ni–Fe catalyst for the decomposition of C_2H_4 and CO in the temperature range 600–725 °C, and they proved that increasing the ambient temperature improves the decomposition of C_2H_4 while the Boudouard reaction is favored thermodynamically by temperatures of around 550 °C.

It has also been found that the crystallographic orientation of the metal atoms plays an important role in the ability of the catalyst to decompose the reactive gases [44]. Zhu et al. [45] showed that the Ni (111) plane adsorbs ethylene and acetylene dissociatively, which was proven in this work by SAED results where the Ni (111) facet was found on the metal particle on the top of CNF (Figure 25). Their calculations show that the rate of diffusion of carbon on the Ni (110) plane is the fastest step. However, the carbon deposited on the facet (110) is poorly crystallized because the distance of this plane does not correspond to that required for forming a graphite network. To form a good crystalline carbon structure, the atoms resulted from the decomposition of HC must first diffuse through Ni to dissolve and then precipitate onto the adequate Fe (110) facet (Table 9), which is required to form a graphite network. The energy difference between poor and well-crystallized carbon is the driving force that leads to transfer from one side of a metal to another [44].

Table 9. Indexation of D-spacing measured by SAED.

Measured D-Spacing (Å)	Indexation [43]	Theoretical d-Spacing (Å) [43]
	(111) Ni	2.03
	(110) Fe	2.04
2.05	(102) Fe$_3$C	2.07
	(220) Fe$_3$C	2.03
	(031) Fe$_3$C	2.01
3.40	(200) carbon	3.35

2.5.2. Influence of Catalyst Particle Sizes on the CNF Diameter

Rodriguez [46] has studied the interaction between a metallic surface and carbon. Figure 29 is a schematic representation of the forces involved in the interaction of a metal catalyst particle with a graphite support in the presence of a gaseous environment. The contact angle θ is determined by the surface energy of the graphite support (Y_{SG}), the surface energy of the metal (Y_{MG}), and the metal-graphite interfacial energy (Y_{MS}), and is expressed in terms of Young's Equation:

$$Y_{SG} = Y_{MS} + Y_{MG} \cos\theta$$

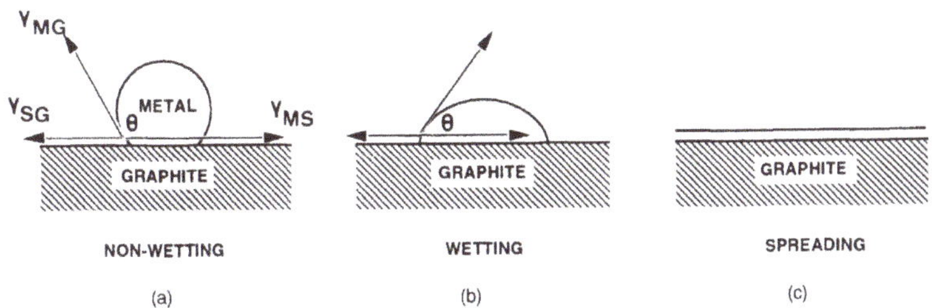

Figure 29. Interaction between surface metal and graphite [46].

It presents changes in the shape of the metal particles as a function of the catalyst wetting degree on the graphite:

(a) When weak forces occur between the metal and the graphite, the resulting contact angle is higher than 90° and there is no wetting;

(b) When strong forces occur between the two components, the contact angle is lower than 90° and wetting occurs;

(c) When the adhesion exceeds the cohesion inside the particle ($Y_{MS} > Y_{SG} + Y_{MG}$), the metal spreads over the graphite support surface [46].

Two forms of catalyst particles associated with nanotubular carbon products, which are clearly different from one another, are commonly observed and presented in the literature [47]: one is conical and the other one is spherical. The conical particles are usually found at the end of the nanofilaments, as it was proven in this work (Figure 25), and the almost spherical particles are observed at the end of the nanotubes [48]. For conical particles, the adhesion exceeds the cohesion inside the particle, which leads the metal to spread on the graphite surface and, after precipitation, the carbon takes the form of piled up stacked cones from the particle, determining the shape of the particle's bottom (Figure 30). In addition, when weak forces occur between the metal and the graphite and the contact angle is >90°, nanotubes are formed (Figure 30).

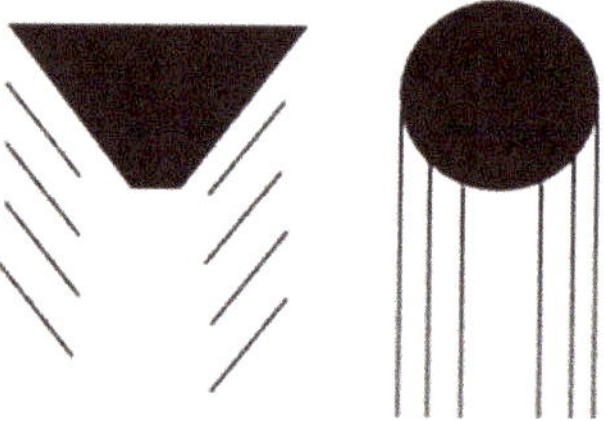

Figure 30. Conical and spherical metal particles on the top of CNF and CNT, respectively.

A sequence of "stop-action" images [48] shows that after a few seconds of initial growth, the particle is pushed upward by the carbon flux and lengthens. As growth continues, the surface in contact with the carbon begins to tilt upward until it forms a conical or tear-shaped form, the tip of the cone being oriented toward the growing carbon nanostructure and pointing in the direction of carbon diffusion. This observation leads to the conclusion that the commonly accepted belief that the catalyst particle determines the size and shape of the product is false. It is more likely the opposite [49].

2.5.3. Influence of Gas Composition on CNF Growth

The type of CNF formed is different for DR reactions and CC reactions; nevertheless, it also depends on other factors, namely the metal type and temperature. Luo et al. [50] have used Ni–La$_2$O$_3$ in a flow of CH$_4$/N$_2$, CO/N$_2$, and CO$_2$/CH$_4$/N$_2$, and they observed the production of both encapsulating carbon and CNF. The former was mostly formed in a CH$_4$/N$_2$ atmosphere whereas the latter was formed in a CO/N$_2$ or CO$_2$/CH$_4$/N$_2$ atmosphere. These results are in accordance with our findings. When using only C$_2$H$_4$, CNF with irregular forms as well as encapsulated carbon was formed; while, during DR reactions (where CO is present), only fishbone-type CNF were formed.

In fact, the composition of the gas affects the composition of the surface of metal particles because of the preferential segregation behavior of one of these components, which affects the arrangement of the atoms in the crystallographic face. This critical characteristic determines the mode of adsorption and decomposition of the reactive gas [51]. It has been found that when CO is present in the reactive gas, particles tended to have a faceted form, which leads to the formation of fishbone CNF [51].

2.5.4. CNF Precursor

As it appeared on the images of the TEM analysis (Figure 25), the metal is located in the tip of carbon nanofilaments, which indicates that carbon has grown in a crystallographic face of the metal.

Several authors have tried to find which phase is the one responsible for CNF growth. First, Baker et al. [52] report an activation energy that suggests that carbon diffuses through the reduced metal and, therefore, they indicate that the reduced metal is the growth crystal. Subsequently, Oberlin et al. [53] studied CNF growth on iron. They used TEM to identify growth crystals and reported that cementite and alpha iron were the only ones present in their work, which led them to conclude that not only is the active metal responsible for the growth of CNF but that it also contributed to the formation of metal carbides. In other research, in order to confirm which solid phase of iron is the most catalytic for carbon formation, Sacco et al. [36] worked on phase diagrams. They experimented by heating iron foils under a stream of hydrogen at 900 K, then fed hydrocarbon gas mixtures of different compositions into the reactor for each experiment. They had a mass gain that corresponds to carbon formation only in the area where Fe_3C is thermodynamically favored. Mass gain does not occur in α-Fe region, which proves that carbides, at least initially, are needed for carbon formation. In another study, it was shown that Fe_3C supported on graphite and exposed to acetylene did not catalyze carbon formation [27]. There are two assumptions to explain this: Fe_3C does not catalytically break up acetylene, or it is necessary to have a Fe_3C/Fe dual phase metal interface to provide the solubility difference needed for carbon diffusion and thus the growth of the nanofilament. The results found in this work, which confirm the presence of the Fe_3C peaks in the XRD pattern and SAED, confirm the assumption that the Fe_3C is the responsible growth crystal for CNF.

3. Experimental

3.1. Materials

The targeted feedstock was gases derived from plastic pyrolysis, which are largely composed of unsaturated HC. The initial step, which is presented in this work, was the use of ethylene as a representative molecule.

UGSO—upgraded slag oxide—is a residue of the UGS process, which has been developed by RTIT to produce, from ilmenite, the world's richest titanium slag (95% of TiO_2). This mining residue is composed largely of Fe, Al, and Mg oxides as determined by inductively coupled plasma mass spectrometry (ICP-MS) elemental analysis shown in Table 10. To produce a reforming catalyst, the UGSO is doped with Ni [30]. Blank experiments have shown that UGSO itself has no significant catalytic activity [30].

Table 10. Average elemental analysis of the upgraded slag oxide (UGSO) residue [30].

Component	Fe	Mg	Al	Ca	Mn	V	Ti	Cr	Na	Si	K	P	Zr	Zn
(wt.%) [a]	31.26	17.49	5.35	1.07	1.01	0.90	0.60	0.51	0.17	0.08	0.02	0.004	0.01	0.01

[a] The balance is oxygen.

3.2. Set-Up

Reactions (DR, CC, and activation reaction) were conducted in a differential fixed-bed reactor, which is a quartz tube of approximately 25 cm of length and 15 mm of internal diameter, put in an oven with temperature control. The catalyst was deposited at the bottom of the tube between two pieces of quartz wool and placed in the oven. Gases were fed from commercial gas (supplied by Praxair) cylinders: C_2H_4 (99%), CO_2 (99%) for the DR reaction, C_2H_4 (99%), Ar (99%) for CC reaction, and H_2 (99%) and Ar (99%) for the activation reaction. Three AALBORG mass flow meters were used to control the gas flow at the inlet (Figure 31). The flow rate of the products was measured using a bubble flow meter and its composition was analyzed by gas chromatography (GC Varian CP-3800) with a measurement error varying between 1% and 2.5% depending on the analyzed gases (i.e., H_2 had the highest error proven during calibration tests).

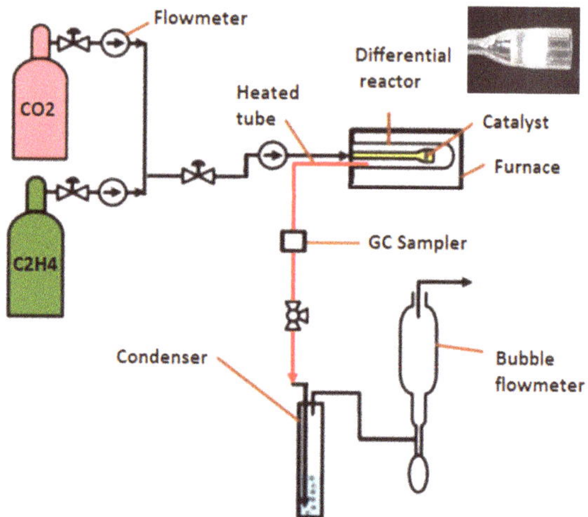

Figure 31. Reaction set-up [30].

3.3. Experimental Methodology

3.3.1. Preparation of Ni-UGSO

Ni-UGSO is prepared via a solid-state reaction developed by Chamoumi et al. [16]. In summary, UGSO was first milled and sieved in a 53 μm sieve, which was the smallest size obtained with our dry powder sieving equipment. Nitrate hexahydrate $Ni(NO_3)_2·6H_2O$ was used as a Ni precursor; the latter was mixed with the UGSO in the targeted proportion. A small quantity of water was added, and the mixture was then milled and homogenized softly in a mortar at ambient temperature. The resulting milled mixture was oven-dried at 105 °C for 4 h and then calcined at 900 °C for 12 h. After calcination, the catalyst was sieved down to 53 μm and was used in the catalytic tests as powder dispersed in the quartz wool placed in the differential reactor.

3.3.2. Activation of Ni-UGSO

The Ni-UGSO was activated by H_2. The evolution of the catalyst structure was studied using TPR, TEM and XRD analysis. The catalyst was activated under a flow of 75% H_2 and 25% Ar for a time-on-stream (TOS) of 2 h. Table 11 shows the activation test conditions.

Table 11. Activation test conditions.

Catalyst	H_2 Flow Rate (mL/min) [a]	Ar Flow Rate (mL/min) [a]	Catalyst Weight (g)	TOS (h)	GHSV (mL·h⁻¹·g⁻¹)	T (°C)
Ni-UGSO (wt.% = 5, 10, 13)	30	10	0.5	2	4800	650

[a] Gases are fed at atmospheric pressure.

3.3.3. Dry Reforming (DR) and Catalytic Cracking (CC) Reactions

Ni-UGSO was used as a catalyst for ethylene DR and CC. The influence of two factors was studied:

- Temperature;
- Weight percentage of Ni in the catalyst (wt.%).

In order to choose the temperature range for the tests, a study of thermodynamic equilibrium was done at temperatures ranging from 350 °C to 850 °C (Section 2.3.1). The wt.% of Ni in the catalyst, 13%, was chosen because the theoretical calculations based on the average UGSO composition show that this Ni content is necessary if all available Fe and Al oxides form spinels with Ni [30]. Nevertheless, since the catalytic activity comes from the local reduction of Ni and Fe into their metallic forms, as well as the concentration of these species at the surface of their support, lower Ni percentage were tested as well. Therefore, 10% and 5% were chosen arbitrarily as intermediate and lower wt.% of Ni.

A 3^2 fully reproduced factorial design of experiments (18 runs) were conducted for each study (nine tests with their duplicates). The experiments were conducted at atmospheric pressure after the activation step. Reaction conditions are summarized in the Tables 12 and 13.

Table 12. Ethylene dry reforming (DR) reaction conditions.

Catalyst	C_2H_4 Flow Rate (mL/min) [a]	CO_2 Flow Rate (mL/min) [a]	Catalyst Weight (g)	TOS (h)	GHSV (mL·h^{-1}·g^{-1})	T (°C)
Ni-UGSO (wt.% = 5, 10, 13)	30	10	0.5	2	4800	550, 650, 750

[a] Gases are fed at atmospheric pressure.

Table 13. Ethylene catalytic cracking (CC) reaction conditions.

Catalyst	C_2H_4 Flow Rate (mL/min) [a]	Ar Flow Rate (mL/min) [a]	Catalyst Weight (g)	TOS (h)	GHSV (mL·h^{-1}·g^{-1})	T (°C)
Ni-UGSO (wt.% = 5, 10, 13)	30	10	0.5	2	4800	550, 650, 750

[a] Gases are fed at atmospheric pressure.

3.4. Characterization Techniques

Several techniques were used to characterize fresh, activated, and spent catalysts. The information derived from the results of these techniques allows the understanding and interpretation of the phenomena that occur during the activation, DR, and CC reactions.

3.4.1. XRD

XRD analysis was used to identify fresh catalyst crystalline structure and to study transformations that might have occurred on this crystalline structure during and after activation, DR reactions, and CC reactions. The diffractometer used was Philips X'Pert PRO equipped with a Cu tube as its X-ray source and a Ni filter that was used to only let through Kα1 radiations from Cu (1.5418 Å) produced at 40 kV and 50 mA. The anti-dispersion slit was set at 1/2 and the diverged slit at 1/4. The analysis was carried out with a scanning angle of 2θ ranging from 15° to 90°.

Crystallite size can be calculated using Scherrer Equation:

$$L_c = \frac{K \times \lambda}{d \times \cos\theta}$$

L_c—crystallite size (nm);
K—0.9;
λ—1.5418 (Å) for Cu Kα1;
d—FHMW (full width at half maximum) calculated using Origin software (nm);
θ—angle (rad).

3.4.2. SEM and EDX

Scanning electron microscopy was used to characterize CNF and to study their morphology. The microscope used was a Hitachi Cold FE SU-823000 characterized by a 0.5 nm resolution at 30 kV and 3 nm resolution at 0.05 kV. It was equipped with a secondary electron (SE) lower detector, an SE/backscattered electron (BSE) upper detector, an SE/BSE top detector with energy filtration of BSE, a five quadrant BSE detector, a STEM (scanning transmission electron microscopy) detector for bright/dark fields, and a drift silicon detector energy dispersive X-ray spectrometry (SDD-EDS) detector, which was used to study the elemental composition and mapping of the sample.

3.4.3. TEM Coupled with EDX and Selected Area Electron Diffraction (SAED)

The microscope used was the Jeol JEM-2100F analytical transmission electron microscope equipped with a field effect gun operating at an acceleration voltage of 200 kV. Capable of imaging resolutions of 0.1 nm, this microscope was also equipped with an EDX spectrometer for chemical analysis. It also allowed the evaluation of crystallography using electron diffraction.

3.4.4. TPR

TPR was performed using a Chemisorb 2750 system (Micrometrics) equipped with a thermal conductivity detector (TCD). Tests were done after calibration of H_2 consumption. Fifteen milligrams of the studied catalyst were put on quartz wool and deposited in the tube reactor. A gas mixture consisting of 10% v/v of H_2 in Ar was fed in the TPR apparatus at a controlled flow rate of 40 mL/min. The sample was heated up to 1000 °C at a ramp rate of 2 °C/min in a temperature-controlled oven. The sensor was put in a cold trap Dewar flask containing isopropanol in liquid nitrogen in order to protect it from H_2O formed during reduction. ChemiSoft TCx software (Micromeritics) was used to calculate the peak area which is proportional to H_2 consumption.

3.4.5. TGA

TGA was done using a Setaram Setsys 24 analyzer under 20% O_2 and 80% Ar in a temperature range between 20 °C and 1000 °C. For carbon deposited quantification, the TGA was carried out at a heating rate of 10 °C/min.

3.4.6. BET

Specific surface area, pore volume, and pore average size were calculated by multipoint BET method. It was performed using an Accelerated Surface Area and Porosimetry System (ASAP 2020 V4.01).

3.5. Reaction Metrics

The performance of the catalyst was evaluated by calculating: C_2H_4 conversions $(X_{C_2H_4})$, H_2 yield $\left(Y_{H_2}\right)$, carbon yield (Y_C), and carbon growth rate $(g_C \cdot g_{cat}^{-1} \cdot h^{-1})$ in accordance with Algorithm 1 below:

Algorithm 1

$$X_{C_2H_4}\ (\%) = \frac{\left(F_{C2H4,in} - F_{C2H4,out}\right)}{F_{C2H4,in}}$$

$$Y_{H_2}\ (\%) = \frac{F_{H_2,}}{2 \times F_{C_2H_4}\,in} \times 100$$

$$Y_C\ (\%) = \frac{m_{C_{deposit}}}{m_{C,in}} \times 100$$

$$m_{C_{deposit}} = m_{catalyst,t0} - m_{catalyst,tf}$$

For DR: $m_{C,in} = ((2{*}\ F_{C2H4,in}{*}\text{time} + F_{CO2,in}{*}\text{time}))\ {*}M_C$

For CC: $m_{C,in} = ((2{*}\ F_{C2H4,in}{*}\text{time}))\ {*}M_C$

Carbon growth rate $(g_C \cdot g_{cat}^{-1} \cdot h^{-1}) = \dfrac{m_{C_{deposit}}}{t \times m_{catalyst,t0}}$

where:

$F_{C2H4,in}$ and $F_{C2H4,out}$ respectively denote the molar flow rates of C_2H_4 at the inlet and the outlet of the reactor,

F_{H_2} the molar flow rates of H_2 at the outlet of the reactor

$m_{C,in}$ the mass of carbon fed.

$m_{C_{deposit}}$ the mass of solid carbon deposit on the catalyst

M_C the molar mass of C = 12.0107 g/mol

---

4. Conclusions

Ni-UGSO was prepared from a mining residue UGSO and then used to produce CNF and H_2 via CC and DR reactions. When mining residues are involved in formulations, there is a concern regarding the variability of its composition. Prior to the present study and the ones published previously [30], four different batches of UGSO have been used to prepare the same catalytic formulations. The elemental analyses of the UGSO have shown that the variations were typically lower than 5% and the subsequent tests demonstrated that the observed conversion and yields deviations were lower than the overall experimental error and, consequently, not statistically significant.

Ni-UGSO 13% at 750 °C for the CC reaction and at 650 °C for the DR reaction exhibited the best performance in terms of H_2 and CNF yields. The literature has already provided the first insight into the factors influencing the formation of CNF and the mechanism of their formation, but the current work reveals the complexity of the latter. Although it is widely thought that the diameter of the CNF depends on the size of the catalyst particles, a more careful literature review along with the results of this work proves that other factors are also important. It has also been proven that carbides are the precursors of CNF and that the CNF-DR have higher structural order than CNF-CC. The type of CNF is also different. TEM images have shown that CNF-DR are fishbone shaped and CNF-CC form into tubular (MWNT) and stacked-cup structures. The results show that Fe is the main precursor of the CNF growth while Ni is more contributing to the split of C–C bonds. In terms of conversion and yield efficiencies, the performance of the catalytic formulations tested is proven at least equivalent to other Ni-based catalyst performances described by the literature. The experiments reported were conducted in a lab scale (g-lab) fixed-bed reactor and serve as a preliminary study. Ongoing work focuses on the production of CNF and H_2 in a kg-lab scale fluidized bed reactor to prove the feasibility at a larger scale towards eventual process commercialization.

Author Contributions: A.A.: PhD student whose topic is directly related to the content of this manuscript. She has written the first draft of the manuscript. E.-H.B. and F.M.: Professors who co-supervise the scientific part of the PhD project. They have read and corrected the manuscript. M.C.: Post-doctoral fellow who has helped and provided guidance during the experimentation. F.G.: Professor who co-supervise the scientific part of the PhD project. He has read and commented the manuscript. N.A.: Professor who co-supervise the PhD work and he is the scientific and technical Director of the Research team and the overall project. He has defined the content of the manuscript, read, corrected and proofread all intermediate steps until final acceptance.

Funding: The authors are indebted to the Natural Sciences and Engineering Research Council (NSERC) of Canada (RDCPJ-500331-16), PRIMA-Quebec (R10-010), and to the industrial partners for providing project's funding. They would also like to thank Gilles L'Espérance, Raynald Gauvin, Nicolas Brodusch, and Jean-Philippe Masse for the assistance provided in terms of TEM analyses. Special thanks are due to Jasmin Blanchard for his scientific and technical contributions to this manuscript, and to all the instrumental specialists at the CCM of the Université de Sherbrooke for the TPR, SEM, and XRD analyses.

Conflicts of Interest: There are no conflicts of interest to declare.

References

1. Baker, R.T.K.; Gadsby, G.R.; Thomas, R.B.; Waite, R.J. The production and properties of filamentous carbon. *Carbon* **1975**, *13*, 211–214. [CrossRef]
2. Baker, R.T.K. Carbon Nanofibers. In *Reference Module in Materials Science and Materials Engineering*; Elsevier: Amsterdam, The Netherlands, 2016; ISBN 978-0-12-803581-8.
3. Albright, L.F.; Baker, R.T.K. (Eds.) *Coke Formation on Metal Surfaces*; ACS Symposium Volume 202; American Chemical Society: Washington, DC, USA, 1983; ISBN 978-0-8412-0745-5.
4. Jankhah, S.; Abatzoglou, N.; Gitzhofer, F.; Blanchard, J.; Oudghiri-Hassani, H. Catalytic properties of carbon nano-filaments produced by iron-catalysed reforming of ethanol. *Chem. Eng. J.* **2008**, *139*, 532–539. [CrossRef]
5. Zeng, J.; Saltysiak, B.; Johnson, W.S.; Schiraldi, D.A.; Kumar, S. Processing and properties of poly(methyl methacrylate)/carbon nano fiber composites. *Compos. Part B Eng.* **2004**, *35*, 173–178. [CrossRef]
6. Tao, X.Y.; Zhang, X.B.; Zhang, L.; Cheng, J.P.; Liu, F.; Luo, J.H.; Luo, Z.Q.; Geise, H.J. Synthesis of multi-branched porous carbon nanofibers and their application in electrochemical double-layer capacitors. *Carbon* **2006**, *44*, 1425–1428. [CrossRef]

7. Ji, L.; Lin, Z.; Medford, A.J.; Zhang, X. Porous carbon nanofibers from electrospun polyacrylonitrile/SiO2 composites as an energy storage material. *Carbon* **2009**, *47*, 3346–3354. [CrossRef]

8. Yoon, S.H.; Park, C.W.; Yang, H.; Korai, Y.; Mochida, I.; Baker, R.T.K.; Rodriguez, N.M. Novel carbon nanofibers of high graphitization as anodic materials for lithium ion secondary batteries. *CARBON* **2004**, *42*, 21–32. [CrossRef]

9. Fauteux-Lefebvre, C.; Abatzoglou, N.; Blais, S.; Braidy, N.; Hu, Y. Iron oxide-functionalized carbon nanofilaments for hydrogen sulfide adsorption: The multiple roles of carbon. *Carbon* **2015**, *95*, 794–801. [CrossRef]

10. Bezemer, G.L.; Radstake, P.B.; Koot, V.; van Dillen, A.J.; Geus, J.W.; de Jong, K.P. Preparation of Fischer–Tropsch cobalt catalysts supported on carbon nanofibers and silica using homogeneous deposition-precipitation. *J. Catal.* **2006**, *237*, 291–302. [CrossRef]

11. Mestl, G.; Maksimova, N.I.; Keller, N.; Roddatis, V.V.; Schlögl, R. Carbon Nanofilaments in Heterogeneous Catalysis: An Industrial Application for New Carbon Materials? *Angew. Chem. Int. Ed.* **2001**, *40*, 2066–2068. [CrossRef]

12. Bockris, J.O.M. The hydrogen economy: Its history. *Int. J. Hydrog. Energy* **2013**, *38*, 2579–2588. [CrossRef]

13. Wei, Z.; Sun, J.; Li, Y.; Datye, A.K.; Wang, Y. Bimetallic catalysts for hydrogen generation. *Chem. Soc. Rev.* **2012**, *41*, 7994–8008. [CrossRef] [PubMed]

14. Blanchard, J.; Oudghiri-Hassani, H.; Abatzoglou, N.; Jankhah, S.; Gitzhofer, F. Synthesis of nanocarbons via ethanol dry reforming over a carbon steel catalyst. *Chem. Eng. J.* **2008**, *143*, 186–194. [CrossRef]

15. Jankhah, S.; Abatzoglou, N.; Gitzhofer, F. Thermal and catalytic dry reforming and cracking of ethanol for hydrogen and carbon nanofilaments' production. *Int. J. Hydrog. Energy* **2008**, *33*, 4769–4779. [CrossRef]

16. Chamoumi, M.; Abatzoglou, N. NiFe2O4 production from α-Fe2O3 via improved solid state reaction: Application as catalyst in CH4 dry reforming. *Can. J. Chem. Eng.* **2016**, *94*, 1801–1808. [CrossRef]

17. Braidy, N.; Bastien, S.; Blanchard, J.; Fauteux-Lefebvre, C.; Achouri, I.E.; Abatzoglou, N. Activation mechanism and microstructural evolution of a YSZ/Ni-alumina catalyst for dry reforming of methane. *Catal. Today* **2017**, *291*, 99–105. [CrossRef]

18. Shah, Y.T.; Gardner, T.H. Dry Reforming of Hydrocarbon Feedstocks. *Catal. Rev.* **2014**, *56*, 476–536. [CrossRef]

19. Pinilla, J.L.; Utrilla, R.; Lázaro, M.J.; Moliner, R.; Suelves, I.; García, A.B. Ni- and Fe-based catalysts for hydrogen and carbon nanofilament production by catalytic decomposition of methane in a rotary bed reactor. *Fuel Process. Technol.* **2011**, *92*, 1480–1488. [CrossRef]

20. Hu, X.; Lu, G. Syngas production by CO2 reforming of ethanol over Ni/Al2O3 catalyst. *Catal. Commun.* **2009**, *13*, 1633–1637. [CrossRef]

21. Takehira, K.; Ohi, T.; Shishido, T.; Kawabata, T.; Takaki, K. Catalytic growth of carbon fibers from methane and ethylene on carbon-supported Ni catalysts. *Appl. Catal. A Gen.* **2005**, *283*, 137–145. [CrossRef]

22. Nakagawa, K.; Nishitani-Gamo, M.; Ogawa, K.; Ando, T. Catalytic growth of carbon nanofilament in liquid hydrocarbon. *Catal Lett* **2005**, *101*, 191–194. [CrossRef]

23. Pinilla, J.L.; de Llobet, S.; Moliner, R.; Suelves, I. Ni-Co bimetallic catalysts for the simultaneous production of carbon nanofibres and syngas through biogas decomposition. *Appl. Catal. B Environ.* **2017**, *200*, 255–264. [CrossRef]

24. Arena, U.; Mastellone, M.L.; Camino, G.; Boccaleri, E. An innovative process for mass production of multi-wall carbon nanotubes by means of low-cost pyrolysis of polyolefins. *Polym. Degrad. Stab.* **2006**, *91*, 763–768. [CrossRef]

25. Svinterikos, E.; Zuburtikudis, I. Carbon nanofibers from renewable bioresources (lignin) and a recycled commodity polymer [poly(ethylene terephthalate)]. *J. Appl. Polym. Sci.* **2016**, *133*. [CrossRef]

26. Krylov, O.V.; Mamedov, A.K.; Mirzabekova, S.R. Oxidation of Hydrocarbons and Alcohols by Carbon Dioxide on Oxide Catalysts. *Ind. Eng. Chem. Res.* **1995**, *34*, 474–482. [CrossRef]

27. Baker, R.T.K.; Alonzo, J.R.; Dumesic, J.A.; Yates, D.J.C.J. Effect of the surface state of iron on filamentous carbon formation. *J. Catal.* **1982**, *77*, 74–84. [CrossRef]

28. Nguyen, H.N.T.; Berguerand, N.; Thunman, H. Mechanism and Kinetic Modeling of Catalytic Upgrading of a Biomass-Derived Raw Gas: An Application with Ilmenite as Catalyst. *Ind. Eng. Chem. Res.* **2016**, *55*, 5843–5853. [CrossRef]

29. Romero, A.; Garrido, A.; Nieto-Márquez, A.; Sánchez, P.; de Lucas, A.; Valverde, J.L. Synthesis and structural characteristics of highly graphitized carbon nanofibers produced from the catalytic decomposition of ethylene: Influence of the active metal (Co, Ni, Fe) and the zeolite type support. *Microporous Mesoporous Mater.* **2008**, *110*, 318–329. [CrossRef]

30. Chamoumi, M.; Abatzoglou, N.; Blanchard, J.; Iliuta, M.-C.; Larachi, F. Dry reforming of methane with a new catalyst derived from a negative value mining residue spinellized with nickel. *Catal. Today* **2017**, *291*, 86–98. [CrossRef]

31. Bali, A.; Blanchard, J.; Chamoumi, M.; Abatzoglou, N. Bio-Oil Steam Reforming over a Mining Residue Functionalized with Ni as Catalyst: Ni-UGSO. *Catalysts (2073-4344)* **2018**, *8*, 1–24.

32. Yu, Z.; Chen, D.; Tøtdal, B.; Holmen, A. Parametric study of carbon nanofiber growth by catalytic ethylene decomposition on hydrotalcite derived catalysts. *Mater. Chem. Phys.* **2005**, *92*, 71–81. [CrossRef]

33. Chesnokov, V.V.; Buyanov, R.A. The formation of carbon filaments upon decomposition of hydrocarbons catalysed by iron subgroup metals and their alloys. *Russ. Chem. Rev.* **2000**, *69*, 623–638. [CrossRef]

34. Díaz, M.C.; Blackman, J.M.; Snape, C.E. Maximising carbon nanofiber and hydrogen production in the catalytic decomposition of ethylene over an unsupported Ni-Cu alloy. *Appl. Catal. A Gen.* **2008**, *339*, 196–208.

35. Ermakova, M.A.; Ermakov, D.Y.; Chuvilin, A.L.; Kuvshinov, G.G. Decomposition of Methane over Iron Catalysts at the Range of Moderate Temperatures: The Influence of Structure of the Catalytic Systems and the Reaction Conditions on the Yield of Carbon and Morphology of Carbon Filaments. *J. Catal.* **2001**, *201*, 183–197. [CrossRef]

36. Sacco, A.; Thacker, P.; Chang, T.N.; Chiang, A.T.S. The initiation and growth of filamentous carbon from α-iron in H2, CH4, H2O, CO2, and CO gas mixtures. *J. Catal.* **1984**, *85*, 224–236. [CrossRef]

37. Pang, L.S.K.; Saxby, J.D.; Chatfield, S.P. Thermogravimetric analysis of carbon nanotubes and nanoparticles. *J. Phys. Chem.* **1993**, *97*, 6941–6942. [CrossRef]

38. Pradhan, D.; Sharon, M. Carbon nanotubes, nanofilaments and nanobeads by thermal chemical vapor deposition process. *Mater. Sci. Eng. B* **2002**, *96*, 24–28. [CrossRef]

39. Sui, Y.C.; Acosta, D.R.; González-León, J.A.; Bermúdez, A.; Feuchtwanger, J.; Cui, B.Z.; Flores, J.O.; Saniger, J.M. Structure, Thermal Stability, and Deformation of Multibranched Carbon Nanotubes Synthesized by CVD in the AAO Template. *J. Phys. Chem. B* **2001**, *105*, 1523–1527. [CrossRef]

40. Serp, P.; Corrias, M.; Kalck, P. Carbon nanotubes and nanofibers in catalysis. *Appl. Catal. A Gen.* **2003**, *253*, 337–358. [CrossRef]

41. Narkiewicz, U.; Podsiadły, M.; Jędrzejewski, R.; Pełech, I. Catalytic decomposition of hydrocarbons on cobalt, nickel and iron catalysts to obtain carbon nanomaterials. *Appl. Catal. A Gen.* **2010**, *384*, 27–35. [CrossRef]

42. Lander, J.J.; Kern, H.E.; Beach, A.L. Solubility and Diffusion Coefficient of Carbon in Nickel: Reaction Rates of Nickel-Carbon Alloys with Barium Oxide. *J. Appl. Phys.* **1952**, *23*, 1305–1309. [CrossRef]

43. Park, C.; Baker, R.T.K. Carbon Deposition on Iron–Nickel During Interaction with Ethylene–Carbon Monoxide–Hydrogen Mixtures. *J. Catal.* **2000**, *190*, 104–117. [CrossRef]

44. Koestner, R.J.; Frost, J.C.; Stair, P.C.; Van Hove, M.A.; Somorjai, G.A. Evidence for the formation of stable alkylidyne structures from C3 and C4 unsaturated hydrocarbons adsorbed on the pt(111) single crystal surface. *Surf. Sci.* **1982**, *116*, 85–103. [CrossRef]

45. Zhu, X.-Y.; White, J.M. Interaction of ethylene and acetylene with Ni(111): A SSIMS study. *Surf. Sci.* **1989**, *214*, 240–256. [CrossRef]

46. Rodriguez, N.M. A review of catalytically grown carbon nanofibers. *J. Mater. Res.* **1993**, *8*, 3233–3250. [CrossRef]

47. Chitrapu, P.; Lund, C.R.F.; Tsamopoulos, J.A. A model for the catalytic growth of carbon filaments. *Carbon* **1992**, *30*, 285–293. [CrossRef]

48. Audier, M.; Coulon, M.; Oberlin, A. Relative crystallographic orientations of carbon and metal in a filamentous catalytic carbon. *Carbon* **1980**, *18*, 73–76. [CrossRef]

49. Nolan, P.E.; Lynch, D.C.; Cutler, A.H. Carbon Deposition and Hydrocarbon Formation on Group VIII Metal Catalysts. *J. Phys. Chem. B* **1998**, *102*, 4165–4175. [CrossRef]

50. Luo, J.Z.; Yu, Z.L.; Ng, C.F.; Au, C.T. CO2/CH4 Reforming over Ni-La2O3/5A: An Investigation on Carbon Deposition and Reaction Steps. *J. Catal.* **2000**, *194*, 198–210. [CrossRef]

51. Krishnankutty, N.; Rodriguez, N.M.; Baker, R.T.K. Effect of Copper on the Decomposition of Ethylene over an Iron Catalyst. *J. Catal.* **1996**, *158*, 217–227. [CrossRef]

52. Baker, R.T.K.; Barber, M.A.; Harris, P.S.; Feates, F.S.; Waite, R.J. Nucleation and growth of carbon deposits from the nickel catalyzed decomposition of acetylene. *J. Catal.* **1972**, *26*, 51–62. [CrossRef]
53. Oberlin, A.; Endo, M.; Koyama, T. Filamentous growth of carbon through benzene decomposition. *J. Cryst. Growth* **1976**, *32*, 335–349. [CrossRef]

MDPI

St. Alban-Anlage 66

4052 Basel

Switzerland

Tel. +41 61 683 77 34

Fax +41 61 302 89 18

www.mdpi.com

Catalysts Editorial Office

E-mail: catalysts@mdpi.com

www.mdpi.com/journal/catalysts